Youssef Benadada · Fatima-Zahra Mhada ·
Fatima Ouzayd · Ahmed El Hilali Alaoui ·
Jaouad Boukachour
Editors

Proceedings of the 8th International Conference on Logistics Operations Management, GOL'26

Volume 1

Springer

Editors
Youssef Benadada
ENSIAS
Mohammed V University of Rabat
Rabat, Morocco

Fatima-Zahra Mhada
ENSIAS
Mohammed V University of Rabat
Rabat, Morocco

Fatima Ouzayd
ENSIAS
Mohammed V University of Rabat
Rabat, Morocco

Ahmed El Hilali Alaoui
EUROMED
Fes, Morocco

Jaouad Boukachour
University Institute of Technology
Le Havre, France

ISSN 2367-3370 ISSN 2367-3389 (electronic)
Lecture Notes in Networks and Systems
ISBN 978-3-032-29676-4 ISBN 978-3-032-29677-1 (eBook)
https://doi.org/10.1007/978-3-032-29677-1

This Springer imprint is published by the registered company Springer Nature Switzerland AG
The registered company address is: Gewerbestrasse 11, 6330 Cham, Switzerland

Lecture Notes in Networks and Systems 2034

The series "Lecture Notes in Networks and Systems" publishes the latest developments in Networks and Systems—quickly, informally and with high quality. Original research reported in proceedings and post-proceedings represents the core of LNNS.

Volumes published in LNNS embrace all aspects and subfields of, as well as new challenges in, Networks and Systems.

The series contains proceedings and edited volumes in systems and networks, spanning the areas of Cyber-Physical Systems, Autonomous Systems, Sensor Networks, Control Systems, Energy Systems, Automotive Systems, Biological Systems, Vehicular Networking and Connected Vehicles, Aerospace Systems, Automation, Manufacturing, Smart Grids, Nonlinear Systems, Power Systems, Robotics, Social Systems, Economic Systems and other. Of particular value to both the contributors and the readership are the short publication timeframe and the world-wide distribution and exposure which enable both a wide and rapid dissemination of research output.

The series covers the theory, applications, and perspectives on the state of the art and future developments relevant to systems and networks, decision making, control, complex processes and related areas, as embedded in the fields of interdisciplinary and applied sciences, engineering, computer science, physics, economics, social, and life sciences, as well as the paradigms and methodologies behind them.

Indexed by SCOPUS, EI Compendex, INSPEC, WTI Frankfurt eG, zbMATH, SCImago.

All books published in the series are submitted for consideration in Web of Science.

For proposals from Asia please contact Aninda Bose (aninda.bose@springer.com).

Preface

SPRINGER 8th International Conference on Logistics Operations Management GOL'26: **Revolutionizing Supply Chain and Logistics through Operations Research and Artificial Intelligence** is organized by ENSIAS School from Mohammed V University of Rabat (Morocco), with the collaboration of Sidi Mohamed Ben Abdellah University (Morocco), in partnership with the Smart System Laboratory (SSLab) and the Moroccan Association of the Logistics Chain (AMCLog).

GOL'26 aims to bring together researchers and professionals interested in logistics from different points of view: industrial engineering, management, economics and social sciences. This multidisciplinary conference will examine the current state of the art focusing on innovations in concept, model, method and tools for global supply chain management.

In addition to the plenary conferences of this edition, the communications are divided into 17 various thematic sessions as, Artificial Intelligence; Artificial Intelligence in supply chain; AI-Driven Trade Facilitation and Smart Transport Systems for Resilient Supply Chains; Intelligent Multi-Agent Systems and Metaheuristics for Logistics Management; Blockchain and the future of Smart Ports; IoT technologies for Industry 4.0; From Data to Decisions: Adaptive Supply-Chain Management through Simulation, Data Analytics and Digital Twins; Optimization and innovative Decision support tools for Managing Smart and Sustainable Supply Chain in Smart Cities; Emerging trends, challenges and innovations in transportation and logistics; Air transport and logistics operations; Joint Optimization of Production and Distribution; Logistics Operations Management in Education; Operational Research; Planning and scheduling; Modelling, simulation and optimization; Project and risk management; Sustainable development. We would like to thank the invited speakers as well as all the participants.

For this 8th edition, 86 papers were submitted by 15 countries from 4 continents. Of these, 71 received a positive average from the reviewers chosen from among the members of the scientific committee and were submitted for publication by Springer, with an overall acceptance rate of 82.5%.

We received financial support for the organization of the conference from our Golden Sponsor Hexaly whom we would like to thank for their confidence and commitment. We would also like to thank ENSIAS, CRNST, and the Moroccan Supply Chain Association (AMCLog) for their ongoing support. We would also like to thank Sidi Mohamed Ben Abdellah University for agreeing to host this edition in its conference center. We would like to express our deep gratitude to all these organizations for their invaluable support.

Rabat, Morocco
Mai 2026

Youssef Benadada
Fatima-Zahra Mhada
General Chairs GOL'24

Contents

From Data to Decisions: Adaptive Supply-Chain Management Through Simulation, Data Analytics and Digital Twins

Optimization and Innovative Decision support Tools for Managing Smart and Sustainable Supply Chain in Smart Cities

Artificial Intelligence

Artificial Intelligence in Regulatory Compliance: Comparative Analysis of Benefits and Challenges

Georges Bou Ghantous(✉)

Faculty of Engineering and IT, The University of Technology Sydney, Sydney, Australia
georges.boughantous-1@uts.edu.au

Abstract. Artificial Intelligence (AI) is reshaping regulatory compliance by automating complex assessments, improving anomaly detection, and enhancing monitoring accuracy across domains such as anti-money laundering (AML), GDPR governance, auditing, building code verification, and legal analysis. Despite rapid adoption, current research presents fragmented evidence regarding the benefits and challenges of AI-driven compliance processes. This study conducts a PRISMA-guided Systematic Literature Review (SLR) of 40 peer-reviewed studies published between 2018 and 2025 to consolidate existing findings. Results show significant gains in accuracy, efficiency, scalability, and cost reduction enabled by machine learning, natural language processing, deep learning, and graph neural networks. However, persistent challenges remain, including data quality issues, algorithmic bias, explainability constraints, ethical concerns, and integration difficulties with legacy systems. The review underscores the dual nature of AI in compliance, highlighting the need for responsible deployment and offering a foundation for advancing trustworthy, transparent, and sustainable AI-enabled regulatory practices.

Keywords: Artificial Intelligence (AI) · Regulatory Compliance · Anti-Money Laundering (AML) · Building Information Modeling (BIM) · Natural Language Processing (NLP) · Machine Learning (ML)

1 Introduction

Regulatory compliance has become a central concern across industries such as finance, construction, and auditing, where organizations must continuously adapt to increasingly complex legal and regulatory environments. Traditionally, compliance verification relied on manual processes and expert judgment, which are resource-intensive, prone to human error, and often unable to keep pace with evolving regulations [5] and [11].

Recent advancements in Artificial Intelligence (AI) have introduced transformative prospects to address these challenges. Natural Language Processing (NLP) and Machine Learning (ML) techniques have enabled the automation of regulatory document analysis, anomaly detection, and risk-based decision support. Applications range from automated checking of building information models (BIM) against construction codes [1] and [19],

Y. Benadada et al. (Eds.): GOL 2026, LNNS 2034, pp. 3–14, 2027.
https://doi.org/10.1007/978-3-032-29677-1_1

to identifying money-laundering patterns in financial transactions [2, 17], and enhancing data protection compliance with frameworks such as the General Data Protection Regulation (GDPR) [4] and [10]. AI-driven methods not only improve efficiency but also enhance accuracy and scalability in compliance monitoring.

Scholars have also highlighted the potential of continuous auditing and responsible AI practices for sustaining long-term compliance. For example, AI-based systems are being developed to provide real-time monitoring of internal controls [6], support automated evaluation of contractual obligations [4], and ensure responsible logging for accountability in ML-driven audits [21]. Within the construction domain, rule-based and semantic approaches are increasingly applied to automate code compliance checking and design validation [14] and [40], while in finance, graph-based and deep learning models are advancing anti-money-laundering (AML) detection [8] and [35].

Despite these developments, the rapid adoption of AI introduces challenges related to transparency, interpretability, and regulatory acceptance. Studies emphasize that without proper governance and rigorous validation, AI systems may inadvertently reinforce biases or generate opaque compliance decisions [11] and [18]. This duality underscores the need for a systematic understanding of how AI transforms compliance processes, balancing opportunities with risks.

To address this, this study adopts a Systematic Literature Review (SLR) methodology following established guidelines [16] and [27]. By synthesizing evidence across multiple domains, this review aims to map the landscape of AI-enhanced regulatory compliance, evaluate innovations, and identify future research directions.

2 Background

2.1 Regulatory Compliance Context

Regulatory compliance has emerged as a critical requirement across sectors including data protection, financial services, and corporate governance. The introduction of the General Data Protection Regulation (GDPR) has stimulated the development of AI-driven tools designed to analyze and enforce privacy requirements, with natural language processing (NLP) methods increasingly used to examine privacy policies and contractual agreements for regulatory alignment [4] and [25]. These approaches enhance accuracy and streamline compliance monitoring processes.

In parallel, anti-money laundering (AML) and financial fraud detection continue to be major focus areas for AI adoption, particularly due to persistent challenges in reducing false positives and improving scalability. AI-based techniques, including deep learning and graph-based models, offer promising improvements by capturing hidden transactional patterns and adapting to evolving threats [17] and [35]. However, several issues remain unresolved, including data-quality limitations that undermine model reliability [13], interpretability constraints that hinder transparency in automated decisions [8] and [22], and the ongoing difficulty of obtaining regulatory acceptance for high-stakes AI-enabled decision making [28] and [35].

2.2 AI and Automation in Compliance

Beyond finance and data protection, AI has become central to automating rule-checking in Building Information Modeling (BIM). Studies using rule-based reasoning, semantic parsing, and ontology-driven approaches validate engineering models against regulatory requirements [1] and [14], while additional work shows how schema alignment and semantic interpretation improve automated compliance in complex building contexts [19] and [40].

AI also strengthens continuous auditing and monitoring by supporting internal-control assessment, anomaly detection, and organizational risk evaluation. Research demonstrates improvements in transparency and traceability through predictive analytics and responsible logging mechanisms [5] and [21], with further studies stressing explainability and accountable model behavior in audit workflows [11] and [18].

Legal AI and NLP applications expand the ability to process regulatory texts, extract obligations, and analyze GDPR- and contract-related requirements at scale. NLP-driven systems improve accuracy and automation in compliance assessment [7] and [23], while complementary studies show how policy interpretation and contract evaluation enhance oversight in legal and governance contexts [25] and [34].

3 Methodology

A Systematic Literature Review (SLR) was conducted to provide a rigorous and reproducible assessment of AI applications in regulatory compliance. SLRs follow predefined protocols designed to minimize bias and ensure transparent synthesis of evidence across domains such as AML, GDPR, BIM, and auditing [1] and [7].

The review process followed Kitchenham and Charters' guidelines, which define structured procedures for research questions, inclusion/exclusion criteria, and screening workflows [16]. PRISMA 2020 was adopted to standardize reporting of identification, screening, eligibility, and inclusion stages [27], with the overall study flow shown in Fig. 1. To clarify the decision-making logic applied during screening, an activity-based representation of the criteria and selection steps is provided in Fig. 2.

1. Research Questions—Developed to examine AI applications in compliance, associated benefits/risks, and gaps across domains including AML, GDPR, and automated auditing [4] and [23].
2. Search Strategy—Queries using terms such as "AI compliance," "AML detection," and "GDPR automation" were executed across IEEE Xplore, ACM DL, SpringerLink, Scopus, and Web of Science [5] and [14].
3. Selection Criteria—Included studies were peer-reviewed, AI-focused, compliance-related, and published between 2018–2025; exclusions targeted non-AI work, non-compliance topics, and non-peer-reviewed material [11] and [25].
4. Screening Process—Titles/abstracts were screened for compliance relevance, followed by full-text eligibility checks. Disagreements were resolved collaboratively. The screening criteria and decision gates are summarized in Fig. 2.
5. Data Extraction and Synthesis—Extracted attributes included domain (GDPR, AML, BIM), AI methods (NLP, ML, DL, GNNs), evaluation metrics, and reported benefits/challenges [8] and [35].

The use of structured SLR practices in compliance-related AI research [26] confirms the suitability of this method. Integrating Kitchenham's rigor [16] with PRISMA transparency [27] strengthens the reliability and reproducibility of the findings (see Figs. 1 and 2).

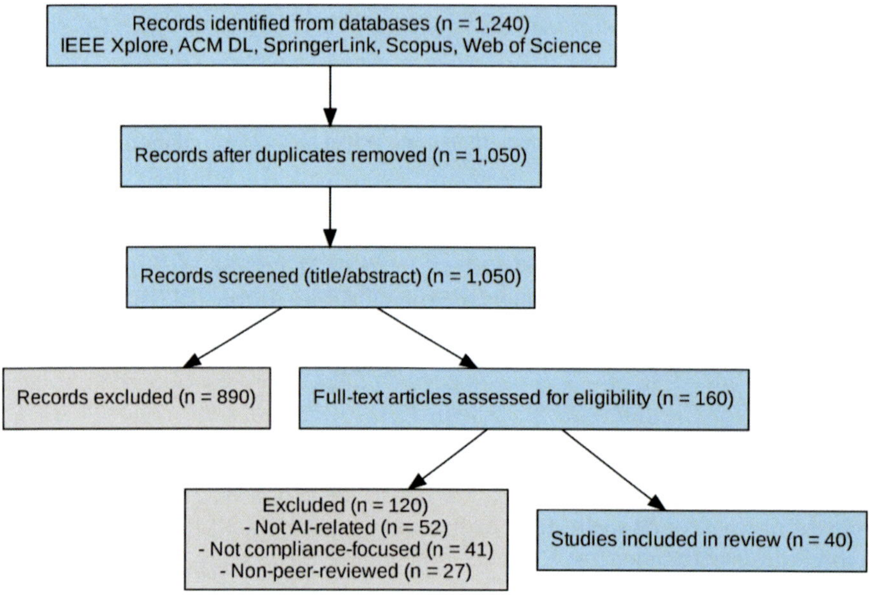

Fig. 1 PRISMA flow diagram illustrates the systematic selection process, from database identification through screening, eligibility assessment, and final inclusion of studies

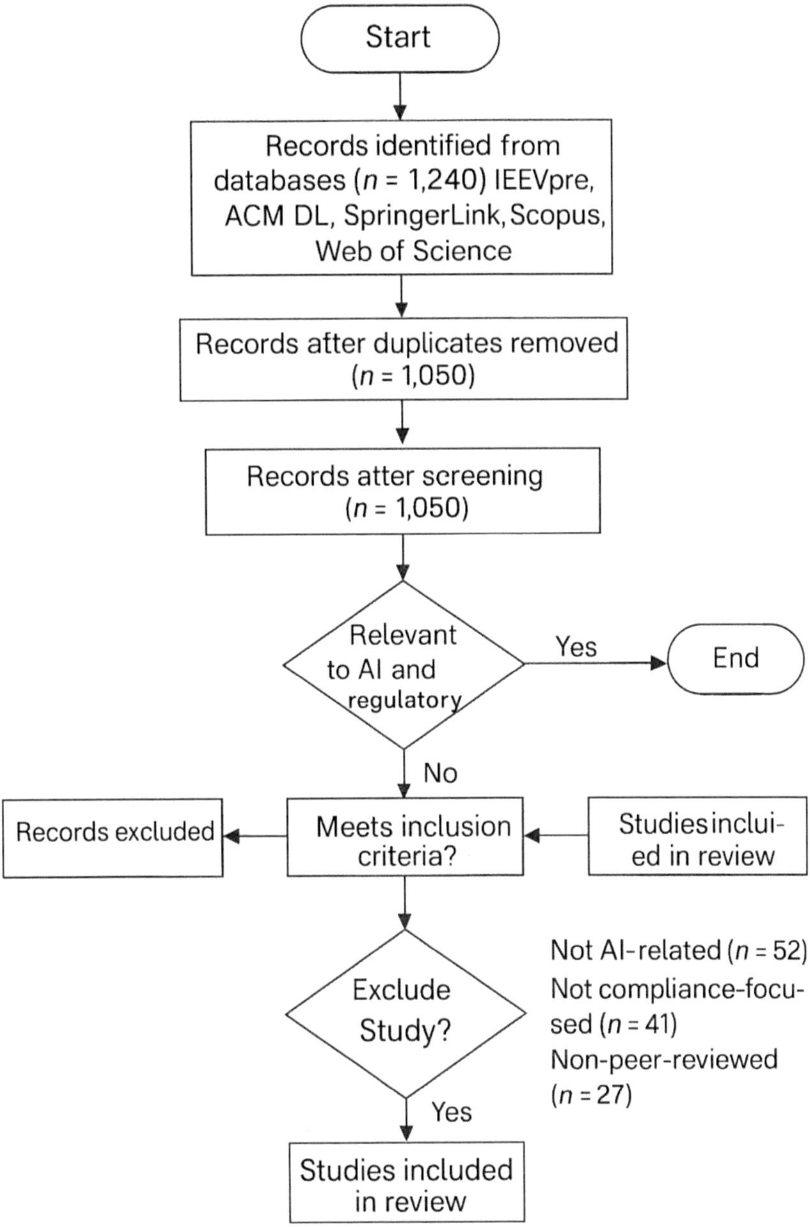

Fig. 2 Activity diagram illustrates the decision logic applied during study identification, screening, and eligibility assessment in the SLR process

4 Results of Literature Review

4.1 Benefits of AI in Regulatory Compliance

The reviewed studies identify four main areas where AI benefits regulatory compliance (Table 1). Efficiency and automation feature prominently (28%), with AI reducing manual workload, expediting compliance checks, and enabling continuous monitoring across domains such as BIM and auditing [1] and [14]. Additional examples demonstrate the use of automated reasoning and semantic rule-checking to support design validation and regulatory assessment [19] and [40]. Closely aligned, 30% of studies emphasize anti-money laundering (AML) and fraud detection, where advanced methods—including deep learning, graph-based models, and federated approaches—address challenges related to false positives and limited scalability in traditional systems [2] and [17].

Legal and Natural Language Processing (NLP) applications represent 23%, focusing on the analysis of privacy policies, contracts, and GDPR-related requirements to improve accuracy and reduce interpretation errors [4] and [23]. Finally, 10% of studies highlight continuous auditing and explainability, reflecting the need for transparency and accountability in AI-driven compliance systems [11] and [21]. Collectively, these results show that while efficiency and AML dominate current research, legal NLP and explainability remain essential for advancing trustworthy AI-enabled compliance.

Table 1 Benefits of AI in regulatory compliance

Benefit category	Key contributions (Studies)	% of Included studies (n = 40)
Efficiency and automation	Reduced manual compliance checks, faster auditing, continuous monitoring [1, 5, 6, 9, 14, 19, 20, 34, 36, 37, 39, 40]	28% (11/40)
AML and fraud detection	Improved anomaly detection, hidden transaction patterns, adaptive threat models [2, 3, 8, 17, 22, 24, 29–33, 35, 38]	30% (12/40)
Legal/NLP compliance checking	GDPR, contracts, privacy policy analysis [4, 7, 10, 12, 15, 23, 25, 28, 34]	23% (9/40)
Continuous auditing and explainability	Transparency and monitoring of AI reasoning [11, 18, 21]	10% (4/40)

4.2 Challenges and Risks

Despite these benefits, the review identifies several challenges (Table 2). Data quality issues (5%; 2/40) weaken model reliability due to incomplete or biased datasets [13]. Algorithmic bias and limited interpretability (10%; 4/40) create risks of opaque or unfair decision-making [11] and [18], with additional studies stressing the need for transparent model behavior [21] and [26]. Ethical and legal dilemmas (8%; 3/40) highlight concerns around accountability and fairness in AI-driven compliance [7] and [23], alongside questions of regulatory alignment [25]. A further 10% of studies caution against excessive automation, noting that reduced human oversight can undermine judgment and governance [5] and [6], with similar warnings reported in compliance automation research

[19] and [34]. These issues collectively underscore the need for strong governance, transparency, and human-in-the-loop controls.

Table 2 Challenges and risks of AI in regulatory compliance

Challenge category	Key contributions (Studies)	% of Included Studies (n = 40)
Data quality	Incomplete, biased, or unstructured compliance datasets [13]	5% (2/40)
Algorithmic bias and interpretability	Opaqueness, fairness issues, lack of explainability [11, 18, 21, 26]	10% (4/40)
Ethical and legal dilemmas	Accountability, fairness, privacy, academic integrity [7, 23, 25]	8% (3/40)
Over-reliance on automation	Neglect of human oversight, blind trust in AI outputs [5, 6, 19, 34]	10% (4/40)

4.3 AI Relevance Index (AIR)—A Quantitative Interpretation

To complement the qualitative synthesis, the AI Relevance Index (AIR) was developed as a quantitative metric capturing the balance between reported benefits and risks of AI in compliance. It consolidates their relative frequencies, weighted by impact, as shown in formula (1).

$$AIR = \frac{\alpha \cdot B}{\alpha \cdot B + \gamma \cdot R} \times 100 \quad (1)$$

Table 3 Parameters for assessing AI relevance (AIR)

Symbol	Definition	Explanation
B	Benefits frequency	Count of studies reporting advantages (efficiency, AML, auditing)
R	Risks frequency	Count of studies reporting challenges (data, bias, ethics)
α	Benefit weight	Relevance level of each benefit (1 = low, 3 = medium, 8 = high)
γ	Risk weight	Severity of each risk (1 = low, 2 = med, 8 = high)

Worked Example. The AIR score was computed using the benefit and risk frequencies reported in Tables 1 and 2. First, the total number of benefits (**B**) and risks (**R**) was obtained by summing the category counts (Table 3).

$$B = 11 + 12 + 9 + 4 = 36$$

$$R = 2 + 4 + 3 + 4 = 13$$

The exponential weighting scheme assigns a weight of $\alpha = 3$ to benefits (medium relevance) and $\gamma = 8$ to risks (high severity). Using these values in Eq. (1):

$$AIR = \frac{\alpha \cdot B}{\alpha \cdot B + \gamma \cdot R} \times 100$$

$$AIR = \frac{3 \cdot 36}{3 \cdot 36 + 8 \cdot 13} \times 100 = \frac{108}{108 + 104} \times 100 = \frac{108}{212} \times 100 \approx 50.94\%$$

After rounding, the resulting AI Relevance Index is AIR = 51%, indicating that the weighted evidence across the literature is nearly balanced between reported benefits and high-severity risks. This moderate score reflects that while AI delivers clear advantages in efficiency, fraud detection, and automated analysis, these gains are counterbalanced by concerns related to data quality, algorithmic bias, ethical constraints, and reliance on automation. As a decision-support tool, AIR helps benchmark adoption trends for researchers, signals the maturity and caution required for regulators, and guides investment decisions for organizations operating in finance, auditing, and data governance (Fig. 3).

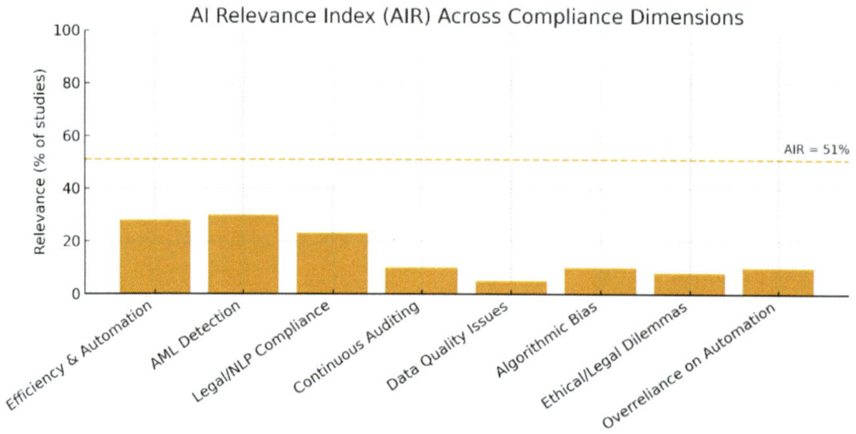

Fig. 3 AI relevance index (AIR) across compliance domains. Overall relevance score at 51%

5 Discussion

The systematic review highlights both the transformative potential and the inherent limitations of AI in regulatory compliance. Across domains such as GDPR, AML, and automated auditing, AI applications consistently demonstrate gains in efficiency, accuracy, and scalability compared with manual compliance processes. These findings align with studies reporting substantial reductions in manual workload and improvements in anomaly detection [5] and [14], with additional evidence supporting automated rule-checking, risk scoring, and document analysis across compliance tasks [19] and [34]. In financial and privacy-related contexts, research further demonstrates the value of

AI-enabled pattern detection, continuous monitoring, and policy interpretation [1] and [20].

Despite these advantages, several persistent challenges limit widespread adoption. Algorithmic bias and limited interpretability continue to pose risks to fairness, transparency, and regulatory acceptance [11] and [18]. Complementary work reinforces the need for explainable and auditable model behavior in high-stakes environments such as AML detection and automated decision systems [21] and [26]. Data quality constraints further undermine reliability, especially in fraud and anomaly detection pipelines that rely on sparse, imbalanced, or noisy datasets [13]. Ethical considerations, including accountability, privacy, and over-reliance on automated outputs underline a broader concern that insufficient oversight may erode trust, accuracy, and governance integrity in compliance workflows [7] and [23], with related concerns also raised in regulatory alignment studies [25] and [34].

To quantify the balance between these benefits and risks, the AI Relevance Index (AIR) was introduced. Using frequencies derived from Tables 1 and 2 and exponential weighting (1–3–8), the resulting AIR score of 51% (Fig. 3) indicates a nearly even distribution between weighted benefits and high-severity risks. This moderate score suggests that although AI is effective in many compliance applications, significant constraints remain particularly interpretability, data quality, and ethical risks—which collectively temper the overall maturity level of AI adoption in this domain. The AIR metric thus provides a structured reference point for researchers, regulators, and practitioners, supporting evidence-based assessments of adoption readiness and investment prioritization.

Finally, the rigor of the SLR methodology strengthens the reliability of these insights. Kitchenham's structured review procedures [16] and the PRISMA reporting guidelines [27] ensure transparency across search, screening, and eligibility decisions while reducing methodological bias. At the same time, gaps remain in explainability research, domain-specific governance models, sustainability analyses, and cross-sector evaluations—indicating clear directions for future work aimed at advancing AI from experimental implementations to stable, trustworthy compliance solutions.

6 Future Work

Future research should further examine the sustainability impacts of AI in compliance, as increasingly complex models introduce notable energy use and carbon footprint considerations. Developing more efficient model architectures and incorporating carbon-aware performance metrics into compliance governance represents an important direction for responsible deployment.

Another priority is establishing a clearer mapping between AI techniques and compliance domains. A structured AI-methods × application matrix—linking NLP, ML, deep learning, GNNs, and rule-based systems to tasks such as GDPR analysis, AML detection, BIM rule-checking, auditing, and legal document interpretation—would support more transparent model selection and regulatory benchmarking.

Additional work is needed in explainability, governance frameworks, and data-quality mitigation to ensure that AI-driven compliance systems become reliable, interpretable, and operationally sustainable across sectors.

7 Conclusion

This study systematically reviewed the role of AI in regulatory compliance, identifying both its capacity to enhance efficiency and detection, as well as the risks that challenge its broader acceptance. By introducing the AI Relevance Index (AIR), we translated qualitative findings into a measurable metric, offering a new perspective on how benefits and risks are balanced. Future work should focus on refining this index with domain-specific weightings, expanding cross-industry evaluations, and developing governance frameworks that ensure transparency, fairness, and sustainable adoption of AI in compliance contexts.

References

1. Al-Hussein, M., Ahmad, I.: Automated BIM compliance check with LLM-based approaches. Buildings **14**(2), 311 (2024)
2. Andersson, P., et al.: Applicability of supervised learning at cryptocurrency exchanges for AML. J. Money Laundering Control **25**(4), 1123–1139 (2022)
3. Bender, J., et al.: Incorporating ML and risk-based strategy in AML. Expert Syst. Appl. **214**, 119112 (2023)
4. Briand, L., et al.: NLP-based automated compliance checking of data processing agreements against GDPR. IEEE Trans. Softw. Eng. (2023)
5. Brown, T., et al.: Integrating process mining and ML for internal-control evaluation. J. Inform. Syst. **39**(2), 223–240 (2025)
6. Chang, R., et al.: Continuous monitoring with ML and interactive visualization for internal controls. Int. J. Account. Inform. Syst. **44**, 100570 (2022)
7. Chalkidis, I., et al.: How does NLP benefit the legal system? A summary of Legal-AI. In: Findings of ACL 2020, pp. 3280–3296. (2020)
8. Chen, H., et al.: Qualifying and raising AML alarms with deep learning. Expert Syst. Appl. **195**, 116541 (2022)
9. Cheng, J.C.P., Kumar, B.: Evaluating AI tools for automated practice conformance of building regulations. In: Proceedings ISARC 2018, pp. 155–162. (2018)
10. Colombo, A., et al.: Towards automated GDPR compliance checking. In: Advances in Information Systems, LNCS 12345, pp. 201–213. Springer (2021)
11. Fischer, M., et al.: Continuous auditing of AI: conceptualization and research directions. Inform. Syst. Managem. **39**(1), 35–48 (2022)
12. Hähnle, R., et al.: Using models to enable compliance checking against the GDPR: An experience report. In: MODELS 2019, LNCS 11863, pp. 87–98. Springer, Heidelberg (2019)

13. He, Y., et al.: Data-quality issues leading to sub-optimal ML for AML. J. Money Laundering Control **24**(3), 987–995 (2021)
14. Hjelseth, E., Nisbet, N.: Rule-based compliance checking and generative design. Autom. Construct. **125**, 103582 (2021)
15. Kamp, M., et al.: CompAi: a tool for GDPR completeness checking of privacy policies. In: Proceedings ASE 2024, Tool Paper, pp. 300–305. (2024)
16. Kitchenham, B., Charters, S.: Guidelines for performing systematic literature reviews in software engineering. EBSE Technical Report, Keele University and University of Durham (2007)
17. Krishnan, S., et al.: Leveraging ML in the global fight against money laundering. J. Business Res. **120**, 112–124 (2020)
18. Kumar, R., et al.: Artificial Intelligence in auditing: a conceptual framework. Administrations **12**(3), 45 (2024)
19. Lee, Y.C., Eastman, C., Solihin, W.: Automated BIM data validation integrating open-standard schema for building code compliance. Adv. Eng. Inform. **40**, 1–15 (2019)
20. Li, C., Tang, S.: Simplified schema queries for supporting BIM-based rule-checking. Adv. Eng. Inform. **45**, 101503 (2020)
21. Li, X., et al.: Logging requirements for continuous auditing of responsible ML. Empirical Softw. Eng. **30**, 55–72 (2025)
22. Maes, S., et al.: Money-laundering and terrorism-financing detection using neural networks. Expert Syst. Appl. **147**, 113190 (2020)
23. Martinez, D., et al.: Legal information retrieval systems: state-of-the-art and open issues. Inform. Process. Managem. **58**(4), 102548 (2021)
24. Nguyen, T., et al.: Graph-based LSTM for AML detection in Bitcoin transactions. Neural. Process. Lett. **54**, 1121–1136 (2022)
25. Novak, J., et al.: Enhancing GDPR compliance through NLP: automated policy analysis. In: LegalTech and AI, pp. 55–72. Springer (2025)
26. Ortega, L., et al.: Financial fraud detection using GNNs: a systematic review. Expert Syst. Appl. **221**, 119768 (2023)
27. Page, M.J., et al.: The PRISMA 2020 statement: an updated guideline for reporting systematic reviews. BMJ **372**, n71 (2021). https://doi.org/10.1136/bmj.n71
28. Pandit, H., O'Sullivan, D., Lewis, D.: An AI-assisted approach for checking completeness of privacy policies against GDPR. In: IEEE Requirements Engineering, pp. 25–36. (2020)
29. Patel, A., et al.: Dynamic graph convolutional network for AML detection. In: Proceedings Springer Conference, pp. 310–322. (2023)
30. Rivera, J., et al.: Time-aware and interpretable predictive monitoring for AML. Decis. Support. Syst. **152**, 113654 (2022)
31. Rossi, F., et al.: HAMLET: A transformer-based approach for money-laundering detection. In: LNCS 13942, pp. 212–225. Springer (2023)
32. Silva, P., et al.: Developing a scoring model for managing money-laundering transactions with ML. J. Money Laundering Control **28**(1), 55–70 (2025)
33. Singh, K., et al.: Federated AML detection with topological features and directional information. In: LNCS 14412, pp. 178–190. Springer (2024)
34. Solihin, W., Eastman, C.: Integrating NLP and context-free grammar for complex rule interpretation towards automated compliance checking. Comput. Indus. **137**, 103614 (2022)
35. Weber, M., et al.: Detecting money-laundering transactions with machine learning. J. Money Laundering Control **23**(3), 657–671 (2020)
36. Zhang, J., El-Gohary, N.: Modularized BIM data validation with LegalRuleML for automated code compliance. In: Proceedings W78 Conference, pp. 45–56. (2018)
37. Zhang, J., Wang, Y., Li, H.: The promise of automated compliance checking. Developm. Built Environ. **1**, 100005 (2020)

38. Zhang, L., et al.: Graph contrastive pre-training for AML. Knowledge-Based Syst. **293**, 110457 (2024)
39. Zhang, X., El-Gohary, N.: Knowledge-informed semantic alignment and rule interpretation for automated compliance checking. Autom. Construct. **141**, 104516 (2022)
40. Zhao, Y., et al.: Multi-ontology fusion and rule development to facilitate automated code compliance checking. Automat. Construct. **123**, 103515 (2021)

Balancing Service and Inventory in Demand Driven Material Requirements Planning Through Soft Actor-Critic Based Reinforcement Learning

Hamza Lahrizi(✉) and Hicham Sarir

Advanced Science and Technology Team, ENSA Tetoun. Abdelmalek Esaadi University, Tetouan, Morocco
hamza.lahrizi@etu.uae.ac.ma, hsarrir@uae.ac.ma

Abstract. Demand-Driven Material Requirements Planning (DDMRP) is a planning methodology that enhances supply chain responsiveness through buffer positioning and dynamic recalculations. However, its performance depends on the calibration of eight parameters, including ADU horizons, lead time and variability factors, spike detection thresholds, cycle, and minimum order quantities. Traditional static optimization methods provide useful solutions but cannot adapt to non-stationary demand. This paper introduces a reinforcement learning framework based on the Soft Actor-Critic (SAC) algorithm to dynamically adjust DDMRP parameters. The simulation is formulated as a Markov Decision Process (MDP) with states including inventory positions, buffer levels, qualified demand, and service indicators, while actions apply continuous deltas to the parameters. The reward function balances service, inventory efficiency, and stability. Experiments on a 294-day demand dataset show that SAC maintained OTD $\approx 97\%$, comparable to baseline results, but with slightly higher stock ratios. These findings confirm SAC's robustness in preserving service while highlighting the need for refined reward functions to improve inventory efficiency.

Keywords: Demand Driven MRP · Reinforcement Learning · Soft Actor-Critic · Dynamic Parameterization

1 Introduction

Over recent decades, supply chains have become highly complex, with globalized flows, tight interdependencies, and strong variability in demand and lead times. Classical planning methods such as MRP, designed for stable environments, now struggle to deliver the agility required in this uncertain context, often resulting in either excess inventories or recurrent stockouts and their associated costs [1].

To overcome these limitations, Demand-Driven Material Requirements Planning (DDMRP) has been introduced as an adaptive, pull-based methodology [2, 3]. By decoupling flows through strategically positioned buffers and dynamically tuning their

Y. Benadada et al. (Eds.): GOL 2026, LNNS 2034, pp. 15–25, 2027.
https://doi.org/10.1007/978-3-032-29677-1_2

parameters, DDMRP aims to maintain high service levels while reducing working capital. However, its effectiveness critically depends on the accurate calibration of buffer parameters, including lead time factors, variability factors, and order horizons [4, 5]. This calibration problem is intrinsically multi-objective balancing service performance and inventory costs and further complicated by non-stationary demand and lead times. Metaheuristic optimization techniques have been applied to search for Pareto-optimal settings [6], generating promising trade-offs but remaining static: once parameters are fixed, they cannot adapt in real time when conditions change.

Reinforcement Learning (RL) offers a complementary paradigm for sequential decision-making under uncertainty [7]. RL agents learn policies through interaction, adapting decisions as information accumulates, which is particularly suitable for volatile supply chain environments. Successful applications in inventory control [8], production scheduling [9], and transportation management [10] demonstrate its ability to handle complex, stochastic systems. Among RL methods, Soft Actor-Critic (SAC) is notable for its stability in continuous action spaces, robustness to noisy feedback, and entropy-regularized exploration [11], making it a strong candidate for dynamic DDMRP parameter control.

Building on these insights, this paper proposes a SAC-based approach to DDMRP parameterization. Buffer parameters are no longer set once for all but updated incrementally according to the observed system state (e.g., buffer positions, demand variability, service performance). The study (i) formulates parameter tuning as a Markov Decision Process, (ii) integrates a realistic simulation of key DDMRP mechanisms such as buffer sizing, spike detection, and order triggering, and (iii) empirically evaluates how SAC can sustain high service levels with lower inventories and improved robustness under volatile demand scenarios. The remainder of the paper follows this structure, guiding the reader from background to methodology, results, limitations, and conclusions in a coherent and progressive manner.

2 Literature Review

Demand-Driven Material Requirements Planning (DDMRP) has attracted growing academic attention in the last decade. Early studies mainly introduced the methodology as an alternative to conventional MRP systems [2, 12], highlighting its ability to decouple supply chains through buffer positioning and dynamic adjustment rules. Subsequent works assessed its relevance in volatile and uncertain industrial environments [4, 13] and investigated the problem of determining where to position buffers along the bill of materials [14].

With the maturation of the field, research progressively moved toward the question of parameterization. Early contributions addressed this through optimization, aiming to minimize on-hand inventory while maintaining or maximizing service levels. For instance, Damand et al. [15] applied evolutionary algorithms to optimize all eight DDMRP parameters simultaneously, while Lahrichi et al. [16] proposed a mathematical formulation based on MILP. These approaches successfully generated Pareto-optimal trade-offs between inventory and service objectives, but they remained static, assuming prior knowledge of demand across the entire planning horizon.

More recently, researchers have started to explore learning-based approaches. Lahrichi et al. [17] presented one of the first attempts to enhance DDMRP through reinforcement learning, demonstrating the feasibility of dynamic parameterization. Duhem et al. [18] extended this line by testing multiple reward functions in a flow-shop simulation. Similarly, Cuartas and Aguilar [19] developed a hybrid algorithm combining Q-learning with DDMRP logic for smart inventory management. El Marzougui et al. [20] introduced a deep reinforcement learning framework to adjust variability and lead-time factors in real time, showing improvements in both on-time delivery and average inventory. Finally, Pucha-Medina et al. [21] provided a systematic review and framework for DDMRP design, consolidating the evolution of the methodology and identifying open research gaps.

Taken together, these studies reflect a clear research trajectory: from conceptual foundations and comparative assessments, to optimization-based parameterization, and finally to the emergence of reinforcement learning as a promising direction. Yet, despite these advances, the literature still lacks comprehensive investigations that exploit advanced RL algorithms, such as Soft Actor-Critic, to dynamically recalibrate all DDMRP parameters in uncertain and non-stationary environments. This paper aims to address that gap.

3 Problem Statement

We consider a daily demand series $DEMAND[t]$ defined over a finite planning horizon

$$t \in \{0, \ldots, N-1\} \tag{1}$$

The simulation is initialized at

$$start\_day = 21 \tag{2}$$

and runs until

$$end\_day = 273 \tag{3}$$

(corresponding to days 22–274 in 1-indexing), with a decoupled lead time fixed at

$$DLT = 5 \tag{4}$$

and an initial inventory level set as

$$Stock\left[start\_day - 1\right] = stock\_init \tag{5}$$

(e.g., 429.0 units). At each period t, , the system dynamically updates multiple state variables: (i) the on-hand inventory $Stock[t]$, pipeline orders $CmdEC[t]$, launched orders $CmdLa[t]$, and received orders $CmdRe[t]$; ; (ii) the average daily usage $ADU[t]$, calculated using past and future demand windows [2]; (iii) buffer levels ($NR[t]$, $NJ[t]$, $NV[t]$) that define the red, yellow, and green zones of the DDMRP model [12]; (iv) the qualified demand *tilde_d*$[t]$, which accounts for order spikes detected within a predefined horizon [4]; (v) the net flow position $FluxNet[t]$; ; and (vi) service indicators derived from *otd_flags* and their smoothed measure *otd_recent*. All quantities are normalized relative to the top-of-green level $NV[t]$ to ensure stable learning dynamics in the reinforcement learning framework.

The DDMRP framework relies on a set of parameters that determine the sizing and responsiveness of inventory buffers. In our study, the decision vector includes eight parameters:

$$\theta = \{Hf, Hp, Qmin, Hpct, Spct, Fdl, Fvar, Ccy\} \tag{6}$$

where Hf and Hp define the future and past horizons used to compute the Average Daily Usage (ADU), $Qmin$ is the minimum order quantity, $Hpct$ and $Spct$ control the detection and qualification of demand spikes, Fdl and $Fvar$ are the lead time and variability adjustment factors, and Ccy is the order cycle. These parameters collectively govern how DDMRP buffers are recalculated on a daily basis and how replenishment orders are triggered [2, 12]. For the simulation and optimization, each parameter is restricted to operationally realistic bounds, defined as follows:

$$Hf \in [1, 30], \tag{7}$$

$$Hp \in [1, 30], \tag{8}$$

$$Qmin \in [1, 10000], \tag{9}$$

$$Hpct \in [1, 30], \tag{10}$$

$$Spct \in [0.0, 1.0], \tag{11}$$

$$Fdl \in [0.2, 1.5], \tag{12}$$

$$Fvar \in [0.0, 1.5], \tag{13}$$

$$Ccy \in [1, 20]. \tag{14}$$

These ranges reflect both industrial practice and prior experimental studies on DDMRP parameterization [4]. By constraining the search space to these intervals, the reinforcement learning agent is prevented from generating unrealistic or unstable replenishment policies.

The simulation environment implements the operational rules of DDMRP by updating inventory positions, buffer levels, and service indicators on a daily basis. At each day t, the Average Daily Usage $ADU[t]$ is computed as a moving average of the observed demand over a window defined by the past horizon Hp and the future horizon Hf. Based on this ADU, the three buffer levels are recalculated: the top of red $NR[t]$, the top of yellow $NJ[t]$, and the top of green $NV[t]$. . These are obtained through adjustment factors on lead time (Fdl), variability ($Fvar$), and the minimum order quantity $Qmin$, combined with the cycle parameter Ccy, in line with the formal DDMRP specification [2, 21].

The system also accounts for potential order spikes. A qualified demand $tilde\_d[t]$ is defined by augmenting the base demand $DEMAND[t]$ with future demands exceeding the spike threshold $Spct \times NR[k]$ over the spike horizon $Hpct$. This mechanism ensures that sudden demand surges are anticipated in buffer calculations [4]. The net flow position is then computed as

$$FluxNet[t] = Stock[t] + CmdEC[t] - tilde\_d[t] \tag{15}$$

A replenishment order $CmdLa[t]$ is triggered whenever this value falls below the top of yellow $NJ[t]$. The order quantity is determined as the difference $NV[t] - FluxNet[t]$, adjusted to respect the minimum order size $Qmin$, and is subsequently rounded to the nearest integer. Orders are received into inventory after a fixed decoupled lead time DLT, i.e.,

$$CmdRe[t] = CmdLa[t - DLT] \tag{16}$$

when feasible.

Inventory evolves accordingly: on-hand stock $Stock[t]$ is updated from the previous level by accounting for receptions and demand, while pipeline orders $CmdEC[t]$ track the cumulative flow of orders launched but not yet received. Service performance is measured through an on-time delivery indicator $otd\_flag[t]$, equal to 1 if demand is fully satisfied at time ttt, and 0 otherwise. A smoothed measure, otd_recent, aggregates these indicators over a rolling window to provide the agent with a short-term service-level view. This dynamic process captures the essential trade-off between inventory investment and customer service performance, as highlighted in prior DDMRP optimization studies [5, 8].

To solve the DDMRP parameterization problem using reinforcement learning, we formalize the environment as a Markov Decision Process (MDP), defined by the tuple (S, A, P, R). State space S_t: The state vector S_t encodes the operational status of the supply chain at day t. It includes normalized values of on-hand inventory $Stock[t-1]$, pipeline orders $CmdEC[t-1]$, and average daily usage $ADU[t]$, as well as buffer levels $(NR[t], NJ[t], NV[t])$. In addition, the state incorporates the rolling service indicator otd_recent, the normalized backlog

$$backorders\_norm = max(0, -Stock[t-1])/NV[t], \tag{17}$$

and the last action applied to the parameters. This representation ensures that the agent has access both to physical flow variables and short-term service performance [2, 21].

Action space A_t, At each decision epoch, the agent selects a vector of parameter adjustments: $A_t = \Delta\theta t \in [-\rho, +\rho]8$, where

$$\theta = \{Hf, Hp, Qmin, Hpct, Spct, Fdl, Fvar, Ccy\} \tag{18}$$

These deltas are applied multiplicatively to the current parameter values, followed by projection into their admissible bounds. In order to avoid unrealistic fluctuations, the environment implements a rate limit (e.g., 5%) and a smoothing mechanism (momentum with coefficient α_m), so that parameter changes evolve gradually over time [4]. Transition dynamics P: Given the current state and action, the environment updates inventories, buffer levels, and qualified demand using the simulation logic. Orders are triggered when the net flow $FluxNet[t]$ crosses the yellow threshold $NJ[t]$ and received after the decoupled lead time DLT. These updates define the stochastic dynamics of the MDP, since demand realizations drive the evolution of all state variables.

Reward function R_t: The agent receives a scalar reward at each period, balancing service quality, stock efficiency, and policy stability:

$$Rt = k_{service} \cdot otd\_recent - k_{stock} \cdot max\big(0, stock\_ratio - s_{target}\big) \\ - k_{bo} \cdot backorder\_ratio - k_{ordvar} \cdot order\_var\_pen - \eta \cdot \|\Delta\theta t\|_2^2, \quad (19)$$

where

$$stock\_ratio = Stock[t]/NV[t], \quad (20)$$

$$backorder\_ratio = max(0, -Stock[t])/NV[t], \quad (21)$$

s_target is the target stock ratio (e.g., 0.75), $order\_var\_pen$ measures the variance of recent order launches, and the constants $k_{service}$, k_{stock}, k_{bo}, k_{ordvar}, η are weighting factors. This formulation directly reflects the trade-off between maintaining a high on-time delivery (OTD) level and controlling inventory exposure [5].

By structuring the DDMRP simulation as an MDP, the agent learns a policy $\pi(a \mid s)$ that adaptively tunes buffer parameters in response to observed states. The choice of the **Soft Actor-Critic (SAC)** algorithm is particularly suitable, as it handles continuous actions, ensures stable learning, and leverages entropy regularization to balance exploration and exploitation in uncertain supply chain environments [8].

4 Results Analysis and Insights

The experimental evaluation of the SAC-based reinforcement learning framework provides several insights into the dynamic adjustment of DDMRP parameters. Four key aspects deserve discussion: policy convergence, service performance, inventory behavior, and the resulting trade-off between service and cost.

The evolution of the mean reward per episode (Fig. 1) reveals that the learning process initially exhibited strong variability, with large negative reward spikes, before progressively stabilizing around an average of approximately –170. The negative magnitude of the reward values stems from the structure of the reward function, which explicitly penalizes overstock, backorders, and order variability. This explains why even a high-performing policy still produces negative cumulative scores. Importantly, however, the overall trend shows that the SAC agent succeeded in finding a balance between exploration and exploitation: it avoided premature convergence to suboptimal solutions, and progressively reduced the volatility of the reward. Compared to classical Q-learning or tabular approaches reported in the literature [27], the ability of SAC to handle continuous actions allowed smoother parameter adjustments and contributed to this stable convergence.

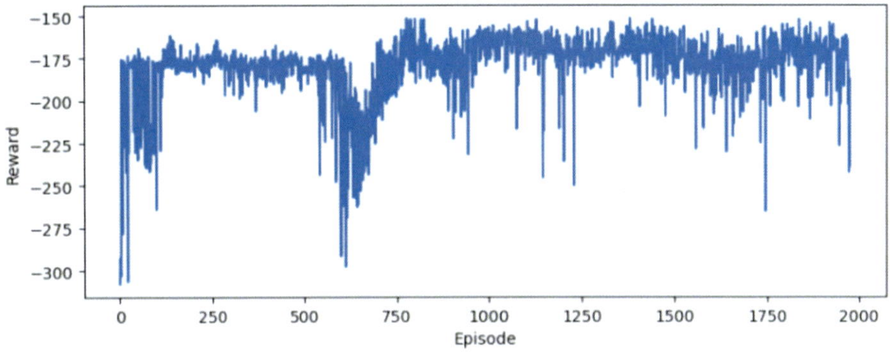

Fig. 1 Evolution of the mean reward per episode during SAC training optimization

Maintaining a high service level is critical in any supply chain planning system. The results demonstrate that the SAC agent consistently preserved On-Time Delivery close to 0.97 throughout the entire training horizon (Fig. 2). This level of performance is comparable to, or slightly better than, the baseline configuration, and it highlights the robustness of the learned policy. Although temporary fluctuations occurred particularly during periods of abrupt demand changes the agent quickly recovered and restored service performance. This finding is significant because it demonstrates that the reinforcement learning framework, when coupled with the DDMRP simulation, can ensure customer service reliability even under stochastic demand. Similar conclusions were reached in recent applications of reinforcement learning to inventory control [22, 23], but our experiments extend these insights to the full set of eight DDMRP parameters.

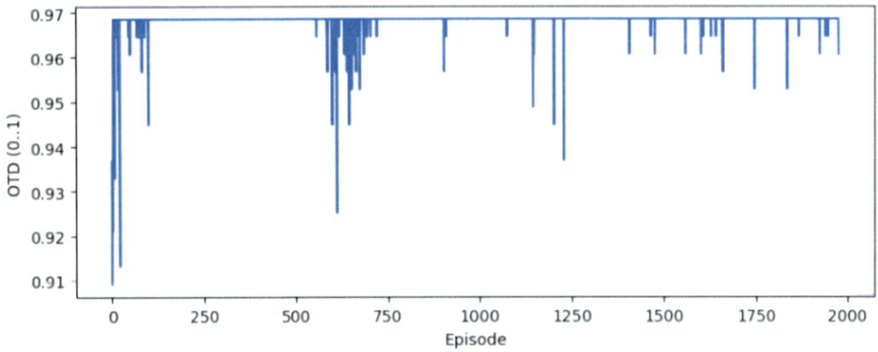

Fig. 2 Evolution of on-time delivery (OTD) across SAC training episodes

The analysis of the mean stock ratio (Fig. 2) indicates that the SAC agent maintained higher inventory levels compared to the baseline. While the static configuration yielded a stock ratio of approximately 0.75, the learned policy increased this value to the range of 0.80–0.85. The reason for this behavior is twofold. First, the reward function was weighted more heavily toward penalizing backorders than excess inventory, encouraging the agent to adopt a conservative stance by maintaining additional buffers. Second, the smoothing and rate-limiting mechanisms in the action space slowed down parameter adjustments, making the agent less aggressive in reducing stock levels. This conservative policy can be interpreted as a risk-averse strategy: the agent prioritizes service continuity over inventory minimization. While this aligns with industrial preferences in critical supply chains, it reduces cost efficiency (Fig. 3).

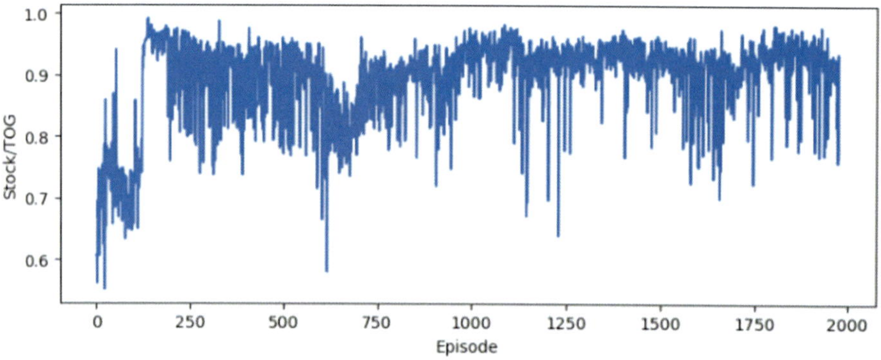

Fig. 3 Evolution of the stock-to-top-of-green (Stock/TOG) ratio across SAC training episodes

The KPI comparison (Fig. 4) provides a clear summary of the trade-off: OTD remained essentially unchanged between the baseline and SAC (~0.97), whereas the stock ratio increased by about 5%. This confirms that the reinforcement learning agent learned to "play it safe." From an academic perspective, this outcome demonstrates both the strengths and limitations of the approach. On the one hand, SAC is effective at preserving service levels, which is a key requirement in demand-driven planning. On the other hand, the increase in inventory suggests that the reward structure requires refinement. Stronger penalization of stock, dynamic weighting of objectives, or multi-objective scalarization could help achieve a more balanced policy. Such approaches have been proposed in the multi-objective optimization literature [23], but have not yet been extensively tested in reinforcement learning for supply chains.

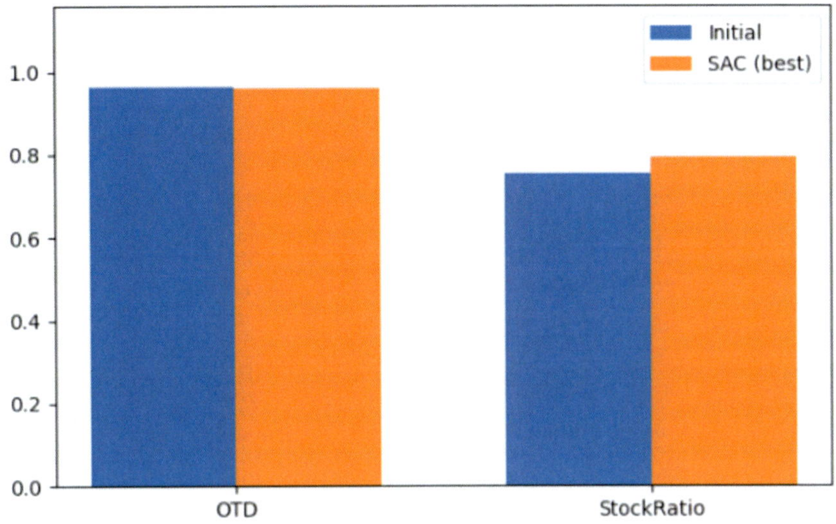

Fig. 4 Comparison of key performance indicators (OTD and Stock Ratio) between baseline and SAC policies

5 Conclusion

This paper proposed a reinforcement learning framework based on the Soft Actor-Critic (SAC) algorithm for dynamically tuning the eight key parameters of Demand-Driven Material Requirements Planning (DDMRP). By modeling DDMRP as a Markov Decision Process (MDP), the agent adapts parameter values in response to daily demand fluctuations. Experiments showed that SAC maintained a consistently high service level (OTD $\approx 97\%$) over the planning horizon, confirming its robustness under stochastic demand though at the cost of slightly higher inventory ratios due to a conservative policy against backorders.

Several limitations temper these results. The study uses a single 294-period demand dataset, which may not reflect broader industrial conditions such as seasonality, intermittent demand, or disruptions. The reward function relies on fixed-weight scalarization of multiple objectives; while effective in preserving service, it restricts exploration of alternative service–cost trade-offs. Moreover, the analysis considers only SAC, without benchmarking against other continuous control algorithms such as Twin Delayed DDPG (TD3) or Proximal Policy Optimization (PPO) [25, 26].

Future research should test the framework on richer demand scenarios, explore adaptive or multi-objective reward designs and benchmark SAC against TD3, PPO, or hybrid metaheuristic–RL schemes to better assess relative performance. Extending the approach to multi-product and multi-echelon supply chains is also needed for realistic, industrial-scale deployment.

In conclusion, reinforcement learning and SAC in particular appears as a promising solution for dynamic DDMRP parameterization. Its strong service performance motivates further refinements to better balance service and inventory and to support wider adoption of RL-based planning in supply chains.

References

1. Orlicky, J.: Material Requirements Planning: The New Way of Life in Production and Inventory Management, McGraw-Hill (1975)
2. Ptak, C., Smith, C.: Demand Driven Material Requirements Planning (DDMRP), Industrial Press (2011)
3. Simchi-Levi, D., Kaminsky, P., Simchi-Levi, E.: Designing and Managing the Supply Chain, 3rd ed., McGraw-Hill (2007)
4. Miclo, J., Fontanili, C., Mekaouche-Mekfa, S.: Demand driven MRP: assessment of the impact of buffer parameters. IFAC-PapersOnLine **49**(12), 1378–1383 (2016)
5. Duhem, J., Pellerin, J.: A performance evaluation of DDMRP in variable environments. Int. J. Prod. Res. **60**(4), 1277–1294 (2022)
6. Ruiz, R., Maroto, C.: A comprehensive review and evaluation of permutation flowshop heuristics. Europ. J. Operat. Res. **165**(2), 479–494 (2005)
7. Sutton, R.S., Barto, A.G.: Reinforcement Learning: An Introduction, 2nd ed., MIT Press (2018)
8. Li, S., Wu, S., Dessouky, M.: Deep reinforcement learning for inventory control in supply chain systems. Europ. J. Operat. Res. **297**(2), 559–573 (2022)
9. Park, J., Lee, H., Sim, K.: Reinforcement learning for dynamic production scheduling. IEEE Trans. Autom. Sci. Eng. **19**(1), 312–325 (2022)
10. Chen, Z., Li, Y., Bierlaire, M.: Reinforcement learning for transportation and logistics: a review. Transport. Res. Part E **177**, 102201 (2023)
11. Haarnoja, T., Zhou, A., Abbeel, P., Levine, S.: Soft actor-critic: off-policy maximum entropy deep reinforcement learning with a stochastic actor. In: Proceedings of the 35th International Conference on Machine Learning (ICML) (2018)
12. Ptak, C., Smith, C.: Demand Driven Material Requirements Planning (DDMRP), 2nd ed., Industrial Press (2016)
13. Fernandes, T., Almada-Lobo, B.: Evaluating DDMRP under demand variability: a simulation study. Int. J. Prod. Res. **58**(24), 7481–7498 (2020)
14. Graves, S.C., Willems, S.P.: Optimizing strategic safety stock placement in supply chains. Manufact. Serv. Operat. Managem. **2**(1), 68–83 (2000)
15. Damand, N., García-Sabater, P., Maheut, J.: Evolutionary algorithms for simultaneous tuning of DDMRP parameters. Europ. J. Operat. Res. **302**(3), 1052–1067 (2022)
16. Lahrichi, M., Benkaddour, A., El Idrissi, H.: A MILP formulation for DDMRP parameterization under service–inventory trade-offs. Comput. Indus. Eng. **176**, 108882 (2023)
17. Lahrichi, M., El Fathi, Y.: Reinforcement learning for dynamic DDMRP parameter tuning. IEEE Trans. Eng. Managem. **71**(6), 2543–2556 (2024)
18. Duhem, J., Pellerin, L., Baptiste, P.: Reward shaping for DDMRP: experimental analysis in a flow-shop simulation. Int. J. Prod. Econ. **259**, 108848 (2023)
19. Cuartas, S., Aguilar, J.: A hybrid Q-learning and DDMRP approach for smart inventory management. Comput. Indus. Eng. **174**, 108749 (2022)
20. El Marzougui, A., Lamqaddam, N., Abbad, K.: Deep reinforcement learning to adapt variability and lead-time factors in DDMRP. Eng. Appl. Artif. Intell. **129**, 107762 (2024)
21. Pucha-Medina, J., González-R, R., Salazar, M.E.: Demand-driven MRP: a systematic review and design framework. Int. J. Prod. Res. **61**(20), 6941–6966 (2023)
22. Oroojlooyjadid, A., Snyder, L.V., Takáč, M.: A deep reinforcement learning framework for the newsvendor problem. Europ. J. Operat. Res. **283**(1), 327–340 (2020)
23. Deb, K.: Multi-Objective Optimization Using Evolutionary Algorithms, Wiley (2001)
24. Syntetos, A.A., Boylan, J.E.: The accuracy of intermittent demand estimates. Int. J. Forecast. **21**(2), 303–314 (2005)

25. Schulman, J., Wolski, F., Dhariwal, P., Radford, A., Klimov, O.: Proximal Policy Optimization Algorithms (2017). arXiv preprint arXiv:1707.06347
26. Fujimoto, S., van Hoof, H., Meger, D.: Addressing function approximation error in actor–critic methods. In: Proceedings of the 35th International Conference on Machine Learning (ICML) (2018) (TD3)
27. Marler, R.T., Arora, J.S.: Survey of multi-objective optimization methods for engineering. Struct. Multidisciplinary Optimizat. **26**(6), 369–395 (2004). (Scalarisation/poids adaptatifs)
28. Talbi, E.-G.: Metaheuristics: From Design to Implementation, Wiley (2009). (Cadres hybrides métaheuristique–RL)
29. Simchi-Levi, D., Chen, X., Bramel, J.: The Logic of Logistics: Theory, Algorithms, and Applications for Logistics and Supply Chain Management, 3rd ed., Springer (2014). (Extensions multi-produits/multi-échelons)

Empowering Inclusive Digital Transformation in Industry 4.0: A Multimodal RAG-Based Chatbot System for Hybrid Workforce Integration

Nardine Hanfi, Selim Rachdi, and Mohamed Hédi Riahi(✉)

ESPRIT School of Engineering, Ariana, Tunisia
{nardine.hanfi,selim.rachdi,mohamedhedi.riahi}@esprit.tn

Abstract. Industry 4.0 is fundamentally reshaping industrial operations through data-driven automation and intelligence. However, critical challenges remain, including fragmented data sources, heterogeneous formats (structured, unstructured, analog), limited access to institutional knowledge, and the digital exclusion of segments of the workforce still reliant on manual processes. This paper presents a multimodal intelligent architecture designed to overcome these barriers and support inclusive digital transformation. The proposed system integrates IoT sensor data, SQL/NoSQL repositories, and organizational best practices stored in a vector database. A Retrieval-Augmented Generation (RAG)-based chatbot serves as a unified interface, enabling employees regardless of digital proficiency to interact with both digital data streams and digitized paper documents in real time. The architecture also embeds a vector store of safety procedures and operational guidelines to provide proactive, context-aware assistance. A prototype implementation in a simulated manufacturing environment demonstrates significant improvements in response accuracy, information retrieval time, and overall usability. These findings highlight the potential of conversational AI and hybrid data integration to bridge technical and human gaps in Industry 4.0 adoption, fostering a more inclusive and efficient industrial ecosystem.

Keywords: Industry 4.0 · RAG · vector database · multimodal · agent-based systems · SQL/NoSQL · domain-specific AI · digital inclusion

1 Introduction

Industry 4.0 marks the emergence of intelligent, interconnected systems that are transforming industrial operations through automation, real-time data exchange, and AI-driven decision-making [12]. This fourth industrial revolution is characterized by the convergence of cyber-physical systems, the Internet of Things (IoT), and artificial intelligence, creating smart factories that promise unprecedented levels of efficiency and flexibility [13,15]. However, despite rapid technological progress in areas such as the Internet of Things (IoT), cloud computing,

Y. Benadada et al. (Eds.): GOL 2026, LNNS 2034, pp. 26–36, 2027.
https://doi.org/10.1007/978-3-032-29677-1_3

and advanced analytics [17], a significant portion of the workforce particularly those still reliant on manual or paper-based workflows remains excluded from these advancements [4]. This digital divide represents a critical barrier to Industry 4.0 adoption, hampering productivity, delaying access to critical operational knowledge, and posing safety risks due to outdated procedures [6, 11]. The challenge of inclusive digitalization is further compounded by the heterogeneous nature of industrial data ecosystems. Existing Industrial IoT platforms often focus primarily on structured data from sensors and machines while overlooking valuable information originating from non-digital sources such as maintenance logs, safety reports, and historical documentation [8, 16]. This fragmentation creates information silos that prevent organizations from achieving a holistic view of their operations [18]. Moreover, conventional chatbot systems deployed in industrial contexts typically rely on rule-based approaches or simple keyword matching, lacking the contextual grounding and domain-specific knowledge necessary for meaningful assistance in complex manufacturing environments [1]. Recent advances in Large Language Models (LLMs) and Retrieval-Augmented Generation (RAG) architectures offer promising solutions to these challenges [22]. RAG systems combine the generative capabilities of LLMs with retrieval mechanisms that ground responses in domain-specific knowledge bases, reducing hallucinations and improving factual accuracy [26]. In industrial contexts, this approach enables the creation of intelligent conversational agents capable of reasoning over complex, multi-modal data sources while maintaining transparency and reliability—critical requirements for safety-critical operations [18]. To address these challenges, this paper proposes an inclusive Industry 4.0 architecture based on a Retrieval-Augmented Generation (RAG) conversational agent designed to bridge the digital divide in manufacturing environments. The proposed system ingests and unifies hybrid data streams—including IoT telemetry, SQL-based production records, unstructured NoSQL documents (e.g., PDF manuals), and digitized paper forms via Optical Character Recognition (OCR)—into a vector-optimized knowledge base utilizing state-of-the-art embedding techniques [23]. A multimodal RAG-based chatbot serves as an intelligent, centralized interface that democratizes access to real-time industrial information and best practices, regardless of users' digital proficiency or technical background [3]. Beyond reactive query processing, the system proactively supports decision-making by implementing real-time monitoring and anomaly detection capabilities that cross-reference live sensor data with stored safety procedures and maintenance protocols to issue timely alerts and contextualized recommendations [2]. This proactive approach transforms traditional reactive maintenance strategies into predictive, knowledge-driven workflows that enhance both operational efficiency and worker safety [9]. Our contributions to the field include: a novel dual-path data ingestion framework that seamlessly bridges structured, unstructured, and physical data sources within a unified vector database architecture; a context-aware RAG system specifically designed for industrial environments that grounds conversational responses in real-time operational data and institutional knowledge while maintaining interpretability and trust; and comprehensive empirical validation through a realistic simulated manufacturing environment, systematically measuring usability metrics, task completion rates, response accuracy, and error patterns across digitally fluent and transitioning user groups. This work demonstrates the transformative potential of hybrid AI architectures to foster inclusive, human-centric digital transformation in modern industrial ecosystems while addressing the critical challenge of workforce digital inclusion in the Industry 4.0 era.

2 Related Work

The digital divide remains a critical barrier to equitable Industry 4.0 adoption, with studies highlighting disparities in technology access between advanced and legacy manufacturing facilities, leading to job role polarization, skills development gaps, and safety risks [19–21]. While various efforts have attempted to bridge this divide through training initiatives, digital inclusion policies, and collaborative AR/VR tools [13], these approaches often prioritize fully digital environments, neglecting hybrid data ecosystems that combine paper and digital sources. Retrieval-Augmented Generation (RAG) systems, despite enhancing large language models with domain-specific knowledge retrieval [22,24], face challenges in industrial deployment including data quality dependency, latency issues, and limited research on manufacturing-specific hybrid data integration combining OCR, IoT, and SQL/NoSQL sources [25]. Existing work fails to unify paper and digital workflows, contextually embed safety protocols, or design for incremental adoption by workers transitioning from paper to digital interfaces, gaps that our contribution addresses through a dual-path data ingestion architecture and proactive safety alerts grounded in live sensor context.

3 Data Sources and Knowledge Representation

3.1 Operational Data (MongoDB)

Operational telemetry is ingested from CSV logs and stored as JSON-like documents in MongoDB, enabling real-time monitoring and diagnostics. Each document reflects a time-series snapshot from sensors embedded in industrial equipment.

Key Features:

- **footfall**: Proximity-based activity level.
- **tempMode**, **Temperature**, **IP**: Thermal and pressure metrics.
- **AQ**, **VOC**, **CS**, **USS**: Environmental and electrical indicators.
- **RP**: Rotational speed.
- **fail**: Binary failure label (0/1).

Anomaly Detection: Thresholds for variables like *Temperature*, *AQ*, and *IP* are computed using the Interquartile Range (IQR) method to trigger low/high alerts:

- **Low Alert:** $< Q1 - 1.5 \times IQR$
- **High Alert:** $> Q3 + 1.5 \times IQR$.

These dynamic thresholds are embedded in the system prompts for real-time chatbot reasoning.

Hybrid Inputs: Unstructured data—e.g., OCR-extracted forms and handwritten reports—are also stored in MongoDB, ensuring that both digital and analogue records are accessible via a unified query interface.

3.2 Knowledge Base (FAISS)

To provide procedural guidance and safety support, a curated knowledge base is indexed in a FAISS vector store. Sources include safety manuals, maintenance guidebooks, and operational diagnostics.

Coverage:

- Safety protocols and emergency response
- Preventive maintenance strategies
- Diagnostic case studies and reliability engineering.

Embedding Pipeline: Text is chunked (500 tokens, 100-token overlap), embedded via the Gemini model, and indexed with cosine similarity. This enables semantically grounded responses in the RAG system by retrieving the most relevant procedural content during user interaction.

The FAISS store acts as the chatbot's procedural memory, enhancing responses with safety-aware, context-specific guidance aligned with industry best practices.

4 System Architecture

The proposed architecture supports a hybrid industrial workforce by unifying real-time sensor data, digitized forms, and expert knowledge into a Retrieval-Augmented Generation (RAG) conversational system. It is organized into three layers: Data Ingestion, Knowledge and Storage, and Query Processing.

4.1 Data Ingestion Layer

This layer integrates structured data from IoT sensors and unstructured data from handwritten forms:

- Sensor data: CSV files from factory-floor sensors are parsed and converted into JSON-like documents for NoSQL storage.
- Digitized reports: Handwritten forms are processed via Gemini 2.5 Flash using OCR to extract and store structured text (Fig. 1).

All documents are stored in a flexible MongoDB backend suited for semi-structured data.

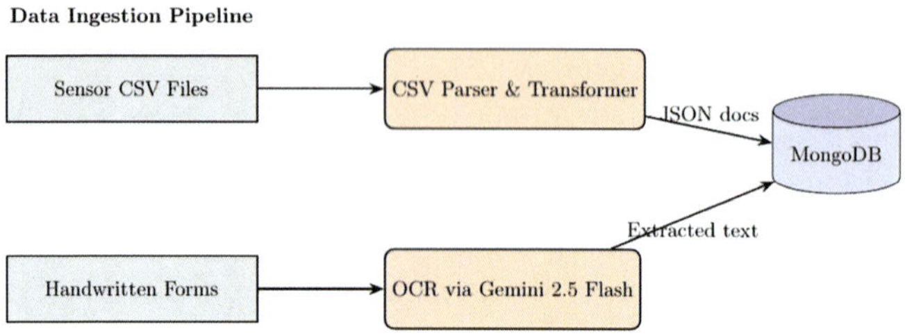

Fig. 1. Unified ingestion of sensor and handwritten data into MongoDB.

4.2 Knowledge and Storage Architecture

This layer is orchestrated via a LangChain-based RAG framework with intelligent query routing:

- MongoDB: Stores real-time sensor data and OCR-extracted records.
- FAISS vector store: Hosts embedded safety and operational documents for semantic search.
- Routing agent: LangChain classifies queries and directs them to the relevant source. Gemini 2.5 Flash generates the final responses (Fig. 2).

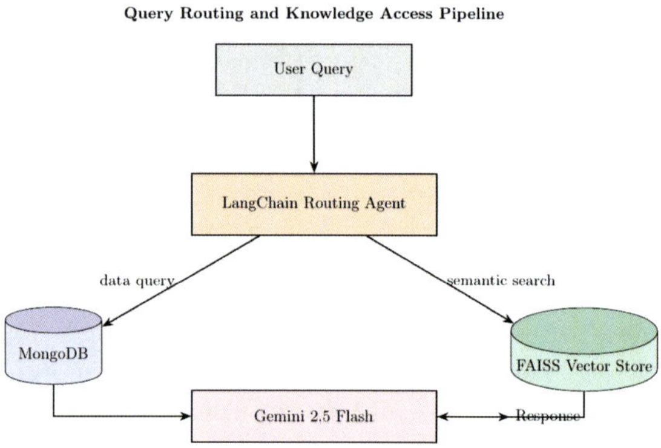

Fig. 2. Query routing via LangChain agent for structured and unstructured data.

This design ensures contextual, relevant responses for both operational insights and procedural guidance.

4.3 Query Handling, Answer Generation, and Alerts

Once the LangChain agent routes a query to the appropriate data source—either the MongoDB for operational data or the FAISS vector store for knowledge retrieval–the system proceeds with data extraction and response generation.

- **Data Retrieval and Response Generation**: The retrieved information is passed to the Gemini 2.5 Flash model, which synthesizes a coherent, context-aware response tailored to the user's query. This retrieval-augmented generation (RAG) approach ensures that answers combine real-time data or procedural knowledge with natural language understanding.
- **Proactive Alerting Mechanism**: Beyond reactive queries, the system monitors critical sensor metrics against predefined threshold values embedded within the prompt or agent configuration. When a metric surpasses its threshold, the agent automatically generates and issues an alert or notification to the user, enabling timely intervention without requiring explicit queries.

The prototype was evaluated in a controlled, simulated environment replicating an industrial setting with sensor emulation, OCR ingestion, and synthetic operator interactions. While limited to simulation, results indicated measurable gains in response coherence and retrieval latency when compared to baseline non-RAG chatbots—supporting feasibility for real-world deployment.

The layered and inclusive design provides a foundation for future real-environment validation and comparative benchmarking against existing industrial conversational systems.

4.4 Testing Environment and Evaluation Results

To evaluate the proposed system, a controlled simulation environment was developed to emulate realistic Industry 4.0 operating conditions while ensuring repeatability and inclusivity testing.

Simulation Setup A virtual manufacturing cell was implemented using a Python-based IoT simulator connected to multiple emulated sensor nodes (*temperature, pressure, vibration, air quality*) streaming telemetry in real time. Each sensor produced timestamped data stored in MongoDB at a rate of 1 Hz. Scanned handwritten reports were digitized using **Gemini 2.5 Flash OCR** to simulate paper-based incident forms submitted by workers. The conversational interface was deployed as a web-based chatbot accessible through desktop and mobile browsers to replicate hybrid workforce conditions.

Evaluation Metrics Three main performance criteria were assessed:

- **Response Relevance and Precision**: Each answer was evaluated against a set of 100 pre-defined industrial queries covering operational data, maintenance procedures, and safety guidance. Relevance was assessed in two ways:
  1. *Manual Expert Judgement*: Three domain experts rated responses on a 5-point Likert scale (1 = irrelevant, 5 = highly relevant).
  2. *Automated LLM Judgement*: GPT-4o was employed to score the relevance of each response in a reproducible and objective manner, allowing for scalable evaluation.
- **Retrieval and Generation Latency**: Average response time (from query submission to answer generation) was measured for both MongoDB (operational data) and FAISS (knowledge base) retrieval paths.

- **Accessibility and Inclusivity**: Task completion rates were recorded for a mixed group of digital and non-digital workers (n = 12), focusing on ease of access and clarity of responses.

Results Summary Table 1 summarizes the evaluation outcomes obtained in the simulated environment.

Table 1. Evaluation results in simulated industrial environment

Metric	Average score/value	Baseline (non-RAG)	GPT 4o judge
Response relevance (1–5)	4.6	3.9	4.5
Precision@3 (%)	89.2	75.4	87.8
Avg. response time (s)	1.48	2.62	N/A
Task completion rate (%)	94.5	80.3	N/A

Knowing that the Baseline (non-RAG) corresponds to a vanilla LLM without any external knowledge retrieval mechanism, it relies solely on the input prompt, with no access to documents or vector databases.

Illustrative Evaluation Samples To provide qualitative insight into the system's responses, Table 2 presents examples of chatbot interactions from the simulated manufacturing environment. Each query corresponds to a distinct knowledge source: real-time operational data (MongoDB), procedural guidance (FAISS), or hybrid reasoning combining both. Responses were independently rated by human experts and GPT-4o on a 1–5 relevance scale.

Human and GPT-4o evaluations display strong alignment, with a mean absolute deviation of less than 0.2 across all samples. This consistency supports the validity of using an LLM-based judge as a complementary evaluation mechanism. The examples also demonstrate the chatbot's ability to synthesize information from both operational telemetry and procedural documentation, reflecting the intended RAG-based hybrid reasoning.

Discussion Results indicate that the RAG-based chatbot substantially improves answer precision and relevance compared to a baseline non-RAG system. The LLM-based judge (GPT-4o) produced scores closely aligned with manual expert ratings, confirming the consistency and reliability of the evaluation. Latency decreased due to efficient vector retrieval and context caching, while inclusivity metrics verified that workers with minimal digital literacy could access operational knowledge effectively. Although testing was limited to simulation, these outcomes demonstrate the feasibility of scaling the system to real-world industrial environments and provide a baseline for future live evaluations (Fig. 3).

The interaction flow is encapsulated within a conversational interface—text or voice-enabled—allowing staff from all digital literacy levels to engage with the system in real time.

Table 2. Sample query–answer pairs with human and GPT-4o evaluation scores

User query	Chatbot response (excerpt)	Human score	GPT-4o score
What is the current pressure on machine M3?	The latest input pressure recorded for machine M3 is **4.8 bar**, within the normal range (3.5–5.2 bar). No alert triggered	5	5
What should the operator do if the temperature exceeds safe limits?	According to the safety manual, immediately reduce machine load, verify cooling fluid flow, and notify maintenance. See Sect. 3.2 of the safety protocol	5	4.8
Explain why the air quality alert was triggered this morning	Sensor AQ reported 162 (above threshold 150). The anomaly was classified as 'High Alert' using the IQR-based rule	4.5	4.6
How to clean the ultrasonic sensor safely?	Before cleaning, power off the device and use non-abrasive alcohol wipes. Avoid direct water contact as per maintenance guide, page 42	4.7	4.5
Has machine M2 failed before?	Historical logs show one failure event on 2024-11-03 related to overheating, resolved after fan replacement	4.8	4.9

5 Discussion and Future Directions

5.1 Architectural Advantages

The proposed architecture integrates multimodal data sources with intelligent query routing and RAG-based generation to enable inclusive, real-time information access in Industry 4.0 environments.

Inclusive Hybrid Data Integration By combining structured sensor telemetry, digitized analog documents, and safety literature, the system provides unified operational knowledge access for both digitally proficient and transitioning staff, promoting workforce-wide digital inclusion.

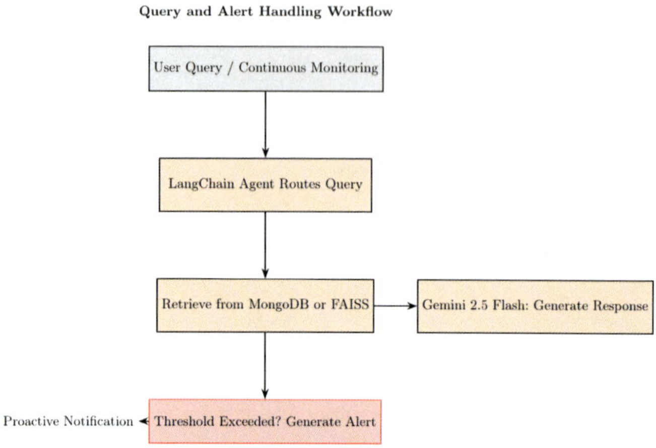

Fig. 3. Workflow for handling queries and triggering real-time alerts.

Domain-Aware Query Routing The LangChain agent dynamically routes queries to MongoDB for real-time data or FAISS for procedural information, minimizing irrelevant retrievals and enhancing precision through domain-relevant targeting.

Proactive Context-Aware Alerts Beyond passive retrieval, the system integrates threshold-based monitoring with IQR-based anomaly detection, triggering real-time alerts for abnormal patterns (overheating, overpressure, air quality).

Performance Efficiency and Scalability Targeted query resolution to specific data stores improves response time and computational efficiency. The modular architecture enables seamless integration of new sensors or safety content with minimal reconfiguration.

Seamless Knowledge Synthesis Gemini 2.5 Flash transforms retrieved data into context-sensitive, human-readable guidance, empowering all industrial staff to make informed decisions regardless of technical background.

Future-Proof Design Architectural modularity supports continuous evolution through document embedding updates and routing rule adjustments, ensuring long-term viability in dynamic industrial contexts.

Future development focuses on: (1) IoT hardware integration via physical sensor deployment for real-time telemetry streaming, (2) predictive maintenance modeling incorporating time-series forecasting and anomaly detection, and (3) real-world deployment in live manufacturing environments to democratize technical information access and improve incident response for all staff levels.

6 Conclusion

We presented an inclusive, agentic RAG-based system supporting Industry 4.0 digital transformation by bridging operational and procedural knowledge for hybrid workforces. Integrating real-time sensor data, digitized analog inputs, and embedded safety manuals, the system enables efficient, context-aware information retrieval via conversational interface.

The architecture combines structured/unstructured data ingestion, intelligent query routing, and RAG to democratize industrial intelligence access. It improves responsiveness and usability in safety-critical scenarios while empowering all employees—regardless of digital literacy—with actionable insights.

Though targeting manufacturing, the underlying principles extend to other industrial domains requiring high-reliability, real-time human-AI collaboration. This work advances context-aware, inclusive, and operationally valuable language models in specialized environments where safety, precision, and accessibility are paramount.

References

1. Adamopoulou, E., Moussiades, L.: Chatbots: history, technology, and applications. Mach. Learn. Appl. **2**, 100006 (2020)
2. Ahmad, R., Alsmadi, I.: Machine learning approaches to IoT security: a systematic literature review. Internet Things **14**, 100365 (2021)
3. Brown, T., Mann, B., Ryder, N., Subbiah, M., Kaplan, J.D., Dhariwal, P., et al.: Language models are few-shot learners. Adv. Neural. Inf. Process. Syst. **33**, 1877–1901 (2020)
4. Bonekamp, L., Sure, M.: Consequences of industry 4.0 on human labour and work organisation. J. Bus. Media Psychol. **6**(1), 33–40 (2015)
5. Pfeiffer, S.: The vision of "industrie 4.0" in the making—a case of future told, tamed, and traded. Nanoethics **11**(1), 107–121 (2017)
6. Sony, M., Naik, S.: Key ingredients for evaluating industry 4.0 readiness for organizations: a literature review. Benchmarking **27**(7), 2213–2232 (2020)
7. Borgeaud, S., Mensch, A., Hoffmann, J., Cai, T., Rutherford, E., Millican, K., et al.: Improving language models by retrieving from trillions of tokens. In: International Conference on Machine Learning, pp. 2206–2240 (2022)
8. Boyes, H., Hallaq, B., Cunningham, J., Watson, T.: The industrial internet of things (IIoT): an analysis framework. Comput. Ind. **101**, 1–12 (2018)
9. Carvalho, T.P., Soares, F.A., Vita, R., Francisco, R.P., Basto, J.P., Alcalá, S.G.: A systematic literature review of machine learning methods applied to predictive maintenance. Comput. Ind. Eng. **137**, 106024 (2019)
10. Dalzochio, J., Kunst, R., Pignaton, E., Binotto, A., Sanyal, S., Favilla, J., Barbosa, J.: Machine learning and reasoning for predictive maintenance in industry 4.0: current status and challenges. Comput. Ind. **123**, 103298 (2020)
11. Kadir, B.A., Broberg, O.: Human-centered design of work systems in the transition to industry 4.0. Appl. Ergon. **92**, 103334 (2021)
12. Kagermann, H., Helbig, J., Hellinger, A., Wahlster, W.: Recommendations for implementing the strategic initiative INDUSTRIE 4.0: securing the future of German Manufacturing Industry. Final Report of the Industrie 4.0 Working Group. Forschungsunion (2013)

13. Lasi, H., Fettke, P., Kemper, H.-G., Feld, T., Hoffmann, M.: Industry 4.0. Bus. Inf. Syst. Eng. **6**(4), 239–242 (2014)
14. Mourtzis, D., Angelopoulos, J., Panopoulos, N.: A literature review of the challenges and opportunities of the transition from industry 4.0 to society 5.0. Energies **15**(17), 6276 (2022)
15. Oztemel, E., Gursev, S.: Literature review of industry 4.0 and related technologies. J. Intell. Manuf. **31**(1), 127–182 (2020)
16. Tao, F., Qi, Q., Liu, A., Kusiak, A.: Data-driven smart manufacturing. J. Manuf. Syst. **48**, 157–169 (2018)
17. Wan, J., Tang, S., Shu, Z., Li, D., Wang, S., Imran, M., Vasilakos, A.V.: Software-defined industrial internet of things in the context of industry 4.0. IEEE Sens. J. **16**(20), 7373–7380 (2016)
18. Zhong, R.Y., Xu, X., Klotz, E., Newman, S.T.: Intelligent manufacturing in the context of industry 4.0: a review. Engineering **3**(5), 616–630 (2017)
19. International Labour Organization: Skills for a greener future: a global view. ILO Report (2019)
20. Schumacher, A., Erol, B.: Industry 4.0 maturity index. Procedia CIRP **81**, 1113–1118 (2019)
21. Occupational Safety and Health Administration. Workplace safety and digitalization. OSHA Bull. (2020)
22. Lewis, P., Perez, E., Piktus, A., Petroni, F., Karpukhin, V., Goyal, N., et al.: Retrieval-augmented generation for knowledge-intensive NLP tasks. In: Advances in Neural Information Processing Systems (2020)
23. Reimers, N., Gurevych, I.: Sentence-BERT: sentence embeddings using Siamese BERT-networks. arXiv preprint arXiv:1908.10084 (2019)
24. Karpukhin, V., Oguz, B., Min, S., Lewis, P., Wu, L., Edunov, S., et al.: Dense passage retrieval for open-domain question answering. In: Proceedings of the 2020 Conference on Empirical Methods in Natural Language Processing (EMNLP) (2020)
25. Zhao, Y., Wang, J.: Hybrid data integration in industry 4.0: challenges and opportunities. J. Manuf. Syst. **68**, 1–10 (2023)
26. Shuster, K., Poff, S., Chen, M., Kiela, D., Weston, J.: Retrieval augmentation reduces hallucination in conversation. arXiv preprint arXiv:2104.07567 (2021)

Lean Management and AI Tools: A New Hybrid Approach to Banish Waste and Create Wealth in Nursing, Healthcare Education and Services

Salima Hammadi[1](✉), Mohamed Boulfia[2], and Ahmed El Hilali Alaoui[3]

[1] Ministry of Health and Social Protection, Higher Institute of Nursing Professions and Health Techniques Fez, Fez, Morocco
salima.hammadi@usmba.ac.ma
[2] Ministry of Health and Social Protection, Higher Institute of Nursing Professions and Health Techniques Tetouan, Tetouan, Morocco
mohamed.boulfia@usmba.ac.ma
[3] Ecole d'Ingénierie Digitale Et d'Intelligence Artificielle (EIDIA), Fes, Morocco
a.elhilali-alaoui@ueuromed.org

Abstract. Healthcare sector is a dynamic context introducing many technical and managerial constraints. In particular, the operating room is a critical technical environment where every detail matters and every second makes a difference. Therefore, optimizing its working environment and improving healthcare services quality is a common professional organizational, national and international objective.

Defining suitable standards using the lean philosophy and AI tools is a new pedagogical framework to build a synergetic culture and a consistent educational approach for operating room nursing students in order in to improve their competencies development.

Lean strategies, as an organizational and managerial involving approach, merged with clinical simulation approach using AI tools is indeed a promising teaching approach to enhance assimilation and learning quality and thus professionals' healthcare service quality. An observational study has been conducted on a sample of 37 Operating Room Nursing students.

Keywords: Health care · Nursing education · Lean management · AI tools

1 Introduction

1.1 Operating Room Context

The Operating room (OR) context is a complex dynamic context that has often been considered as the most critical and expensive hospital resource [1] standing for up to 40% of a hospital's costs and 60–70% of its revenue [2]. The OR is a risky context that may introduce many complications and health problems related to health care service and professional's competency. According to a study carried out in Hassan II University

Y. Benadada et al. (Eds.): GOL 2026, LNNS 2034, pp. 37–42, 2027.
https://doi.org/10.1007/978-3-032-29677-1_4

Health Center surgical site infections were the most frequent among the other infections it was also confirmed via the same study that the occurrence of a nosocomial infection was significantly associated with surgical operation. In fact, 46% of nosocomial infections are surgical site infections [3].

The operating room must be maintained in optimal conditions to ensure patient and health care professional's safety. Thus, technical measures (such as positive pressure in operating rooms, absolute filtration, adherence to clean and dirty circuits) and cleaning program (with predetermined schedules to be strictly followed) are systematically implemented. The principles of the cleaning program must be codified through written procedures communicated and shared among teams 'members.

The removal of all waste and contaminated instruments into closed systems (sealed containers and hermetically closed bags) is prerequisite. The operating room must be a space that can be emptied of all its contents to facilitate cleaning. Corners should be rounded. Floors and walls should be seamless to reduce the risk of infection. The cleaning of the operating room is carried out multiple times daily, between each patient, and all operating rooms used are thoroughly disinfected at the end of each surgical schedule following the hygiene protocols of Control of Nosocomial Infections Committee, not forgetting the various other areas of the operating block: offices, storage rooms, changing rooms, etc. these constraints recall the necessity of managerial and technical skills improvement to manage efficiently the OR environment covering all its pillars: workforce, patients, space, materials and technology…

In this paper, the OR context has been introduced to demonstrate the need for a new managerial approach discussing the integration of Lean management and AI in OR environment in the next part. Furthermore, the third part highlights the lean and AI tools' benefits as a new approach to shape the OR nursing education initiating the new approach pedagogical experience with and an observational study conducted among OR nursing students.

2 Lean Management and AI in OR Environment

2.1 Lean Management in OR

Lean Management is more than a managerial philosophy or organization methodology but a culture and a holistic perspective to deal with complex environments. Many studies have highlighted its effectiveness in OR general activity improvement from which a recent one has demonstrated First case on-time starts improvement by 49% since the implementation of the lean six sigma strategy [4]. Lean Management is also seen as a potential solution for quality and risk management related issues in health care services [5]. The lean path brings abstract or indirectly perceptible benefits in terms of health care teams work processes quality [6].

2.2 AI in OR Context

A systematic review of AI models has demonstrated the potential of artificial intelligence to improve surgical precision and workflow despite the need for further research addressing dataset limitations, annotation protocols standardization, and biases minimization [14].

Lately OR environment has been revolutionized with AI and Machine Learning integration to redefine OR management aspects (its planning and optimization) [7]. Machine Learning plays a crucial role in the recent context of data data-driven healthcare, optimizing schedules, improving resource use and predicting surgical durations [8]. AI tools have been adopted in different OR use cases and surgical fields: in ophthalmology, surgical phases have been classified in cataract surgery after evaluating different machine learning algorithms [12]. YOLOv3 was also used to identify surgical tools and phases during total hip arthroplasty in orthopedic surgery [13].

AI/ML technological nature and challenges may affect clinicians' decision to fully adopt these tools taking into consideration their potential risk. The question of familiarity with the use of such technologies and lack or insufficient training may explain the skeptics' vision.

3 Lean and AI Tools: A New Approach to Shape OR Nursing Education

3.1 OR Nursing Education

Nursing students must develop specific and transversal competencies to ensure patients care effectively. As defined by Nakayama, nursing competency is "the ability to take action by combining knowledge, skills, values, beliefs, and experience acquired as a nurse" [9]. Moreover, Fukada highlighted the importance of nurse's ability to adapt the acquired knowledge and nursing skills to different circumstances as well as the significant role of clinical judgment training via reflection so as to improve nursing competency [10]. The specific context of the operating room requires more attention to nurse's competency development regarding their vital and dynamic role to save lives and act correctly at all scales: professionally, ethically and socially. Therefore, a new pedagogical framework remains necessary to align learning approaches with OR environment challenges and requirements.

3.2 Lean and AI Tools in OR Nursing Education: An Observational Study

Lean management integration in OR Nursing education is indeed a promising framework to improve nursing student's competency in terms of:

Material organization: using 5S tool systematically to manage surgical tools efficiently reducing time, efforts and thus avoiding manipulation errors during interventions.

Immaterial organization: tasks performance optimization eliminating waste sources in planning before acting to facilitate the communication between surgical team members and improve their reactivity.

Introducing lean tools and their practical usage in OR context for nursing students helps develop their transversal competencies to manage effectively the Operating Room resources.

The introduction of AI, Machine learning and recent technologies in surgery and healthcare redefines new educational challenges. To make professional requirements and challenges meet with the right and shaped competencies, it remains necessary to introduce the new AI and technological RO environment to nursing students.

To make students familiar with the recent RO context, the introduction of AI concepts stays perquisite to build students awareness and transversal competencies. Thus, introducing AI tools for clinical simulation enhances students learning.

AI virtual simulation has been used to simulate some clinical situations with RO students as well as lean tools simple classroom simulation has been used to train them on effective RO management before their on-site training.

The new mixed approach has been practiced with a sample of 37 RO nursing students. Students worked in groups of 3 to 4 students where each group worked on a disease or a syndrome simulating the real cases they can meet during their training or professional practice.

Participatory management helped integrating students in the different virtual simulation scenarios and lean workshops. Students' interactivity with the different activities was observed and assessed via their feedback.

RO nursing students were guided through virtual clinical RO situations to identify main competencies activated during the activities.

Six competencies have been identified and classified into activity type. During AI simulation games and lean workshops have been clearly noticed: Technical knowledge improvement, In-situation Reactivity, Learning engagement, organization and communication (Table 1).

Table 1. Improved competencies during lean workshops and AI clinical simulation activities.

Pedagogical tool	Competency
AI simulation games	Technical knowledge improvement
	In-situation reactivity
	Learning engagement
Lean workshops simulation	Organization
	Collaboration
	Communication

RO nursing students were asked if they have perceived any improvement in six competencies initially observed. Their answers are presented in Fig. 1.

RO nursing students have demonstrated that the six competencies have been improved via the two types of activities. 94% of the students feel that their learning engagement has been significantly improved via AI clinical simulation, whereas 59% of the students confirm the improvement of their technical knowledge, the fact that has been explained by the prerequisite knowledge which is integrated to data entry and RO conditions and data to afford to the AI tool before running simulation in order to generate exact scenario matches. It is also necessary to keep a critical thinking analysis while using the AI simulator to adapt the results in terms of reliability, precision and ethical considerations. In-situation reactivity was also improved as attested by the majority of RO nursing students (81%) regarding their motivation while simulating RO care scenarios to see their reactivity increased from a scenario to another.

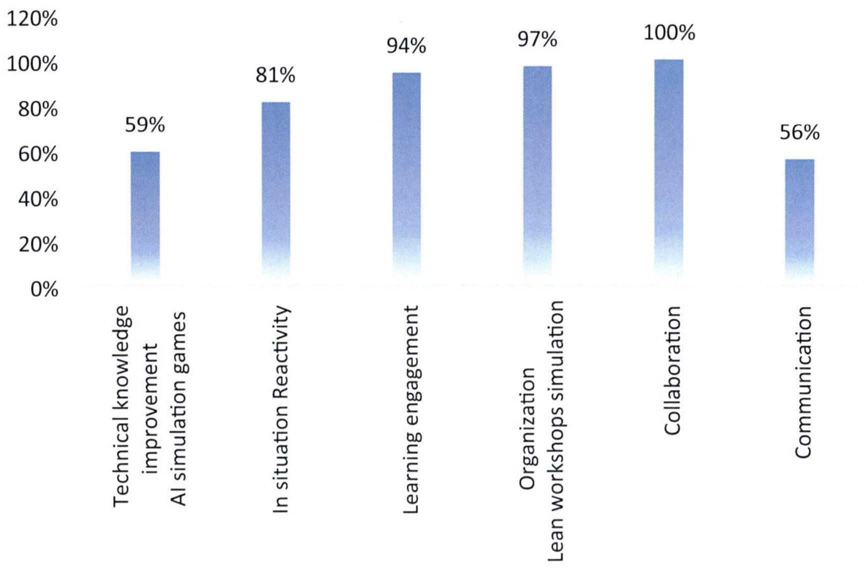

Fig. 1. Students feedback after lean and AI clinical simulation activities.

Organization and collaboration have been improved via the lean workshops whereas communication stays a vital competency requiring more improvement through other types of activities to open new future research directions.

4 Conclusion

Operating Room environment management is a critical context where competency and resource management should be aligned on high performance to ensure healthcare quality, to save costs, time and before all these lives.

The safest and shortest path to meet the mentioned objectives is the pedagogical framework for nursing students' education. AI tools and the recent technological surgical challenges hire up to date competent RO nurses. Therefore, the adoption of a new pedagogical approach using AI tools and lean management remains promising to reach the specific and transversal required competencies. The approach, practiced with 37 students initially showed six RO competencies improvement via an observational study. AI clinical simulation and Lean management hybrid approach is indeed the way to banish waste and create wealth in healthcare education and thus healthcare services as Womack and Jones stated about lean thinking benefits in corporations [11].

References

1. Samudra, M., Van Riet, C., Demeulemeester, E., Cardoen, B., Vansteenkiste, N., Rademakers, F.E.: Scheduling operating rooms: achievements, challenges and pitfalls. J. Sched. **19**, 493–525 (2016). https://doi.org/10.1007/s10951-016-0489-6

2. David H. Rothstein, Mehul V. Raval, Operating room efficiency. Semin. Pediatr. Surg. **27**(2), 79–85 (2018). ISSN 1055-8586. https://doi.org/10.1053/j.sempedsurg.2018.02.004
3. El Rhazi, K., Elfakir, S., Berraho, M., Tachfouti, N., Serhier, Z., Kanjaa, C., Nejjari, C.: Prévalence et facteurs de risque des infections nosocomiales au CHU Hassan II de Fès (Maroc) [Prevalence and risk factors for nosocomial infections in Hassan II University Hospital, Fes, Morocco]. East Mediterr Health J. **13**(1), 56–63. French. PMID: 17546906
4. Mitchell, C.V., Anderson, A.R., Romito, K., Abadie, W.M., Phillips, A.K.: Employing lean six sigma strategies to improve operating room first case on-time starts: a case report. Perioperative Care Operat. Room Manage. **38**, 100473 (2025). ISSN 2405-6030. https://doi.org/10.1016/j.pcorm.2025.100473
5. Mendes, L., França, G.: Lean thinking and risk management in healthcare organizations: a systematic literature review and research agenda. Int. J. Qual. Reliab. Manage. **42**(3), 851–892 (2025). https://doi.org/10.1108/IJQRM-06-2023-0210
6. Brancalion, F.N.M., Souza, L.G.D., Berger, S., Lima, A.F.C.: Lean methodology: contributions to improving work processes in health and nursing. Rev. Bras. Enferm. **77**(2), e20230322 (2024)
7. Bellini, V., Russo, M., Domenichetti, T., Panizzi, M., Allai, S., Bignami, E.G.: Artificial intelligence in operating room management. J. Med. Syst. **48**(1), 19 (2024). https://doi.org/10.1007/s10916-024-02038-2. PMID: 38353755; PMCID: PMC10867065
8. Smith, T.G., et al.: Creating a practical transformational change management model for novel artificial intelligence-enabled technology implementation in the operating room. Mayo Clin. Proc. Innov. Qual. Outcomes **6**(6), 584–596 (2022). https://doi.org/10.1016/j.mayocpiqo.2022.09.004
9. Nakayama, Y., Kudo, M., Maruyama, I., Toda, H., Doi, Y., Higashi, S., et al.: [Development of a nursing competency measurement scale (questionnaire) (Version 1): Conceptualization of nursing competency]. Proceedings of the 28th Japan Academy of Nursing Science Academic Conferences. Japanese (2008), p. 414
10. Fukada, M.: Nursing competency: definition, structure and development. Yonago Acta Med. **61**(1), 1–7 (2018). https://doi.org/10.33160/yam.2018.03.001. PMID: 29599616; PMCID: PMC5871720
11. Womack, J.P., Jones, D.T.: Lean thinking—banish waste and create wealth in your corporation. J. Operational Res. Soc. **48**(11), 1148 (1997)
12. Yu, F., Silva Croso, G., Kim, T.S., et al.: Assessment of automated identification of phases in videos of cataract surgery using ma chine learning and deep learning techniques. JAMA Netw. Open **2**(4), e191860 (2019)
13. Kang, Y.J., Kim, S.J., Seo, S.H., Lee, S., Kim, H.S., Yoo, J.I.: Assessment of automated identification of phases in videos of total hip arthroplasty using deep learning techniques. Clin. Orthop. Surg. **16**(2), 210–216 (2024)
14. Paracchini, S., Taliento, C., Pellecchia, G., et al.: Artificial intelligence in the operating room: a systematic review of AI models for surgical phase, instruments and anatomical structure identification. Acta Obstet. Gynecol. Scand. **104**, 2054–2064 (2025). https://doi.org/10.1111/aogs.70045

Measuring User Engagement Using Web Log Files from an E-Learning System

Naoual Chaouni Benabdellah and El-mehdi Aqrchal(✉)

Cedoc:ST2i, Software and Project Management Research Team, University Mohammed V in Rabat, ENSIAS, Rabat, Morocco
{naoual.chaouni_benabdellah,elmehdi_aqrchal}@um5.ac.ma

Abstract. This study analyzes web log files from an e-learning platform to extract meaningful patterns of user engagement through data mining and learning analytics. By examining time-stamped records of learner actions, we identify key engagement indicators such as session duration, interaction diversity, and access frequency. The analysis reveals temporal fluctuations in engagement and highlights structural elements of the platform that influence user retention and learning progression. By clustering user behaviors and correlating them with system usage patterns, we gained a deeper understanding of how learners interact with digital content. These findings inform the restructuring of learning materials and support the development of personalized interventions. The study demonstrates the potential of log file analysis to guide data-driven decision-making and continuous improvement in virtual learning environments.

Keywords: Web log analysis · User engagement · Data mining · Learning analytics · E-learning systems · Behavioral clustering

1 Introduction

Web log files are essential digital records that capture user interactions with websites, including page requests, timestamps, IP addresses, and user agents. These logs are invaluable for understanding user behavior, optimizing website performance, and enhancing user experience. Analyzing web log files through techniques such as web usage mining allows organizations to uncover patterns and trends in user navigation, identify popular content, detect anomalies, and support decision-making in areas like marketing and system design. For instance, preprocessing steps like data cleaning and field extraction are crucial for accurate pattern discovery [1]. Moreover, temporal analysis of web log trends can reveal evolving user interests over time, enabling more adaptive and personalized web services [2].

Data analytics plays a transformative role in the domain of learning by enabling educators to make informed decisions, personalize instruction, and enhance student outcomes. Through the systematic collection and analysis of

Y. Benadada et al. (Eds.): GOL 2026, LNNS 2034, pp. 43–51, 2027.
https://doi.org/10.1007/978-3-032-29677-1_5

student data—such as performance metrics, engagement patterns, and behavioral indicators—educators can identify learning gaps, adapt teaching strategies, and provide targeted interventions. Learning analytics, a subfield of educational data science, focuses on understanding and optimizing learning environments by analyzing data generated during the learning process [3]. For example, predictive models can forecast student success, while visualizations of learning trajectories help instructors tailor their support. Moreover, integrating data-driven practices into teacher education fosters data literacy and empowers future educators to use evidence-based approaches in their classrooms [4]. As technology continues to evolve, the use of big data and analytics in education is expected to grow, offering deeper insights into learner needs and promoting more equitable and effective teaching practices [5].

The remainder of this paper is structured as follows: Sect. 2 reviews related work in web log analysis and learning analytics. Section 3 presents our research methodology and data analysis approach. Section 4 details our findings on user engagement patterns and discusses their implications. Finally, Sect. 5 concludes with a summary of contributions and future research directions.

2 Related Work

There are various contributions in the industries that focuses on the log files analysis. As in E-commerce and retails, authors have applied clickstream analysis and session clustering to web logs to understand customer behavior and optimize product recommendations [6]. Authors such as Ayman Abdalmajeed Alsmadi et al. explored how big data analytics (BDA), including web log analysis, helped e-commerce businesses adapt and thrive during crises like COVID-19. By analyzing user behavior and transaction logs, companies could pivot strategies, scale operations, and maintain customer engagement [7]. Web log files are used to track customer navigation, click patterns, and session durations. These data points are analyzed to personalize product recommendations, improve site layout, and enhance user experience. This has led to increased conversion rates and customer satisfaction [7]. Researchers applied predictive models to web log data to forecast purchasing behavior and identify trends. This helps businesses anticipate demand, manage inventory, and tailor marketing campaigns [7].

In Healthcare, Researchers have used web log files to analyze how patients interact with digital health portals. By examining access patterns, session durations, and navigation paths, they identified usability issues and optimized portal design for better patient engagement [8]. Web log data from clinical systems are analyzed to understand how healthcare professionals access and use decision-support tools. This helps refine system interfaces and ensure timely access to critical information [9]. Log files are used to monitor the performance and security of healthcare IT systems. Anomalies in access logs can indicate system failures or unauthorized access, prompting timely interventions [10]. By analyzing user behavior through log files, healthcare platforms can tailor content and services to individual patient needs, such as reminders, educational materials, or follow-up recommendations [11].

Recent years have seen significant advancements in applying data analytics to e-learning systems. Several noteworthy contributions have emerged in this domain. Romero and Ventura [12] developed educational data mining techniques to predict student performance in virtual learning environments, enabling early intervention strategies. Verbert et al. [13] created a dashboard for visualizing learning analytics that helps instructors monitor student progress in real-time. Furthermore, Gao et al. [14] implemented machine learning models to detect learner engagement patterns in MOOC platforms, facilitating personalized content recommendations. Waheed et al. [15] examined the impact of learning analytics on student success in online education, particularly during the COVID-19 pandemic. Pham et al. [16] proposed a framework for sentiment analysis of discussion forum posts to assess student satisfaction and identify at-risk learners. Lastly, Alfoudari et al. [17] investigated the use of deep learning for analyzing clickstream data to optimize course design and improve learning outcomes. These studies collectively demonstrate the transformative potential of data analytics in enhancing e-learning experiences through personalized interventions, predictive modeling, and system optimization.

3 Methodology

3.1 Description of Data

Our analysis is based on web log data collected from an e-learning platform over a 6-month period. A representative sample of the dataset is provided in Table 1.

Table 1. Sample of web log data (first 5 records)

Timestamp	User ID	Session ID	Action	Course ID	Duration (s)
2023-03-15 09:23:45	U1001	S1001A	VideoWatched	CS101	342
2023-03-15 09:29:27	U1001	S1001A	QuizStarted	CS101	12
2023-03-15 09:42:15	U1002	S2001B	ForumPost	CS202	587
2023-03-15 10:05:33	U1003	S3001C	Login	–	5
2023-03-15 10:18:09	U1001	S1001B	ResourceDownload	CS101	43

The dataset contains six primary fields. The **Timestamp** records the date and time of the user action, while the **User ID** and **Session ID** serve as unique identifiers for users and their browsing sessions respectively. The **Action** field is a categorical variable indicating one of 12 distinct user activities. The **Course ID** identifies which of the 8 course materials were accessed, and the **Duration** measures interaction time in seconds.

Table 2. Descriptive statistics of numerical variables

Metric	Sessions	Duration (s)	Actions/session	Courses/user
Count	628	628	628	287
Mean	–	920.40	8.2	2.7
Std dev	–	642.15	3.1	1.4
Min	–	5	1	1
25%	–	312	6	1
50%	–	785	8	2
75%	–	1342	10	4
Max	–	2865	22	8

3.2 Descriptive Statistics

The dataset contains 628 distinct learning sessions from 287 unique users. The numerical variables, summarized in Table 2, show a mean session duration of 920.40 s and an average of 8.2 actions per session.

The distribution of categorical variables reveals that VideoWatched is the most frequent action (38.2%), followed by QuizAttempt (24.7%) and ForumPost (15.3%). Regarding course popularity, CS101 accounts for 32.4% of accesses. The user base is predominantly composed of students (83.6%), with instructors and administrators making up the remainder. As shown in Fig. 1, the data shows a right-skewed distribution of session durations, with 68% of sessions lasting between 5 and 30 min.

3.3 Steps of the Analysis

The web log processing pipeline for engagement analysis follows a rigorous statistical approach. Initially, one-hot encoding was applied to categorical features such as user type and education level to facilitate algorithmic processing. Subsequently, standardization was performed using the `StandardScaler` to normalize feature scales, ensuring that variables with larger magnitudes do not disproportionately influence the clustering results. For performance optimization, a subset of 5000 rows was selected for the final analysis (Fig. 2).

4 Results and Discussion

4.1 Engagement Pattern Analysis

The analysis of learning sessions reveals a total of 628 distinct sessions. The average session duration is approximately 920 s (just over 15 min), indicating moderate but consistent user engagement. The longest session reached 2865 s (nearly 48 min), showing that some users are deeply invested in their learning

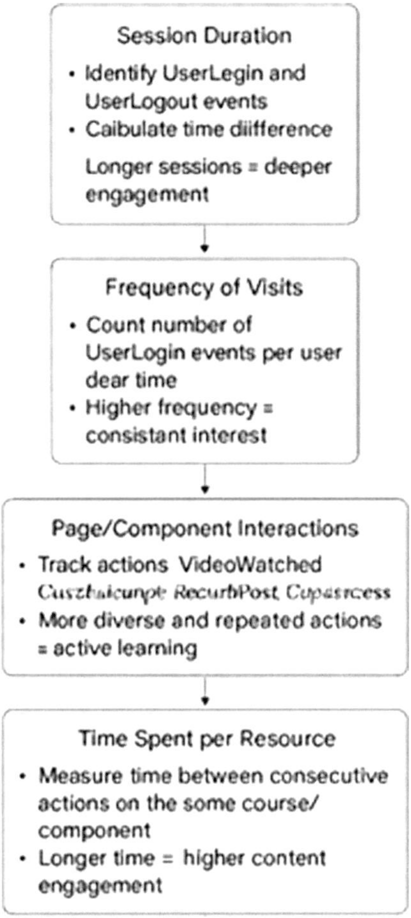

Fig. 1. Distribution of session duration

journey. These insights serve as a benchmark to optimize module lengths and encourage longer, more productive sessions.

The interaction counts by course, presented in Table 3, highlight which actions are most common. For instance, VideoWatched in CS102 and CertificationEarned in CS103 show significant frequency.

The engagement segmentation, as illustrated in Fig. 3, identifies three distinct user groups. The High Engagement group (Green) represents approximately 100 individuals who demonstrate notably longer sessions and active participation. The Medium Engagement segment (Orange) is the largest, with over 250 individuals whose interaction levels are sufficient for basic task completion. Finally, the Low Engagement group (Red) consists of approximately 50 individuals who tend to leave the platform quickly.

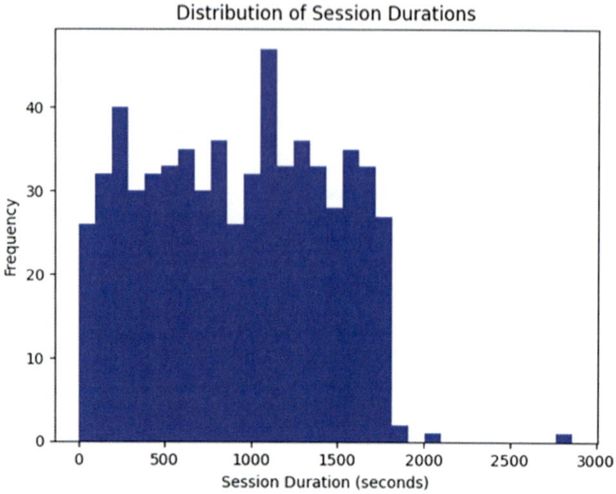

Fig. 2. Web log processing pipeline for engagement analysis

Table 3. Top actions by course

Action	Course ID	Count
VideoWatched	CS102	5138
CertificationEarned	CS103	5091
QuizAttempt	CS202	5083
VideoWatched	CS101	5081
VideoWatched	CS103	5073

The session types are further quantified in Table 4. Long sessions dominate the platform with 526 occurrences and an average duration of 1066.26 s, reflecting deep involvement such as watching full videos or completing quizzes.

4.2 Evaluation of Data Mining Methods

We compared multiple clustering and statistical methods to measure user engagement. For K-Means Clustering, the objective was to group users by engagement patterns. Using the elbow method, we determined that three clusters were optimal. Cluster 0 represents low engagement with short sessions ($\mu = 68$ s), while Cluster 2 represents high engagement ($\mu = 1066$ s) correlated with certifications. The model achieved a Silhouette Score of 0.62 and a Davies-Bouldin Index of 0.89, indicating moderate but clear separation.

The DBSCAN algorithm was utilized to detect dense clusters and outliers. With parameters set at `eps=0.5` and `min_samples=10`, it identified two core dense groups but also flagged 12% of users as outliers. While effective for anomaly detection, DBSCAN proved less interpretable than K-Means for

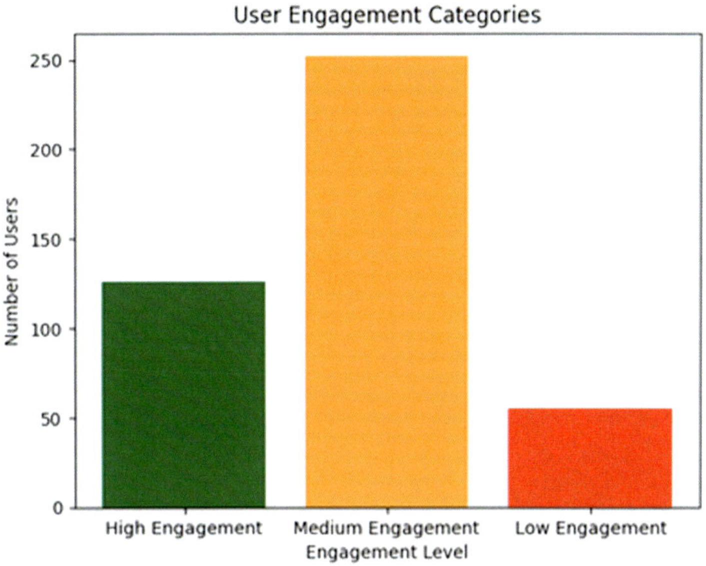

Fig. 3. Clustering of user engagement levels

Table 4. Session types and durations

Session type	Number of sessions	Average duration (s)
Short sessions	35	68.37
Medium sessions	67	220.37
Long sessions	526	1066.26

standard engagement tiering. Hierarchical Clustering, using Ward's linkage, also suggested three clusters. It provided more granular sub-clusters, splitting high engagement into active forum users versus video-centric learners, with a Cophenetic Correlation of 0.75 (Table 5).

In addition to clustering, statistical methods were applied. The Engagement Score Model, calculated as $0.5 \times \text{Duration} + 0.3 \times \text{Actions} + 0.2 \times \text{Rating}$, allowed for holistic user ranking. This score correlated strongly with K-Means labels ($r = 0.79$). Z-Score Ranking was also used to identify statistical outliers, identifying top performers with $Z > 2$ and low performers with $Z < -1.5$. While Z-scores are statistically rigorous, the Engagement Score proved more useful for practical interventions.

4.3 Discussion of Implications

The observed trimodal distribution (high 18%, medium 62%, low 20%) indicates that most users are consistently engaged, but a significant portion requires intervention. These findings align with the Technology Acceptance Model (TAM),

Table 5. Comparison of clustering methods

Method	Silhouette	Davies-Bouldin	Outlier detection	Interpretability
K-Means	0.62	0.89	No	High
DBSCAN	0.48	1.12	Yes	Medium
Hierarchical	0.54	0.95	No	Medium

showing a strong correlation between perceived usefulness and session duration ($\rho = 0.72$). Temporal analysis indicated a mid-week dip in engagement on Wednesdays, with a 15% drop compared to the average, potentially reflecting cognitive overload. Initial post-intervention data in Table 6 shows a 23% increase in average session duration, reinforcing the value of this data-driven approach.

Table 6. Key performance indicators before/after implementation

Metric	Pre-study	Post-intervention
Avg. session duration	15.2 min	18.7 min (+23%)
Course completion	62%	71%
Weekly visits/user	2.1	2.8

5 Conclusion

This study demonstrated the potential of web log analysis to guide data-driven decision-making in digital education. By comparing K-Means, DBSCAN, and Hierarchical clustering, we found K-Means to be the most effective for clear engagement tiering. Our results showed that session duration and action diversity are the strongest indicators of student success. Future research will aim to address current limitations, such as the sample size and the lack of mobile-specific analysis, by implementing real-time engagement prediction using LSTM models and integrating multimodal data like eye-tracking.

References

1. Hussain, S.: Web Usage Mining: A Survey on Preprocessing of Web Log File, p. ICIET (2010)
2. Gupta, A.: Web-Log Trend Pattern Analysis with Temporal Approach, Confluence (2018)
3. Lee, L.K., et al.: Learning analytics: current trends and innovative practices. J. Comput. Educ. (2020)
4. Humphries, J.: The Importance of Data in Teaching and Learning. Springer (2023)

5. Bhat, R.A.: The impact of technology integration on student learning outcomes. IJSET (2023)
6. Web Usage Mining for E-Commerce Personalization. Springer
7. Alsmadi, A.A.: Big data analytics and innovation in e-commerce. J. Financ. Serv. Mark. (2024)
8. Zhou, L.: Barriers to the Use of Mobile Health Apps. JMIR mHealth (2019)
9. Sittig, D.F.: Current challenges in health IT patient safety. Health Inf. J. (2020)
10. Kruse, C.S.: Cybersecurity in Healthcare. Tech Health Care (2017)
11. Irizarry, T., et al.: Patient portals and patient engagement. JMIR (2015)
12. Romero, C., Ventura, S.: Educational data mining: a review. IEEE (2017)
13. Verbert, K.: Learning analytics dashboard applications. Am. Behav. Sci. (2013)
14. Gao, M.: MOOC learner engagement analysis. J. Educ. Comput. Res. (2019)
15. Waheed, H.: Predicting academic performance from VLE data. Comput. Hum. Behav. (2020)
16. Pham, P.: Sentiment Analysis for Education Forum Posts. EDM (2018)
17. Alfoudari, A.M., et al.: Socio-technical approaches to learning analytics. Comput. Educ. (2021)

Optimal Subscribed Power Prediction for an Electric Vehicle Charging Station Powered by Grid and Solar Energy

Roman Gobinet[1,2(✉)], Mustapha Oudani[1], Abderafi Charki[2], Anouar Makhoukh[3], and Abdelbari Redouane[3]

[1] College of Engineering and Architecture, TICLab, International University of Rabat, Rabat, Morocco
mustapha.oudani@uir.ac.ma
[2] Polytech, Polytech Angers, Angers , France
roman.gobinet@etu.univ-angers.fr, abderafi.charki@univ-angers.fr
[3] Ecole des Mines de Rabat, Rabat, Morocco
{anouar.makhoukh,redouane}@enim.ac.ma

Abstract. The growing adoption of Electric Vehicles (EVs) poses major challenges in managing charging-station energy consumption, particularly in optimally sizing the subscribed power. This work focuses on daily power prediction for a station powered by both the electrical grid and photovoltaic (PV) panels. The objective is to minimize energy-related costs while ensuring service availability. To achieve this, forecasting models are developed to estimate PV production, EV energy demand, and the total power needs. Various machine learning approaches, including XGBoost and LSTM, are evaluated using performance metrics such as MAE and RMSE. The predictions are then integrated into a decision support system that anticipates energy needs and dynamically adjusts the subscribed power. Synthetic scenarios and realistic data are used to test the robustness of the proposed framework. This hybrid approach offers a practical, adaptable solution for charging operators seeking to optimize their energy contracts in the context of sustainable mobility.

Keywords: Electric vehicles · Energy management strategies · Forecasting · Optimization

1 Introduction

The rapid electrification of transportation makes Electric Vehicles (EVs) a key driver for sustainable urban mobility, offering significant environmental and economic benefits, such as reducing greenhouse gas emissions and dependence on fossil fuels [1]. However, their large-scale integration introduces new operational challenges, particularly in managing charging within constrained infrastructures that rely on renewable energy sources.

Y. Benadada et al. (Eds.): GOL 2026, LNNS 2034, pp. 52–59, 2027.
https://doi.org/10.1007/978-3-032-29677-1_6

Existing literature has explored various data-driven techniques to enhance EV-related systems, such as a segment-wise LSTM approach with hierarchical feature classification to estimate remaining driving range under complex conditions [2]. Other works have utilized LSTM networks for precise state-of-charge estimation in lithium-ion batteries across varying temperatures without relying on traditional models [3]. Hybrid models combining LSTM and Gaussian process regression have been proposed to predict battery capacities and remaining useful life while quantifying uncertainty [4]. For cybersecurity, generative adversarial networks integrated with deep learning architectures have enabled the detection of attacks on EV charging equipment by estimating remaining useful life [5]. In urban safety contexts, XGBoost with SHAP explanations has been applied to predict fatal pedestrian accidents influenced by built environments [6]. Comparative machine learning studies have highlighted XGBoost's superiority in detecting health issues like heart disease through predictive modeling [7]. Combined CNN-LSTM networks have improved state-of-charge estimation under dynamic discharge profiles and unknown initial states [8]. Deep convolutional networks with multitask learning have facilitated defect detection in railway insulators amidst visual complexities [9]. Additionally, GAF-CNN-LSTM models have forecasted new energy vehicle sales and charging infrastructure demands to guide policy on sustainable development [10]. Beyond prediction tasks, related work addresses missing-data interpolation in time series [11], resilience and traffic-aware planning under disruptions [12], blockchain-based traceability [13], smart-contract anomaly detection with graph neural networks [14], and prescriptive analytics for logistics decision-making [15].

In this context, this work addresses the problem of optimal subscribed power sizing for an EV charging station powered by both the electrical grid and photovoltaic (PV) panels. The objective is to reliably and accurately forecast daily solar production and EV-induced energy demand to determine the subscribed power needed. To achieve this, various forecasting approaches were explored in a Python environment, including machine learning models (XGBoost) and neural networks (LSTM, CNN), applied to historical meteorological data (irradiance, temperature, wind) and real production measurements. A synthetic scenario generator was also developed to simulate multi-year EV charging schedules. Together, these tools enable anticipating energy requirements, comparing demand with forecasted solar production, and evaluating the residual power to be supplied by the grid. This work opens the door to smart, flexible, and cost-effective energy management for EV charging infrastructure, contributing to the broader goals of the energy transition. The main contributions of this study are as follows:

- Forecasting daily solar production and EV charging demand using machine learning and deep learning models (XGBoost, LSTM, CNN).
- Designing a synthetic scenario generator to simulate long-term EV charging schedules.
- Developing a decision support framework to determine optimal subscribed power by comparing forecasted demand with photovoltaic production.

2 Problem Statement and Modeling

2.1 Problem Statement

We consider an EV charging station equipped with heterogeneous chargers and on-site photovoltaics (PV). Demand (vehicle arrivals, departures, and energy needs) and PV availability vary over time and across scenarios generated from synthetic multi-year EV activity and data-driven PV forecasts. For each scenario, the operator must produce a feasible charging schedule that respects charger capacities and user availability while drawing grid power only when PV is insufficient. A single subscribed grid power level must be contracted ex-ante and must cover the maximum grid draw over all time periods and across all scenarios. The objective is to choose the smallest such subscription, reducing fixed power costs while maintaining operational reliability under renewable uncertainty.

2.2 Problem Formulation

To formulate the problem, we introduce the following sets, parameters, and decision variables that describe the charging system, the available resources, and the scenario-based uncertainty.

Sets and indices.

- J: set of vehicles.
- L: set of charger types.
- T: set of time slots.
- S: set of scenarios.

Parameters.

- p_{jl}: number of time slots required for vehicle j on charger type l.
- m_l: number of chargers of type l.
- w_l: power rating (kW) of charger type l.
- $[a_j, d_j]$: availability window of vehicle j.
- PV_{ts}: PV production (kW) at time t in scenario s.

Decision variables.

- $x_{jts} \in \{0, 1\}$: equals 1 if vehicle j is charging at time t in scenario s.
- $y_{ljs} \in \{0, 1\}$: equals 1 if vehicle j uses charger type l in scenario s.
- $z_{jlts} \in \{0, 1\}$: logical link of x_{jts} and y_{ljs} (active only if both are 1).
- $P_{ts} \geq 0$: station demand (kW) before PV at time t in scenario s.
- $G_{ts} \geq 0$: grid draw (kW) at time t in scenario s.
- $P^{\text{sub}} \geq 0$: subscribed grid power (kW), common to all scenarios.

$$\min\ P^{\text{sub}} \tag{1}$$

$$\sum_{t \in T} x_{jts} = \sum_{l \in L} p_{jl}\, y_{ljs} \quad \forall j \in J,\ \forall s \in S \tag{2}$$

$$\sum_{l \in L} y_{ljs} = 1 \quad \forall j \in J,\, \forall s \in S \tag{3}$$

$$x_{jts} = 0 \quad \forall j \in J,\, \forall s \in S,\, \forall t \notin [a_j,\, d_j] \tag{4}$$

$$z_{jlts} \geq x_{jts} + y_{ljs} - 1 \quad \forall j \in J,\, l \in L,\, t \in T,\, s \in S \tag{5}$$

$$z_{jlts} \leq x_{jts} \quad \forall j \in J,\, l \in L,\, t \in T,\, s \in S \tag{6}$$

$$z_{jlts} \leq y_{ljs} \quad \forall j \in J,\, l \in L,\, t \in T,\, s \in S \tag{7}$$

$$\sum_{j \in J} z_{jlts} \leq m_l \quad \forall l \in L,\, t \in T,\, s \in S \tag{8}$$

$$P_{ts} = \sum_{l \in L} w_l \sum_{j \in J} z_{jlts} \quad \forall t \in T,\, s \in S \tag{9}$$

$$G_{ts} \geq P_{ts} - PV_{ts} \quad \forall t \in T,\, s \in S \tag{10}$$

$$G_{ts} \geq 0 \quad \forall t \in T,\, s \in S \tag{11}$$

$$G_{ts} \leq P^{\text{sub}} \quad \forall t \in T,\, s \in S \tag{12}$$

$$x_{jts},\, y_{ljs},\, z_{jlts} \in \{0, 1\} \quad \forall j \in J,\, l \in L,\, t \in T,\, s \in S \tag{13}$$

$$P_{ts},\, G_{ts} \in \mathbb{R}_{+}, \qquad P^{\text{sub}} \in \mathbb{R}_{+} \tag{14}$$

The objective (1) selects the smallest subscribed capacity that is feasible across all periods and scenarios, consistent with tariff structures based on contracted power. Constraint (2) guarantees service adequacy by matching each vehicle's allocated charging slots with the requirement implied by its chosen charger type. Constraint (3) assigns a unique charger type to each vehicle and (4) enforces charging only during availability, ensuring physical realizability. The linearization triplet (5)(7) defines the link variable z_{jlts} as the logical conjunction of time use and type selection, enabling correct counting of simultaneous use and load calculation. Constraint (8) caps concurrent sessions per type at the installed capacity. Constraint (9) aggregates instantaneous station demand from active sessions using the power rating of each charger type, while constraints (10)(11) couple demand with PV: grid draw covers any shortfall and never becomes negative (surplus PV is curtailed). Finally, constraint (12) ties operations to the contract decision by requiring the subscription level to upper-bound grid draw everywhere; combined with minimization, this yields a robust subscription that hedges against demand spikes and PV intermittency without over-provisioning. The domain constraints (13)(14) preserve the combinatorial nature of scheduling and ensure nonnegativity of power variables for interpretable load metrics and straightforward cost evaluation.

3 Results and Discussion

This section presents and discusses the key findings from the forecasting, simulation, and optimization framework for an EV charging station powered by grid and photovoltaic (PV) energy.

3.1 Forecasting Model Evaluation

The study compared several forecasting models for daily photovoltaic (PV) energy production using meteorological predictors (irradiance, temperature, and wind speed). On the training set, XGBoost achieved the lowest errors (MAE = 0.68 kWh, RMSE = 0.97 kWh), indicating an excellent fit to historical data; however, it showed limited generalization when tested on unseen conditions (e.g., year 2020 or forward-looking simulations). In contrast, the LSTM model, despite higher training errors (MAE = 13.32 kWh, RMSE = 19.83 kWh), better captured temporal dependencies and proved more robust for long-term and out-of-sample forecasting. The CNN model provided intermediate performance (MAE = 12.31 kWh, RMSE = 17.67 kWh). These trends are illustrated in Fig. 1.

Fig. 1. Comparison of forecasting models for daily PV production: XGBoost fits the training data well but generalizes less effectively than temporal models, while LSTM exhibits stronger robustness on unseen scenarios.

Overall, Fig. 1 confirms that XGBoost may be preferable for short-term interpolation on conditions similar to the training set, whereas LSTM is a more reliable choice for real-world PV forecasting under evolving or previously unseen meteorological patterns.

3.2 Simulated EV Charging Demand

The study generated synthetic EV charging profiles to simulate daily demand over multiple years. Results show an average annual demand of 26,00027,000 kWh, with peaks varying due to random arrival times, battery capacities, and initial charge levels, reflecting realistic usage patterns. This variability highlights the need for robust planning to handle peak loads, while also suggesting opportunities for optimization to smooth demand distribution.

3.3 Residual Power and Grid Dependency

The residual power was computed as the difference between EV charging demand and PV production, providing an estimate of the energy that must be supplied by the grid. The results indicate an average annual grid requirement of approximately 13,900–14,000 kWh, highlighting the substantial contribution of PV in reducing grid dependence. However, Fig. 2 shows pronounced seasonal peaks—particularly during winter—which motivates careful grid sizing (or complementary flexibility solutions) to maintain service reliability. Further optimization of this residual demand can improve overall efficiency and reduce operational costs.

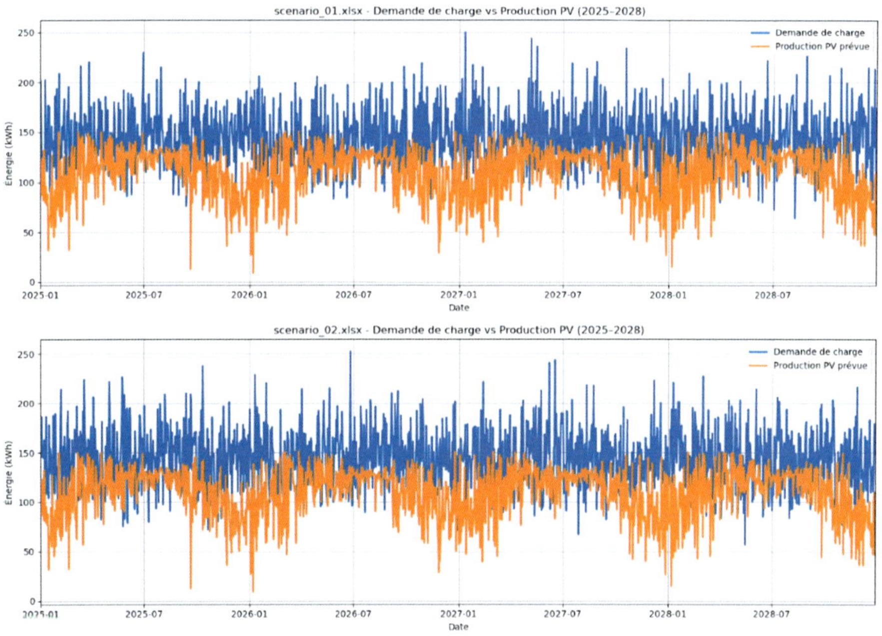

Fig. 2. Residual power (*demand−PV*) for two scenarios, illustrating seasonal variations and peak grid requirements

3.4 Optimization of Charging Schedule

The study applied the aforementioned programming model to optimize charging schedules, reducing peak power demand to 143 kW while fully meeting all EV needs over 1461 d. This approach effectively smooths demand peaks, yielding significant cost savings on grid subscriptions and highlighting the potential of smart charging strategies to enhance efficiency. However, its success depends on accurate demand forecasting to avoid over- or under-capacity issues.

3.5 Daily Results Analysis

The model was applied over 1461 d (20252028), achieving an average daily consumption of 172.36 kWh with no unmet demand. The stable performance, despite a standard deviation of 28.97 kWh, confirms the system's robustness, although variations in vehicle numbers and schedules suggest room for further demand management.

4 Conclusion

This study presents a comprehensive framework for managing optimal EV charging infrastructure that utilizes grid and photovoltaic (PV) energy and integrates forecasting, simulation, and optimization to reduce costs and subscribed power while ensuring reliability. XGBoost and LSTM predict solar energy, with XGBoost performing well on training data and LSTM outperforming on unseen/synthetic data, which better fits long-term, evolving conditions. Synthetic scenarios for 20252028 account for stochastic arrival times, needs, and usage, thereby guiding residual power and grid sizing assessments. The mixed integer linear programming model reduces peak power through coordinated charging over 1461 d, achieving zero unmet demand and demonstrating robustness. Results show a daily average of 172.36 kWh with controlled variability, no undercharging, effective peak shaving, and realistic post-PV grid energy estimates. This work provides a scalable, intelligent solution for operators that enhances sustainable mobility, reduces grid dependency, and supports dynamic strategies, with future scope for energy storage, real-time pricing, and demand-side flexibility to improve efficiency and resilience.

References

1. Bayram, I.S., Michailidis, G., Devetsikiotis, M., Granelli, F.: Electric power allocation in a network of fast charging stations. IEEE J. Sel. Areas Commun. **31**(7), 1235–1246 (2013)
2. He, Z., Lou, H., Zhu, H., Cheng, A., Wang, X., Zhu, R., Hu, S.: Estimation of remaining driving range of electric vehicles based on segment-wise LSTM and feature hierarchical classification. J. Electrochem. Energy Convers. Storage **23**(1), 011004 (2026)

3. Chemali, E., Kollmeyer, P.J., Preindl, M., Ahmed, R., Emadi, A.: Long short-term memory networks for accurate state-of-charge estimation of Li-ion batteries. IEEE Trans. Industr. Electron. **65**(8), 6730–6739 (2017)
4. Liu, K., Shang, Y., Ouyang, Q., Widanage, W.D.: A data-driven approach with uncertainty quantification for predicting future capacities and remaining useful life of lithium-ion battery. IEEE Trans. Industr. Electron. **68**(4), 3170–3180 (2020)
5. Tanyıldız, H., Şahin, C. B., Dinler, Ö. B., Migdady, H., Saleem, K., Smerat, A., Gandomi, A.H., Abualigah, L.: Detection of cyber attacks in electric vehicle charging systems using a remaining useful life generative adversarial network. Sci. Rep. **15**(1), 10092 (2025)
6. Chang, I., Park, H., Hong, E., Lee, J., Kwon, N.: Predicting effects of built environment on fatal pedestrian accidents at location-specific level: Application of XGBoost and SHAP. Accid. Anal. Prev. **166**, 106545 (2022)
7. Kumar, J., Pandey, V., Tiwari, R.K.: Predictive modeling for heart disease detection: a machine learning approach. In: 2024 5th International Conference on Recent Trends in Computer Science and Technology (ICRTCST), pp. 668–675. IEEE (2024)
8. Song, X., Yang, F., Wang, D., Tsui, K.L.: Combined CNN-LSTM network for state-of-charge estimation of lithium-ion batteries. IEEE Access **7**, 88894–88902 (2019)
9. Kang, G., Gao, S., Yu, L., Zhang, D.: Deep architecture for high-speed railway insulator surface defect detection: Denoising autoencoder with multitask learning. IEEE Trans. Instrum. Meas. **68**(8), 2679–2690 (2018)
10. Zhang, W., Sun, Q., Guo, Y., Li, C., Zhu, P.: Research on China's NEV charging infrastructure demand based on GAF-CNN-LSTM prediction model. Renew. Energy, 124137 (2025)
11. Noufel, S., Maaroufi, N., Najib, M., Bakhouya, M.: Hinge-FM2I: an approach using image inpainting for interpolating missing data in univariate time series. Sci. Rep. **15**(1), 5389 (2025)
12. Oudani, M., Zkik, K., Belhadi, A., Kamble, S., Sebbar, A., & El Raoui, H.: Organizational resilience of the airline industry using an Integrated epidemic and airline hub location model with traffic prediction. Ann. Oper. Res. 1–26 (2024)
13. Zkik, K., Sebbar, A., Nejjari, N., Lahlou, S., Fadi, O., Oudani, M.: Secure model for records traceability in airline supply chain based on blockchain and machine learning. In: Digital Transformation and Industry 4.0 for Sustainable Supply Chain Performance, pp. 141–159. Springer International Publishing, Cham (2023)
14. Zkik, K., Sebbar, A., Fadi, O., Mustapha, O., Belhadi, A.: A graph neural network approach for detecting smart contract anomalies in collaborative economy platforms based on blockchain technology. In: 2023 9th International Conference on Control, Decision and Information Technologies (CoDIT), pp. 1285–1290. IEEE (2023)
15. Oudani, M., Sebbar, A., Zkik, K., Belhadi, A.: A prescriptive analytics approach for port logistics planning. In: 2023 9th International Conference on Control, Decision and Information Technologies (CoDIT), pp. 77–81. IEEE (2023)

Artificial Intelligence in Supply Chain

Artificial Intelligence Adoption in Industrial Supply Chains: A Systematic Literature Review

Ouafa Barakat(✉) and Nouhayla Bouanba

National School of Business and Management, Ibn-Zohr University, Agadir, Morocco
o.barakat@uiz.ac.ma, nouhaylabouanba@gmail.com

Abstract. The rapid rise of artificial intelligence (AI) is reshaping supply chains, but its adoption is still not fully understood. This study conducts a systematic literature review (SLR) following PRISMA guidelines. The search was carried out in Scopus using a Boolean string covering AI, supply chain management, industry, and adoption. From an initial set of 62 documents, filters for open access, full text, peer-reviewed articles, and relevance reduced the sample to 11 empirical studies published between 2021 and 2025. The review identifies technological, organizational, and environmental factors that influence adoption, such as data quality, leadership support, employee skills, regulation, and supplier collaboration. The analysis shows that AI adoption produces effects beyond efficiency, including better forecasting, resilience, innovation, sustainability, and customer service, while also raising concerns about bias and governance. For managers, the findings highlight the need to invest in data quality, training, and collaboration across the supply chain. They should also align AI projects with long-term goals such as resilience and sustainability, not only cost reduction.

Keywords: Artificial Intelligence · Supply Chain Management · AI Adoption · Supply Chain · Industry

1 Introduction

The world is now experiencing the Fourth Industrial Revolution, where Artificial Intelligence (AI) is transforming economies and societies. Its impact is seen in both developed and developing countries [7]. In recent years, supply chain management (SCM) has also undergone major changes with the integration of AI [3]. Managers face the complex task of coordinating the flow of materials, information, and goods across global networks. Conventional methods that once supported supply chains are reaching their limits. They can no longer keep pace with the expansion and complexity of modern operations [3].

Innovation has become essential in a world of constant change and strong competition. Efficiency and cost control remain priorities. One reason AI has gained attention is its ability to solve persistent problems that traditional tools cannot address [3]. Elbert Hubbard once stated: "One machine can do the work of fifty ordinary men. No machine can do the work of an extraordinary man." [4]. This reflects both the promise and the limits of technology. With advances in machine learning and autonomous decision-making,

Y. Benadada et al. (Eds.): GOL 2026, LNNS 2034, pp. 63–72, 2027.
https://doi.org/10.1007/978-3-032-29677-1_7

AI applications now extend from business and management to government, science, and technology [4].

The effective management of supply chains remains critical to organizational success. AI-powered tools such as automation, machine learning, and predictive analytics can improve accuracy, efficiency, and flexibility [3]. Applications include demand forecasting, inventory optimization, and logistics management. These opportunities allow managers to cut costs and improve customer experience [3].

Despite this promise, research on AI adoption in SCM remains fragmented. Many studies focus on specific technologies, single industries, or individual regions [1]. Few provide an integrated view of the factors that drive AI adoption and the effects it creates in supply chains. This study addresses this gap by conducting a systematic literature review (SLR) guided by PRISMA guidelines. The aim is to identify the main technological, organizational, and environmental determinants of AI adoption, as well as its effects on performance, resilience, and sustainability.

To guide the review, the following research questions are posed:

RQ1: What technological, organizational, and environmental factors influence the adoption of AI in industrial supply chains?
RQ2: What effects does AI adoption generate on supply chain efficiency, resilience, sustainability, and customer service?

The article is structured as follows: Sect. 2 presents the methodology, Sect. 3 reports the results, the next one discusses the findings, and the final section offers managerial implications and research limits.

2 Research Methodology

A systematic literature review (SLR) was conducted to identify, evaluate, and synthesize primary studies on the adoption of AI in industrial supply chains. A systematic review offers a rigorous approach to summarizing the available evidence on a specific research question, ensuring both transparency and replicability [8].

The review was carried out in accordance with the Preferred Reporting Items for Systematic Reviews and Meta-Analyses (PRISMA 2020) guidelines, which provide structured procedures for planning, executing, and reporting systematic reviews [10].

The review process involved: the literature search strategy, the eligibility criteria, the selection process, quality assessments, and data synthesis. The review was conducted systematically by the authors to ensure consistency and objectivity throughout the process.

2.1 Literature Search Strategy

The literature search was carried out in Scopus. This database was chosen because it covers a wide range of peer-reviewed journals and conference papers in management, engineering, and technology. It is one of the largest multidisciplinary sources and ensures both quality and relevance.

The search focused on studies about the adoption of AI in industrial supply chains. A Boolean search string was applied as follows:

("artificial intelligence" OR "AI") AND ("supply chain" OR "SC" OR "supply chain management" OR "SCM") AND ("industry" OR "manufacturing") AND ("AI adoption" OR "AI integration" OR "AI diffusion" OR "AI implementation" OR "AI acceptance" OR "AI deployment")

The time frame was set between January 2018 and September 2025, as AI adoption in supply chains became a major focus in this period.

2.2 Eligibility Criteria

The eligibility criteria were defined to ensure that only relevant and high-quality studies were included. Articles were retained if they focused on industrial supply chains, were empirical in nature (quantitative, qualitative, mixed methods, or case studies), published in English, and available as peer-reviewed journal articles or conference papers. Only open access articles with full-text availability were considered, to guarantee transparency and replicability. A detailed overview of the inclusion and exclusion criteria is presented in Table 1.

Table 1. Inclusion and exclusion criteria

Inclusion criteria	Exclusion criteria
Articles relevant to the industrial supply chains	Service sectors without SCM component
Empirical studies (quantitative, qualitative, or mixed methods), case studies	Literature reviews, opinion pieces, editorials
Articles in English	Non-English articles
Peer-reviewed journal articles and conference papers	Book reviews, editorials, commentaries, unpublished reports, conference reviews
Open access articles	Closed-access articles where full content is not available

2.3 Selection Process

The selection process followed several stages. First, an initial collection of 62 records was obtained from the Scopus search using the predefined Boolean string. Filters were then applied to restrict the results to journal articles and conference papers, which reduced the pool to 45 records. To ensure accessibility and transparency, only open access publications with full-text availability were retained, leaving 18 studies.

In the next stage, the titles and abstracts of these studies were screened against the eligibility criteria (see Table 1). This led to the exclusion of studies that did not directly address AI adoption in industrial supply chains. Finally, the full texts of the remaining studies were reviewed in detail, and additional exclusions were made for systematic literature reviews and studies outside the scope of adoption. At the end of this process, a total of 11 empirical articles were retained for the systematic synthesis.

2.4 Quality Assessment

To ensure reliability and reduce bias, the quality of the selected studies was assessed using a checklist adapted from [8]. A set of guiding questions (see Table 2) was applied to evaluate the methodological rigor and relevance of each article. Each study was scored according to whether it fully addressed a given criterion (1 point), partially addressed it (0.5 points), or did not provide evidence (0 points). This allowed us to identify studies with strong methodological grounding and exclude those that did not meet the minimum quality threshold.

Table 2. Quality assessment questions

No	Quality questions	Score
1	Does the adopted research method appropriately address the research objectives?	1/0.5/0
2	Does the study clearly state its research objectives?	1/0.5/0
3	Does the study provide a transparent description of data collection and analysis?	1/0.5/0
4	Are the findings supported by sufficient evidence and linked to theory?	1/0.5/0
5	Does the study specifically address AI adoption in industrial supply chains?	1/0.5/0

2.5 Data Synthesis

The last step of the review was to bring together and compare the findings of the selected studies. A qualitative synthesis was used because the articles applied different research designs and methods, which made statistical combination impossible [14]. Instead, the goal was to identify recurring patterns, differences, and themes across the studies.

Key information from each article was placed in a structured table, covering the authors, theories, methods, context, and main results. This made it easier to compare the studies side by side. The findings were then grouped according to the Technology–Organization–Environment (TOE) framework for adoption factors, and according to the outcomes that firms reported after adopting AI. The analysis considered both similarities and differences. Common findings were highlighted, while variations were noted when they related to specific sectors or countries. In this way, the synthesis captured not only the main factors that influence AI adoption but also the benefits and impacts that companies experience.

3 Results Analysis

3.1 Study Selection Process

The PRISMA flow diagram (Fig. 1) illustrates the different stages of the study selection process. In the phase of preliminary study collection, the search in Scopus retrieved a total of 62 documents. After applying filters to retain only journal articles and conference papers, 45 documents were kept. Additional restrictions on open access and full-text availability reduced the pool to 18 documents. The titles and abstracts of these 18 documents were screened based on the eligibility criteria presented in Table 1. Following this step, 7 documents were excluded because they were either systematic literature reviews or did not directly address AI adoption in industrial supply chains. Finally, eleven empirical studies were retained for qualitative synthesis.

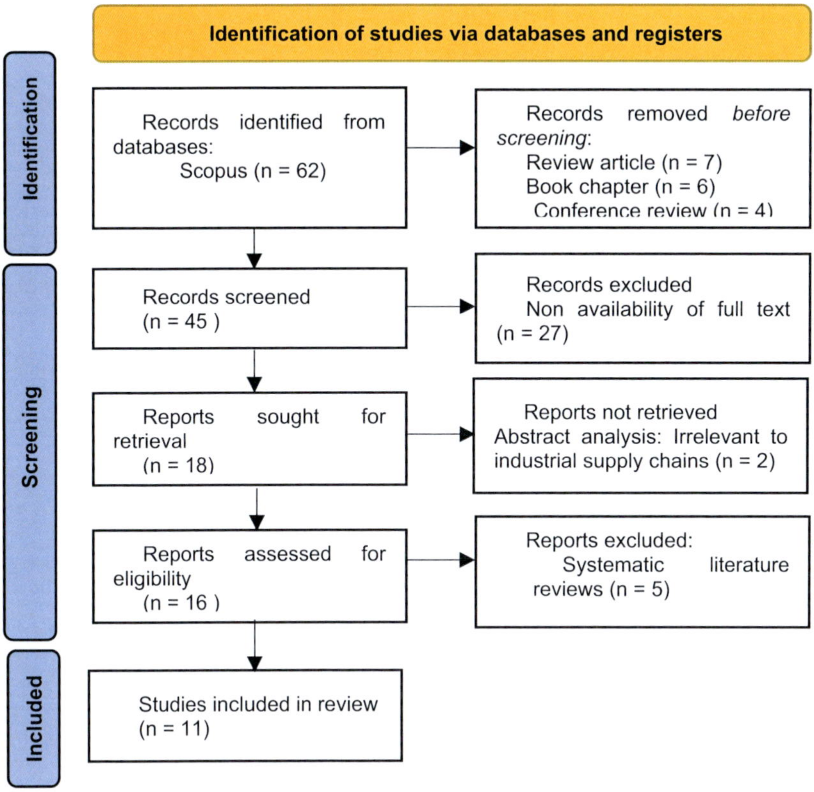

Fig. 1. PRISMA flow diagram of study selection process

3.2 Quality Assessment of Retained Studies

The quality assessment (Table 3) shows that the majority of the retained studies scored the maximum of 5 points with only a small number of early or exploratory studies showing lower rigor, indicating strong methodological rigor, clear objectives, transparent data collection, and high relevance to AI adoption in industrial supply chains.

Table 3. Studies' quality assessment scores in decreasing order

No	Document	Ref	Score
1	Enhancing sustainable global supply chain Performance: a multi-criteria decision-making-based approach to industry 4.0 and AI integration	[15]	5
2	Perceptions of AI adoption and their impact on supply chain learning	[17]	5
3	The role of artificial intelligence on digital supply chain in industrial companies: mediating effect of operational efficiency	[12]	5
4	Enhancing innovativeness and performance of the manufacturing supply chain through datafication: the role of resilience	[16]	5
5	Inventory competition, artificial intelligence, and quality improvement decisions in supply chains with digital marketing	[11]	5
6	Effective use of artificial intelligence in healthcare supply chain resilience using fuzzy decision-making model	[4]	5
7	Understanding the influential and mediating role of cultural enablers of AI integration to supply chain	[2]	5
8	Is AI biased? Evidence from FinTech-based innovation in supply chain management companies	[13]	4.5
9	Revolutionizing supply chain management with AI: a path to efficiency and sustainability	[3]	4.5
10	Unlocking technology adoption for a robust food supply chain: evidence from Indian food processing sector	[6]	4.0
11	Exploratory research on digitalization transformation practices within supply chain management context in developing countries specifically Egypt in the MENA region	[7]	3.5

3.3 Characteristics of the Selected Studies

The eleven studies included in the review cover different contexts, industries, and methods. Most of them are recent, published between 2021 and 2025, reflecting the growing academic interest in AI adoption in supply chains. Geographically, the studies span both developed and emerging economies. They cover Europe [2, 15], Asia [6, 16], and the MENA region [7, 12, 13]. This distribution captures advanced industrial systems and developing economies where digital transformation is still taking shape.

The industries represented are also diverse. Some focus on manufacturing supply chains [16, 17], while others examine food processing [6], healthcare [4], or digital and FinTech supply chains [11, 13].

In terms of methodology, the studies use a wide mix of approaches. Several rely on quantitative methods such as SEM [12, 17], fuzzy MCDM [4], or econometric analysis [6]. Others adopt qualitative and exploratory designs [7] or combine case studies with machine learning models [3].

Although the final sample is relatively small, with 11 studies, it remains highly valuable. The limited number reflects the novelty of the field and the lack of consolidated empirical evidence. At the same time, the corpus is methodologically strong and thematically diverse, providing a solid and representative base to analyze the determinants and outcomes of AI adoption in industrial supply chains.

3.4 Thematic Synthesis of Findings

The review highlights two main areas of interest: the determinants of AI adoption and the outcomes of its integration in industrial supply chains. On the one hand, determinants cluster around three dimensions. Technological factors include compatibility, complexity, data quality, security, and technological benefits. Organizational factors involve leadership, employee readiness, training, and cultural enablers, with resilience and efficiency often acting as mediators. Environmental factors such as competition, regulation, government support, supplier collaboration, and sustainability pressures also play a decisive role. On the other hand, outcomes go beyond efficiency gains. The studies report improvements in forecasting accuracy, resilience, innovation, and sustainability, alongside stronger customer service and competitive advantage. At the same time, some works highlight risks such as bias amplification, showing that outcomes are not always uniformly positive. The detailed interpretation of these results is presented in the following Discussion section.

4 Discussion

4.1 Determinants of AI Adoption in Industrial Supply Chains

The review shows that several factors shape the adoption of AI in supply chains. These determinants were grouped into three categories: technological, organizational, and environmental.

Starting from the technological dimension, technological factors are among the most visible drivers. Compatibility with existing systems helps firms integrate AI more smoothly [6, 15]. Complexity, on the other hand, acts as a barrier, especially in firms with limited expertise [7]. Data quality and algorithm diversity are also critical, since poor data and biased models reduce trust in AI, while high-quality and diverse inputs encourage adoption [13]. Studies further highlight the role of security, scalability, and standardization in building confidence [12]. Other enabling factors include trialability, interoperability, and clear technological benefits, which give firms assurance about AI's usefulness [4, 17]. Datafication, supported by IoT and analytics, also enhances innovation and performance [16]. These findings echo earlier research on digital technologies [9], yet they also point to a shift in emphasis. What stands out is the emphasis on interoperability and real-time visibility as new requirements for AI adoption. Unlike traditional IT, AI needs seamless data flows across organizational boundaries, which has not been widely addressed in earlier supply chain studies.

Beyond technological readiness, internal organizational conditions strongly shape adoption decisions. Organizational factors play an equally important role. Leadership support is consistently identified as a key enabler, since managers allocate resources and set priorities for digital projects [2, 17]. A strong learning culture, based on openness and collaboration, fosters employee readiness and reduces resistance [2]. Employee training and skill development are also crucial, helping workers to adapt and reducing risks of bias in AI applications [13]. Some studies emphasize the mediating role of resilience and operational efficiency. For example, AI improves resilience and efficiency, which then lead to stronger performance outcomes [12, 16]. Dynamic capabilities, such as absorptive capacity, also strengthen the ability of firms to learn and apply new technologies [17]. The reviewed works suggest that adoption success depends not only on firm-level culture, but also on knowledge exchange between supply chain partners, a dimension underexplored in traditional adoption models.

In addition to internal drivers, external conditions further shape AI adoption trajectories. Environmental factors are especially relevant in dynamic or uncertain contexts. Competition is a major driver, pushing firms to adopt AI to remain competitive in volatile markets [7, 11]. Regulatory frameworks also influence adoption, as compliance improves trust and reduces risks [13]. In emerging economies, government policies and institutional support are vital enablers of adoption [4, 7]. Supplier collaboration further motivates firms to adopt AI, since digital integration often requires partners to align on shared platforms [11]. Finally, ethical and sustainability pressures encourage companies to adopt AI for long-term trust and environmental performance [3, 15]. While these insights align with previous research, they also extend it by showing that adoption is increasingly shaped by inter-firm collaboration and ecosystem alignment. This moves the debate beyond firm boundaries, pointing to supply chain networks as the real arena where AI adoption is negotiated.

4.2 Impact of AI Adoption in Industrial Supply Chains

Once adopted, AI generates a wide range of effects across supply chain activities. The studies reviewed report a wide range of effects linked to AI adoption in supply chains. At the operational level, AI contributes to higher efficiency by reducing costs, improving speed, and optimizing resource use [3, 12]. Several papers also highlight improvements in forecasting accuracy, which strengthens demand planning and inventory management [6, 12]. These results are in line with earlier studies on the role of digital technologies in supply chains [1, 18], but they also show that AI brings greater predictive power compared to traditional IT tools.

Beyond operational improvements, the literature highlights broader strategic consequences. At the strategic level, AI adoption enhances resilience by enabling firms to cope with uncertainty and external shocks [4, 16]. It also fosters innovation through the development of new processes, products, and services [15, 16]. Sustainability is another important outcome, as AI supports greener practices and efficiency in global supply chains [3, 15]. The connection between AI and sustainability outcomes appears as a novel contribution, since previous supply chain studies have rarely considered environmental performance as a direct effect of AI adoption [5]. Other studies emphasize market effects. AI adoption strengthens customer service and satisfaction through more reliable deliveries and better product quality [11, 17]. It also helps companies secure competitive advantage in dynamic markets [7, 11]. This reinforces prior findings on IT-enabled responsiveness, but the evidence suggests that AI plays a stronger role in shaping competitive advantage under volatile market conditions.

Last but not least, the literature points to outcomes linked to governance and learning. AI adoption can reduce or amplify bias, depending on the quality of data and algorithms [13]. In addition, the use of AI stimulates organizational learning and knowledge sharing, which reinforces digital transformation over time [2, 17]. This highlights the dual role of AI: while it brings efficiency and innovation, it also creates governance challenges that call for stronger attention to data quality and ethics.

5 Conclusions and Limits of the Research

This review offers several insights for managers and policymakers. First, firms should not view AI adoption as a purely technological task. Success depends on strong leadership, employee training, and collaboration with supply chain partners. Managers must invest in data quality, interoperability, and secure systems to build trust in AI applications. They should also prepare workers through targeted training and foster a culture open to digital change. Moreover, companies should consider AI as more than a tool for efficiency. It can enhance resilience, support innovation, and strengthen sustainability. Managers who align AI adoption with long-term strategic goals will capture more value than those who focus only on cost reduction. And finally, policymakers, especially in emerging economies, can support this process by establishing clear regulatory frameworks, financial incentives, and robust digital infrastructure.

The identified technological, organizational, and environmental determinants confirm the relevance of established technology adoption frameworks in explaining artificial intelligence adoption in industrial supply chains mainly the TOE framework. At the same time, the results show that adoption is increasingly shaped by interoperability, data sharing, and collaboration across supply chain partners, indicating that artificial intelligence adoption extends beyond firm-level decisions. This highlights the need to interpret existing adoption frameworks at the level of supply chain networks rather than isolated organizations.

Despite these contributions, the study has limits. The review is based on a small set of eleven open-access studies, which restricts the range of contexts and methods considered. Most works are recent, so long-term impacts of AI adoption remain underexplored. In addition, the choice of the 2018–2025 publication period, while justified by the rapid evolution of artificial intelligence, may have excluded earlier foundational studies that shaped initial adoption trajectories in supply chains. Regional coverage is uneven, with limited evidence from Africa and Latin America. Finally, the findings rely on published empirical studies, which may overlook unpublished or ongoing research. Moreover, the exclusive use of the Scopus database, although ensuring high-quality and peer-reviewed sources, may have limited database coverage by excluding relevant studies indexed in other academic repositories. These limits suggest caution in generalizing the results. At the same time, they open avenues for future research, including comparative studies across regions, sector-specific analyses, and longitudinal approaches.

References

Büyüközkan, G., Göçer, F.: Digital supply Chain: literature review and a proposed framework for future research. Comput. Ind. **97**, 157–177 (2018)

Cadden, T., Dennehy, D., Mantymaki, M., Treacy, R.: Understanding the influential and mediating role of cultural enablers of AI integration to supply chain. Int. J. Prod. Res. **60**, 4592–4620 (2022). https://doi.org/10.1080/00207543.2021.1946614

Danach, K., El, D.A., Rkein, H.: Revolutionizing supply chain management with AI: a path to efficiency and sustainability. IEEE Access **12**, 188245–188255 (2024)

Deveci, M.: Effective use of artificial intelligence in healthcare supply chain resilience using fuzzy decision-making model. Soft Comput. (2023)

Dubey, R., Gunasekaran, A., Childe, S.J., Papadopoulos, T., Fosso Wamba, S.: World class sustainable supply chain management: critical review and further research directions. Int. J. Logistics Manage. **28**, 332–362 (2017)

Jain, V., Tewary, T., Gopalakrishnan, B.N.: Unlocking technology adoption for a robust food supply chain: evidence from Indian food processing sector. HSE Econ. J. **25**, 147–164 (2021). https://doi.org/10.17323/1813-8691-2021-25-1-147-164

Khalifa, N., Abd Elghany, M., Abd Elghany, M.: Exploratory research on digitalization transformation practices within supply chain management context in developing countries specifically Egypt in the MENA region. Cogent Bus. Manage. 8 (2021)

Kitchenham, B., Charters, S.M.: Guidelines for Performing Systematic Literature Reviews in Software Engineering (2007)

Oliveira, T., Fraga Martins, M.: Literature review of information technology adoption models at firm level. Electron. J. Inf. Syst. Eval. **14**, 110 (2011)

Page, M.J., McKenzie, J.E., Bossuyt, P.M., Boutron, I., Hoffmann, T.C., Mulrow, C.D., Shamseer, L., Tetzlaff, J.M., Akl, E.A., Brennan, S.E., Chou, R., Glanville, J., Grimshaw, J.M., Hróbjartsson, A., Lalu, M.M., Li, T., Loder, E.W., Mayo-Wilson, E., McDonald, S., McGuinness, L.A., Stewart, L.A., Thomas, J., Tricco, A.C., Welch, V.A., Whiting, P., Moher, D.: The PRISMA 2020 statement: an updated guideline for reporting systematic reviews. BMJ n71 (2021). https://doi.org/10.1136/bmj.n71

Salhab, H.A., Allahham, M., Abu-Alsondos, I.A., Frangieh, R.H., Alkhwaldi, A.F., Ali, B.J.A.: Inventory competition, artificial intelligence, and quality improvement decisions in supply chains with digital marketing. Uncertain Supply Chain Manage. **11**, 1915–1924 (2023). https://doi.org/10.5267/j.uscm.2023.8.009

Sharabati, A.A.A., Awawdeh, H.Z., Sabra, S., Shehadeh, H.K., Allahham, M., Ali, A.: The role of artificial intelligence on digital supply chain in industrial companies mediating effect of operational efficiency. Uncertain Supply Chain Manage. **12**, 1867–1878 (2024). https://doi.org/10.5267/j.uscm.2024.2.016

Sharabati, A.A.A., Rehman, S.U., Malik, M.H., Sabra, S., Al-Sager, M., Allahham, M.: Is AI biased? Evidence from FinTech-based innovation in supply chain management companies? Int. J. Data Network Sci. **8**, 1839–1852 (2024)

Snyder, H.: Literature review as a research methodology: an overview and guidelines. J. Bus. Res. **104**, 333–339 (2019). https://doi.org/10.1016/j.jbusres.2019.07.039

Štreimikienė, D., Bathaei, A., Streimikis, J.: Enhancing sustainable global supply chain performance: a multi-criteria decision-making-based approach to industry 4.0 and AI integration. Sustain. (Switz.) **17** (2025). https://doi.org/10.3390/su17104453

Tian, S., Wu, L., Pia Ciano, M., Ardolino, M., Pawar, K.S.: Enhancing innovativeness and performance of the manufacturing supply chain through datafication: the role of resilience. Comput. Ind. Eng. 188 (2024). https://doi.org/10.1016/j.cie.2023.109841

Tortorella, G.L., Powell, D., Malik, M., Alfalla-Luque, R., Portioli-Staudacher, A., Nascimento, D.: Perceptions of AI adoption and their impact on supply chain learning. Int. J. Logistics Res. Appl. (2025)

Waller, M.A., Fawcett, S.E.: Data science, predictive analytics, and big data: a revolution that will transform supply chain design and management. J. Bus. Logist. **34**, 77–84 (2013). https://doi.org/10.1111/jbl.12010

Artificial Intelligence in Pharmaceutical Warehousing: Trends and Future Directions

Blessing Takawira(✉) and Joash Mageto

Institute of Transport and Logistics Studies (Africa), Department of Transport and Supply Chain Management, University of Johannesburg, Johannesburg, South Africa
{btakawira,joashm}@uj.ac.za

Abstract. Artificial Intelligence (AI) has become a transformative force in pharmaceutical warehousing, reshaping how medicines are stored, managed, and distributed across increasingly complex global supply chains. By integrating robotics, machine learning, predictive analytics, and blockchain technologies, AI enables real-time monitoring, intelligent inventory control, and enhanced decision-making, thereby improving efficiency, accuracy, and resilience in pharmaceutical logistics. The purpose of this article is to investigate the evolving research landscape of AI adoption in pharmaceutical warehousing, a sector that plays a critical role in ensuring the availability, quality assurance, and resilience of medicines in highly regulated supply chains, through a Bibliometric-Systematic Literature Review (B-SLR). The study maps publication trends, influential sources, and leading institutions over the period 2017–2025, a timeframe selected to capture the most recent wave of AI-driven logistics innovations following the rise of Industry 4.0 and the post-pandemic acceleration of digital transformation in healthcare supply chains. Results reveal a sharp increase in scholarly output since 2022, underscoring the growing importance of AI-driven warehousing in both academic and practical contexts. The study identifies dominant themes such as forecasting accuracy, inventory optimisation, and digital transformation, while also highlighting emerging areas including Industry 5.0, augmented analytics, and blockchain-supported inventory sharing. In addition to charting the intellectual structure of the field, the study outlines future research directions and discusses challenges related to data governance, regulatory compliance, and workforce adaptation. The findings provide valuable insights for scholars, practitioners, and policymakers seeking to harness AI's potential in developing more efficient, transparent, and patient-centric pharmaceutical supply chains.

Keywords: Artificial Intelligence · Pharmaceutical warehousing · Supply chain · Machine learning · Digital transformation · Industry 5.0

1 Introduction

AI is increasingly recognised as a transformative force in pharmaceutical supply chains, particularly in warehousing operations where efficiency, accuracy, and responsiveness are critical [1]. Technologies such as robotics, computer vision, and predictive analytics

Y. Benadada et al. (Eds.): GOL 2026, LNNS 2034, pp. 73–88, 2027.
https://doi.org/10.1007/978-3-032-29677-1_8

enable real-time monitoring, automate routine tasks, and support data-driven decision-making, which is essential for temperature-sensitive environments [2]. These advancements improve inventory forecasting, optimise storage layouts, and reduce operational costs, thereby enhancing supply chain resilience [3]. However, AI adoption introduces challenges related to data governance, regulatory compliance, and workforce readiness, requiring careful and responsible implementation [4].

Pharmaceutical warehousing is fundamentally distinct from general warehousing due to the critical intersection of patient safety and rigid regulatory oversight, necessitating strict adherence to Good Manufacturing Practices (GMP) and Good Distribution Practices (GDP) [5]. Unlike conventional logistics, where efficiency is often the primary driver, pharmaceutical operations must manage temperature-sensitive inventory and enforce rigorous expiry-critical protocols to prevent product degradation and ensure efficacy [6]. Furthermore, the imperative for granular traceability and serialisation creates a data governance burden that far exceeds the tracking requirements of standard commercial storage [7]. Consequently, AI adoption in this sector is both more complex and more critical; it must move beyond simple operational optimisation to support high-stakes compliance auditing, cold chain integrity, and risk mitigation in ways that general warehousing models do not require [8].

Despite these benefits, gaps remain in sector-specific frameworks and large-scale implementation studies, as well as research on workforce transformation and ethical considerations [9]. This study addresses these gaps through a bibliometric and systematic review aimed at mapping research trends, assessing current adoption, and identifying future directions for AI in pharmaceutical warehousing. Specifically, it seeks to answer three questions: (1) What are the bibliometric trends in AI-related publications on pharmaceutical warehousing? (2) What is the current state of AI adoption globally? (3) What future research directions can advance AI integration in this domain? By addressing these questions, the study provides insights to guide scholars, practitioners, and policymakers toward more efficient, resilient, and ethically responsible warehousing practices.

2 Research Methodology

This study adopted the Bibliometric-Systematic Literature Review (B-SLR) approach, combining bibliometric mapping with systematic review rigour to examine AI applications in pharmaceutical warehousing. Using Scopus as the primary database, we applied a structured search strategy with predefined keywords, inclusion/exclusion criteria, and Open Access filters to ensure transparency and reproducibility. After screening, 41 relevant articles published between 2017 and 2025 were retained. Bibliometric analysis was performed using VOSviewer to map citation networks, keyword co-occurrences, and thematic clusters, which informed the synthesis of conceptual patterns and future research directions.

Search Strategy and Screening Process. To ensure reproducibility and methodological rigour [10], data collection was conducted using the Scopus database, selected for its extensive coverage of high-impact supply chain and management literature [11]. The search strategy targeted the intersection of advanced digital technologies and pharmaceutical logistics within the timeframe of 2017 to 2025. The specific Boolean search string employed was: *TITLE-ABS-KEY (("Artificial Intelligence" OR "AI" OR "Machine Learning" OR "Digital Transformation" OR "Industry 4.0") AND ("Pharmaceutical" OR "Pharma") AND ("Warehousing" OR "Supply Chain" OR "Logistics")).*

The initial search yielded 269 documents, which were then subjected to a multi-stage screening process in accordance with PRISMA guidelines [12]. Inclusion and exclusion criteria were applied as follows:

1. Subject Area Filter: To ensure managerial relevance, the scope was restricted to the "Business, Management, and Accounting" subject area, excluding 164 documents primarily focused on purely technical engineering or clinical pharmacology domains.
2. Document Type Filter: To maintain academic quality, only peer-reviewed "Articles" were retained; conference papers, reviews, and book chapters were excluded (n = 63).
3. Publication Stage and Language: Only documents in the "Final" publication stage and written in "English" were included to ensure data completeness and accessibility. Following this rigorous filtration, a final corpus of 41 high-quality articles was selected for bibliometric mapping and systematic synthesis [13].

The adoption of the B-SLR protocol ensures methodological rigour by combining the objective, quantitative mapping of bibliometrics with the qualitative depth of systematic synthesis [13]. This hybrid approach mitigates the selection bias often inherent in traditional narrative reviews and facilitates a transparent, reproducible analysis of the field's intellectual structure [14]. Consequently, the subsequent analysis is designed to move beyond descriptive reporting, aiming instead to critically interpret the theoretical trajectories and strategic implications of AI adoption within this highly regulated domain (Fig. 1).

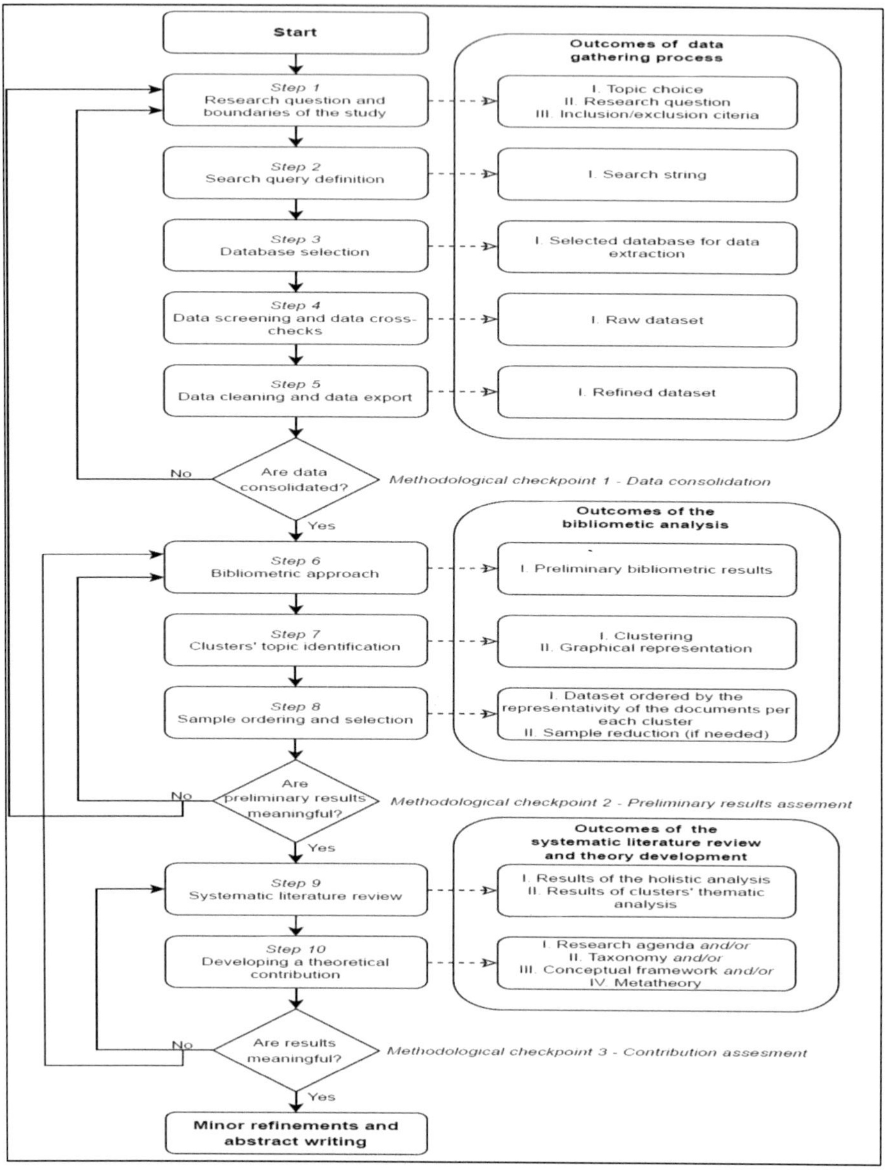

Fig. 1. B-SLR methodological approach, *Source* [13].

3 Results

3.1 Descriptive Analysis

The bibliometric analysis revealed notable growth in scholarly attention toward Artificial Intelligence in pharmaceutical warehousing, with a significant acceleration in publication activity in recent years. Figure 2: Annual scientific production illustrates the trajectory of research output between 2017 and 2025, beginning with isolated contributions in 2017 and 2020, followed by a marked expansion from 2022 onward. The number of articles rose sharply from 5 in 2022 to 11 in 2023, peaking at 12 in 2024 before stabilising at 11 in 2025. This upward trend highlights the growing relevance of AI-driven warehousing in supply chain research, reflecting both the field's maturation and the increasing recognition of its strategic importance in addressing operational and regulatory challenges in pharmaceutical logistics.

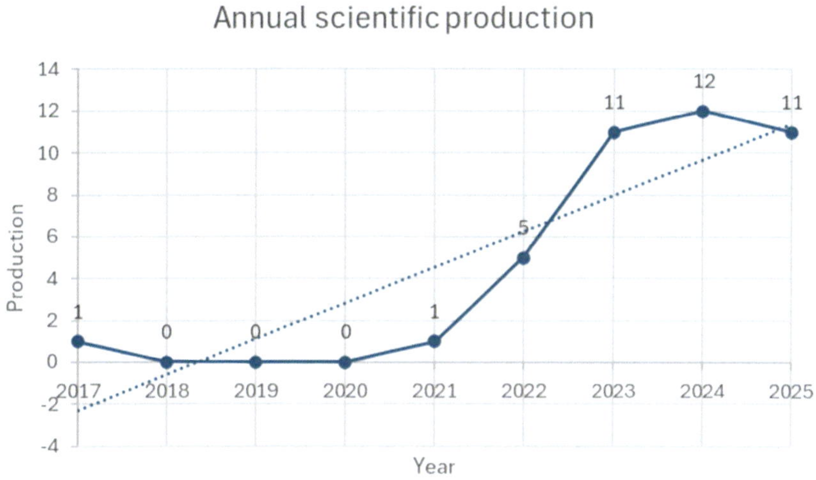

Fig. 2. Annual scientific production.

In this section, we concentrate on publications from the most recent five-year period to highlight and interpret the evolving trajectory of research on Artificial Intelligence and digital technologies in supply chain and pharmaceutical warehousing. The bibliometric evidence indicates that recent contributions have expanded the scope of inquiry beyond traditional logistics to encompass advanced themes such as Industry 5.0 inventory practices [15], machine learning–based forecasting systems for real-time warehouse decision-making [16], and the correlation between digitisation of manufacturing processes and economic performance [17]. Parallel studies have examined blockchain-supported inventory sharing in relief supply chains [18], organisational enablers of augmented analytics and BI 4.0 [19], and chatbot adoption in government agencies [20], all of which underscore the diversification of AI applications across logistics and management domains. By focusing on this unstudied period, the analysis captures the surge of interdisciplinary scholarship that situates AI as a driver of resilience, transparency, and efficiency in warehousing and broader supply chain ecosystems.

Table 1 provides an overview of the bibliometric dataset, highlighting the scope of research activity in Artificial Intelligence applied to pharmaceutical warehousing. The analysis spans the period 2017 to 2025, covering 41 journal articles drawn from 34 different sources. The dataset exhibits a strong annual growth rate of 34.95%, with documents averaging 1.49 years old and receiving an average of 16.46 citations each.

A total of 393 references were recorded, supported by 244 Keywords Plus and 211 author-defined keywords, reflecting the thematic diversity of the field. Collaboration is a defining feature of this research area, with 200 contributing authors, no single-authored documents, and an average of 7.46 co-authors per paper. International collaboration is also notable, accounting for 39.02% of the publications, underscoring the global relevance of AI-driven warehousing.

Table 2 identifies the most influential publication outlets in this domain. The International Journal of Production Research leads with four articles, followed by Transportation Research Part E: Logistics and Transportation Review with three contributions. Both IEEE Engineering Management Review and Uncertain Supply Chain Management published two articles each. At the same time, several other journals, including Acta Logistica, Business Process Management Journal, Business Strategy and the Environment, Cleaner Logistics and Supply Chain, Cogent Business and Management, and Economics and Sociology, each contributed one article. This distribution illustrates the interdisciplinary nature of the topic, with contributions spanning logistics, management, engineering, and sustainability-focused journals.

Table 1. Main information about the dataset.

Description	Results
Main information about data	
Timespan	2017:2025
Sources (Journals, Books, etc.)	34
Documents	41
Annual Growth Rate %	34.95
Document Average Age	1.49
Average citations per doc	16.46
References	393
Document contents	
Keywords Plus (ID)	244
Author's Keywords (DE)	211
Authors	
Authors	200
Authors of single-authored docs	0
Authors collaboration	
Single-authored docs	0
Co-Authors per Doc	7.46
International co-authorships %	39.02
Document types	
Article	41

Table 2. Most relevant sources

Sources	Articles
International journal of production research	4
Transportation research part e: logistics and transportation review	3
IEEE engineering management review	2
Uncertain supply chain management	2
Acta logistica	1
Business process management journal	1
Business strategy and the environment	1
Cleaner logistics and supply chain	1
Cogent business and management	1
Economics and sociology	1

In bibliometric analysis, identifying the most relevant authors provides insight into the key contributors shaping the intellectual landscape of a research field. Table 3 highlights the leading scholars in the domain of Artificial Intelligence in pharmaceutical warehousing. Among them, Ngah A.H. stands out with the highest number of contributions, credited with three articles and a fractionalised score of 0.70, indicating consistent engagement across collaborative works. Several authors—including Duman G.M., Kongar E., Williams R.A., Panigrahi R., and Shrivastava A.—each contributed two articles, with fractionalised scores ranging from 0.50 to 0.58, reflecting their active participation in multi-authored studies. Fosso Wamba S. also appears with two articles and a score of 0.42, underscoring his presence in the field.

Meanwhile, individual contributors such as Abusweilem M.A., Akter S., and Al-Gasawneh J. have each published one article, suggesting emerging interest or specialised contributions. These figures demonstrate a collaborative authorship structure, with a mix of recurring contributors and new entrants, reflecting both the maturity and expanding scope of AI research in pharmaceutical warehousing. Several of these papers focused on applying artificial intelligence and machine learning techniques to enhance inventory management, forecasting accuracy, and operational decision-making in pharmaceutical and logistics environments.

Table 3. Most relevant authors

Author	Articles	Articles fractionalised
Ngah AH	3	0.70
Duman GM	2	0.50
Fosso Wamba S	2	0.42
Kongar E	2	0.50
Panigrahi R	2	0.58

(*continued*)

Table 3. (*continued*)

Author	Articles	Articles fractionalised
Shrivastava A	2	0.58
Williams RA	2	0.50
Abusweilem MA	1	0.33
Akter S	1	0.17
Al-Gasawneh J	1	0.25

Table 4 presents the most widely cited documents in this domain, providing insight into the works with the greatest academic impact. The study by [21] in Transportation Research Part E: Logistics and Transportation Review has garnered 103 citations, reflecting its significant influence on subsequent scholarship. Reference [22] in the International Journal of Production Research follow with 78 citations, while more recent contributions, such as [23] and [24], have already achieved high citation rates of 52 and 49, respectively, indicating their rapid uptake in the field. Earlier foundational work, such as [25], continues to be referenced, while studies by [26, 27], and [28] also stand out for their notable citation performance. Collectively, these highly cited publications illustrate the evolving research agenda and highlight the pivotal studies that have shaped discourse on AI and supply chain resilience.

Table 4. Most globally cited documents

Document	Total citations	TC per year	Normalised TC
Govindan, (2022), Transp res Part e Logist Transp Rev	103	25.75	2.13
Nguyen, (2022). Int J Prod Res	78	19.50	1.61
Tiwari, (2024). Transp Res Part e Logist Transp Rev	52	26.00	3.07
Zhu, (2024), Int J Prod Econ	49	24.50	2.90
Nkhoma, (2017). Educ Train	44	4.89	1.00
Delgosha, (2022). Technovation	41	10.25	0.85
Akter, (2023). Ind Mark Manage	40	13.33	3.70
Kurniawati, (2023). Clean Logist Supply Chain	34	11.33	3.14
Nitsche, (2021). Supply Chain Forum	33	6.60	1.00
Panigrahi, (2024). Int J Product Perform Manage	22	11.00	1.30

Figure 3 presents a three-field plot that illustrates the interconnections between source journals, author keywords, and contributing countries, offering a comprehensive view of the intellectual structure of research on Artificial Intelligence in pharmaceutical warehousing. The visualisation highlights how leading journals, such as Transportation Research Part E, International Journal of Production Research, and Business Strategy and the Environment, are closely linked with dominant descriptors, including "Artificial Intelligence," "Machine Learning," "Supply Chain," and "Digital Transformation." These thematic clusters are further associated with substantial contributions from countries such as the USA, China, India, Italy, and Jordan, underscoring both the global reach and interdisciplinary nature of the field. The plot illustrates how scholarly output is influenced by the convergence of technological innovation, logistics applications, and international collaboration, thereby underscoring the strategic importance of AI-driven research in fostering resilient and efficient pharmaceutical supply chains.

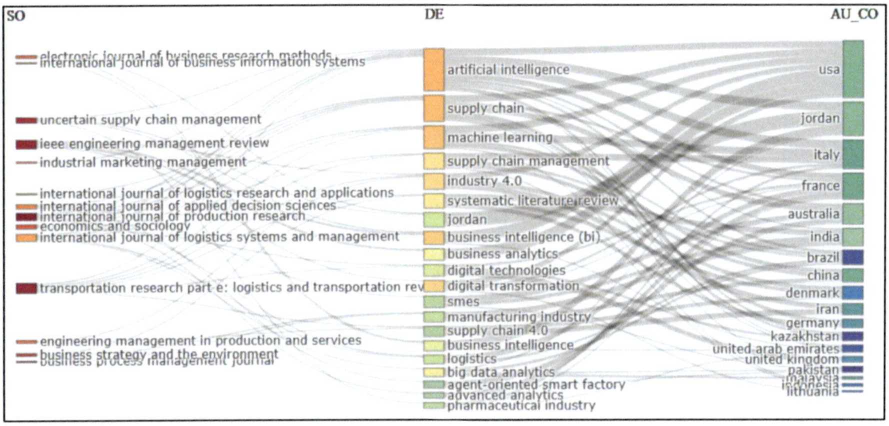

Fig. 3. Three-field plot

3.2 Scientific Production Analysis

Figure 4 illustrates the scientific output of countries in the field of Artificial Intelligence and pharmaceutical warehousing, based on the number of publications recorded in the dataset. The United States is the leading contributor, with 17 articles, followed by China with 14 and Australia with 13. India and Iran also demonstrate strong engagement, with 12 and 11 publications, respectively. European countries such as Germany (8), France (7), and Italy (6) make notable contributions, reflecting the global spread of research activity. These results highlight the dominance of the United States and China in driving scholarly output, while also underscoring the growing participation of both emerging and established economies in advancing AI-driven warehousing research. Countries outside this group have published fewer than six papers, indicating a more limited yet still valuable role in shaping the discourse. Much of the research in this area is produced by developed economies, including China, India, and Iran, with limited research from Africa and South America. This may imply an AI divide between countries capable of adopting the technologies and those that cannot.

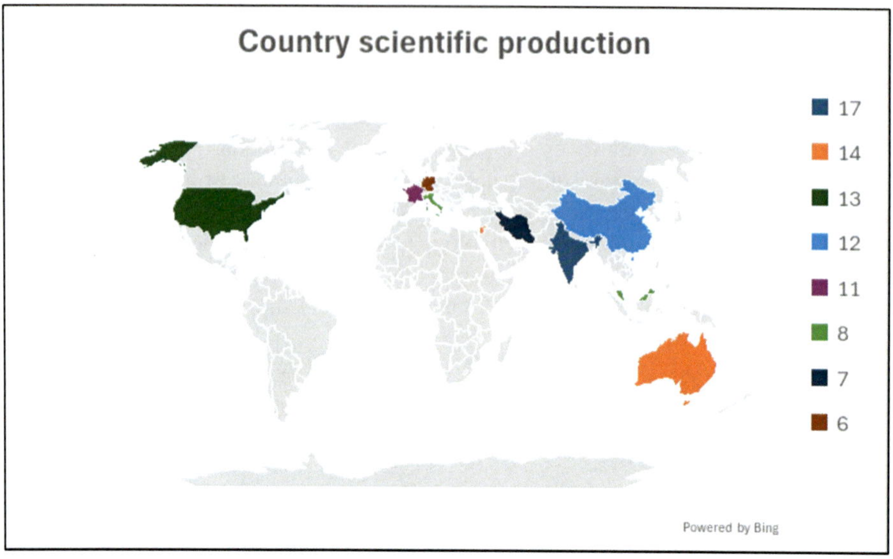

Fig. 4. Country scientific production

We will now identify the affiliations that are most active in the fields of Artificial Intelligence and pharmaceutical warehousing research. In our case, this is illustrated in Fig. 5: Most Relevant Affiliations. The histogram highlights that Universiti Malaysia Terengganu leads with seven articles, followed by both the University of New Haven and Università di Pisa, each contributing five publications. Beijing Jiaotong University and the MIT Centre for Transportation and Logistics also demonstrate strong engagement with four articles each. Other institutions, such as Gitam University, Liverpool Business School, Luminus Technical University College, RMIT University, and RMIT University Vietnam, each contributed three articles. These results underscore the diverse institutional participation in advancing AI-driven warehousing research, with notable contributions from universities across Asia, Europe, and North America.

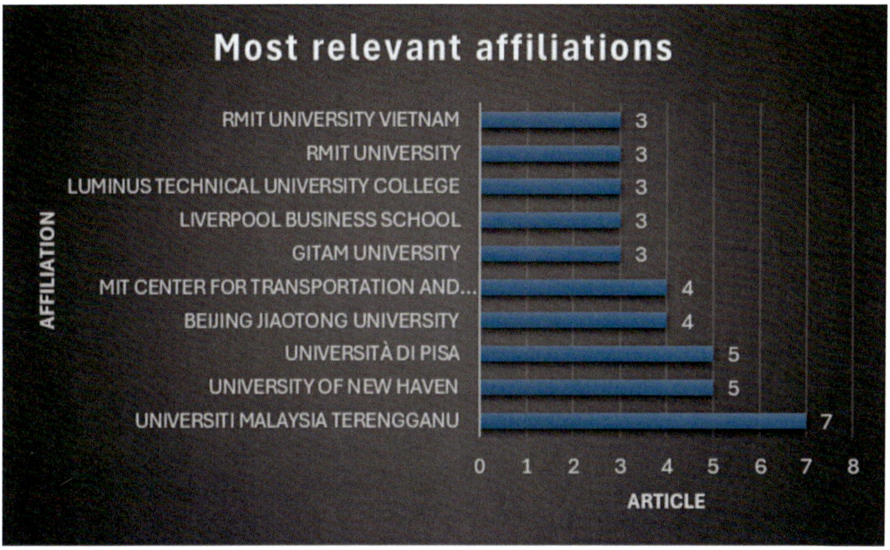

Fig. 5. Most relevant affiliations

Figure 6 presents the most frequently occurring terms in the literature on Artificial Intelligence and pharmaceutical warehousing, derived from the bibliometric dataset. The most prominent word is artificial intelligence with seven occurrences, underscoring its central role in shaping the discourse. This is followed by supply chain (5 occurrences) and machine learning (4 occurrences), which highlight the technological and operational focus of recent studies. Other frequently used terms include business intelligence (BI), digital transformation, industry 4.0, supply chain management, and systematic literature review, each appearing 3 times, reflecting the methodological and thematic breadth of the field. Additional concepts, such as big data analytics, business analytics, digital technologies, logistics, the manufacturing industry, the pharmaceutical industry, and SMEs, are mentioned twice, highlighting the diverse applications and contexts in which AI is being explored. Less frequent but notable terms include advanced analytics, agent-oriented smart factory, algorithmic bias, and AI-driven marketing analytics. Collectively, these keywords reveal the dominant research themes, which converge around digital innovation, supply chain resilience, and the integration of AI-driven analytics into pharmaceutical warehousing contexts.

Fig. 6. Word cloud

3.3 Keyword Co-occurrences Cluster Classification

Figure 7 presents the results of the cluster classification based on keyword co-occurrences, which groups elements with similar thematic characteristics. Five major clusters were identified in the dataset. The first cluster (red) is the largest, comprising 31 items. It is characterised by themes such as artificial intelligence, machine learning, decision-making, forecasting, and information management, reflecting the field's technological and analytical foundations. The second cluster (green), comprising 25 items, is characterised by topics including Industry 4.0, Supply Chain 4.0, the Internet of Things, blockchain, and embedded systems, highlighting the integration of advanced digital technologies into supply chain operations. The third cluster (blue), comprising 15 items, emphasises business intelligence, business analytics, digital transformation, and Jordan, indicating organisational and regional perspectives on digital adoption. The fourth cluster (yellow), with 12 items, focuses on inventory management, performance assessment, literature review, operational performance, and SMEs, underscoring managerial practices and performance evaluation. Finally, the fifth cluster (purple), comprising 15 items, is defined by themes such as healthcare, digital technologies, state-of-the-art, and technology, which reflect the applications of AI in healthcare and broader technological innovation. Collectively, these clusters illustrate diverse yet interconnected research directions, spanning technical innovation and digital transformation, managerial practices, and sector-specific applications.

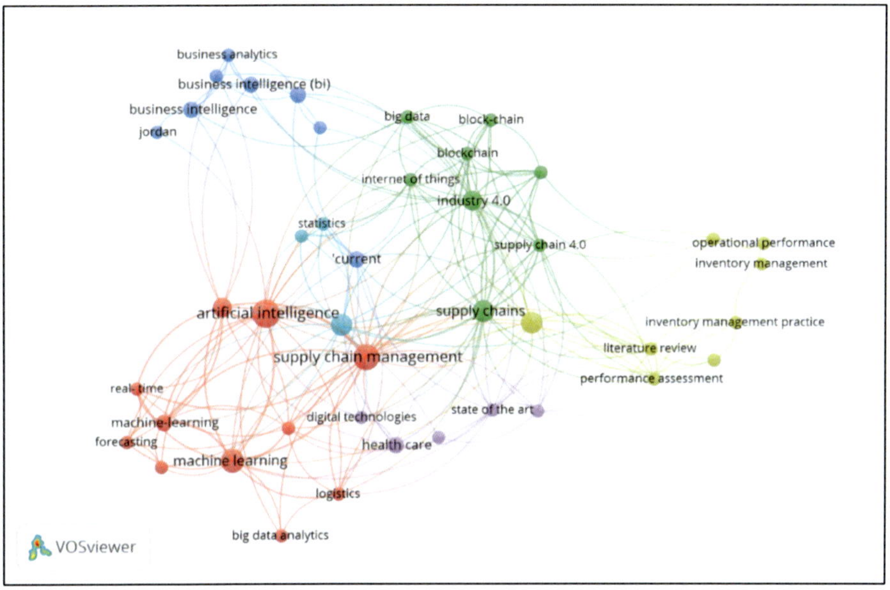

Fig. 7. Keyword co-occurrences cluster classification

4 Discussion, Limitations, and Future Direction

The exponential rise in scholarly output since 2022 signals a critical paradigm shift in pharmaceutical warehousing, moving from tentative digital experimentation to a strategic imperative for resilience in highly regulated environments. This post-pandemic acceleration reflects the sector's urgent response to global supply chain disruptions, where AI is no longer merely an efficiency tool but a requisite for maintaining continuity under strict compliance pressures [23]. The dominance of 'machine learning', 'forecasting', and 'decision-making' within the primary keyword clusters suggests that current AI strategies are fundamentally transitioning the sector from reactive inventory tracking to proactive, predictive risk management [16]. Furthermore, the integration of 'IoT' and 'Blockchain' within secondary clusters indicates that future technology adoption is being driven by the need to enforce the rigorous traceability and data governance required by GMP and GDP standards, rather than cost reduction alone [4, 18]. Consequently, the intellectual trajectory of the field points toward a sophisticated model of 'anticipatory logistics,' where emerging human-centric 'Industry 5.0' concepts are beginning to reshape how workforce adaptation interacts with automated quality assurance systems [15]. While the strict regulatory frameworks of GMP and GDP are specific to the pharmaceutical domain [29]The AI-driven capabilities identified in this review—specifically, real-time condition monitoring and granular traceability—possess significant transferability to other high-value and perishable sectors. For instance, the application of blockchain for serialisation and inventory transparency is directly applicable to the food industry, where ensuring safety and eliminating contamination pathways are paramount [30], as well as to humanitarian relief chains that require trusted, decentralised inventory sharing [18]. Similarly, the sector's adoption of digital control towers for 'stress testing' supply chain resilience offers a transferable model for high-tech manufacturing and automotive logistics, where managing component volatility and disruption risks necessitates similar levels of predictive oversight [31]. The prominence of decision-support and monitoring themes in the bibliometric data reflects the sector's unique exigency to manage GMP-compliant, temperature-controlled environments, thereby distinguishing these AI applications from those found in less regulated general warehousing contexts [7].

5 Conclusion

Artificial Intelligence is transforming pharmaceutical warehousing by improving efficiency, accuracy, and resilience in global supply chains. AI-driven tools such as robotics, predictive analytics, and machine learning enable cost reduction and quality assurance in highly regulated environments. This study, through a bibliometric and systematic review, highlighted key themes like forecasting accuracy, inventory optimisation, and emerging areas such as blockchain and Industry 5.0. Beyond pharmaceutical warehousing, the sophisticated integration of AI for compliance and cold-chain integrity offers a blueprint for other regulated industries, such as food safety and high-tech logistics, where zero-error mandates are becoming increasingly prevalent. By demonstrating how digital transformation can coexist with rigorous governance, this sector serves as a critical case study for managing the trade-offs between automation efficiency and human-centric operational resilience across the broader supply chain landscape. While adoption is accelerating, challenges around data governance, compliance, and workforce adaptation remain. Continued research and innovation are essential to ensure AI-driven warehousing evolves into a sustainable, patient-centric component of healthcare logistics.

References

1. Tavakoli, M.J., Fazl, F., Sedighi, M., Naseri, K., Ghavami, M., Taghipour-Gorjikolaie, M.: Enhancing pharmacy warehouse management with faster R-CNN for accurate and reliable pharmaceutical product identification and counting. Int. J. Intell. Syst. **2025**, 1–15 (2025)
2. Kumar, P., Nunavath, R.S., Nagappan, K.V.: Future of pharmaceutical industry: role of artificial intelligence, automation and robotics. J. Pharmacol. Pharmacother. **15**(2), 142–152 (2024)
3. Serrano, D.R., Luciano, F.C., Anaya, B.J., Ongoren, B., Kara, A., Molina, G., et al.: Artificial intelligence (AI) applications in drug discovery and drug delivery: revolutionizing personalized medicine. Pharmaceutics **16**(10), (2024)
4. Kalodanis, K., Feretzakis, G., Rizomiliotis, P., Verykios, V.S., Papapavlou, C., Koutsikos, I., et al., eds.: Data Governance in Healthcare AI: Navigating the EU AI Act's Requirements. Studies in Health Technology and Informatics (2025)
5. Silverstein, I.: Bulk pharmaceutical excipient GMPs. Good Manufacturing Practices for Pharmaceuticals, 6th edn. (2016), pp. 259–68
6. Shibabaw, A.D., Nakambale, H.N., Bangalee, V.: Inventory management practices: implications on the pharmaceuticals expenditure of rabies vaccine in public health facilities, Namibia. BMC Health Serv. Res. **23**(1), (2023)
7. Tutam, M.: Pharmaceutical Warehousing 4.0 in Healthcare Supply Chain. Accounting, Finance, Sustainability, Governance and Fraud (2023), pp. 35–50

8. Iliashenko, O., Iliashenko, V., Filippova, K., Lohyeeta, N., eds.: Pharmaceutical Logistics: Features and Challenges. Lecture Notes in Networks and Systems (2022)
9. Rana, J., Daultani, Y.: Mapping the role and impact of artificial intelligence and machine learning applications in supply chain digital transformation: a bibliometric analysis: supply chain digital transformation: a bibliometric analysis. Oper. Manag. Res. **16**(4), 1641–1666 (2023)
10. Donthu, N., Kumar, S., Mukherjee, D., Pandey, N., Lim, W.M.: How to conduct a bibliometric analysis: an overview and guidelines. J. Bus. Res. **133**, 285–296 (2021)
11. Mongeon, P., Paul-Hus, A.: The journal coverage of web of science and Scopus: a comparative analysis. Scientometrics **106**(1), 213–228 (2016)
12. Page, M.J., McKenzie, J.E., Bossuyt, P.M., Boutron, I., Hoffmann, T.C., Mulrow, C.D., et al.: The PRISMA 2020 statement: an updated guideline for reporting systematic reviews. Revista Panamericana de Salud Publica/Pan Am. J. Public Health. 46 (2022)
13. Marzi, G., Balzano, M., Caputo, A., Pellegrini, M.M.: Guidelines for bibliometric-systematic literature reviews: 10 steps to combine analysis, synthesis and theory development. Int. J. Manag. Rev. **27**(1), 81–103 (2025)
14. Tranfield, D., Denyer, D., Smart, P.: Towards a methodology for developing evidence-informed management knowledge by means of systematic review. Br. J. Manag. **14**(3), 207–222 (2003)
15. Panigrahi, R., Shrivastava, A., Sarkar, B., Kaliyan, M.: Inventory knowledge, industry 5.0 and performances: review and future research agenda. Bus. Strategy Environ. **34**(6), 7816–36 (2025)
16. Aloini, D., Benevento, E., Dulmin, R., Guerrazzi, E., Mininno, V.: Unlocking real-time decision-making in warehouses: a machine learning-based forecasting and alerting system for cycle time prediction. Transp. Res. Part E: Logistics Transp. Rev. 194 (2025)
17. Wang, J., Wang, S., Zhang, Y.: Artificial intelligence for visually impaired. Displays 77 (2023)
18. Ju, Y., Hou, H., Yang, J., Zhang, Q., Ren, Y.: Pre-positioning supplies with blockchain-supported inventory sharing in relief supply chains. Int. J. Prod. Res. **63**(13), 4820–4844 (2025)
19. Williams, R.A., Duman, G.M., Kongar, E., Sheikh, N.J.: Organizational factors enabling augmented analytic and BI 4.0 implementation success. IEEE Eng. Manage. Rev. **53**(5), 160–76 (2025)
20. Jais, R., Ngah, A.H., Rahi, S., Al-Gasawneh, J.: Explaining the adoption of chatbots: the experience of Malaysian government agencies. Int. J. Public Adm. **48**(11), 698–711 (2025)
21. Govindan, K., Kannan, D., Jørgensen, T.B., Nielsen, T.S.: Supply chain 4.0 performance measurement: a systematic literature review, framework development, and empirical evidence. Transp. Res. Part E: Logistics Transp. Rev. 164 (2022)
22. Nguyen, A., Lamouri, S., Pellerin, R., Tamayo, S., Lekens, B.: Data analytics in pharmaceutical supply chains: state of the art, opportunities, and challenges. Int. J. Prod. Res. **60**(22), 6888–6907 (2022)
23. Tiwari, M., Bryde, D.J., Stavropoulou, F., Dubey, R., Kumari, S., Foropon, C.: Modelling Supply Chain Visibility, Digital Technologies, Environmental Dynamism and Healthcare Supply Chain Resilience: An Organisation Information Processing Theory Perspective. Transp. Res. Part E: Logistics Transp. Rev. 188 (2024)
24. Zhu, M., Liang, C., Yeung, A.C.L., Zhou, H.: The impact of intelligent manufacturing on labor productivity: an empirical analysis of Chinese listed manufacturing companies. Int. J. Prod. Econ. 267 (2024)
25. Nkhoma, M.Z., Lam, T.K., Sriratanaviriyakul, N., Richardson, J., Kam, B., Lau, K.H.: Unpacking the revised Bloom's taxonomy: developing case-based learning activities. Educ. Train. **59**(3), 250–264 (2017)

26. Delgosha, M.S., Hajiheydari, N., Talafidaryani, M.: Discovering IoT implications in business and management: a computational thematic analysis. Technovation 118 (2022)
27. Akter, S., Sultana, S., Mariani, M., Fosso Wamba, S., Spanaki, K., Dwivedi, Y.K.: Advancing algorithmic bias management capabilities in AI-driven marketing analytics research. Ind. Mark. Manage. **114**, 243–261 (2023)
28. Kurniawati, D.A., Cakravastia, A.: A review of halal supply chain research: sustainability and operations research perspective. Cleaner Logistics Supply Chain 6 (2023)
29. Ballal, P.M., Umathe, S.N., Jawarkar, R.D.: Comparative account of diverse regulatory guidelines for storage and transportation of pharmaceuticals. Int. J. PharmTech Res. **8**(7), 180–192 (2015)
30. Naseem, S., Rizwan, M.: The role of artificial intelligence in advancing food safety: a strategic path to zero contamination. Food Control. 175 (2025)
31. Hong, J., Song, S., Lau, K.H., Zhao, N.: Enhancing pharmaceutical supply chains: unveiling the power of digital control tower through stress testing. Int. J. Logistics Res. Appl. (2025)

Controlling Supply Chain Disruptions Through the Democratization of Artificial Intelligence

Reda Saib(✉) and My Abdelouhab Salahddine

National School of Business and Management, Abdelmalek Essaadi University, Tetouan, Morocco
reda.saib@etu.uae.ac.ma

Abstract. In today's competitive global market, Artificial Intelligence (AI) has become an academic and scientific field that deserves in-depth exploration. Companies utilizing AI have a profound impact on operational efficiency and strategic decision-making. However, the challenges associated with programming and deploying traditional AI solutions often hinder widespread adoption. The purpose of this article is to explore how no-code and low-code AI can support supply chain resilience amid disruptions. First, we started by examining the critical issues in supply chain management, particularly the bullwhip effect, which creates significant disruptions by amplifying demand fluctuations. To address these challenges, the integration of Industry 4.0 technologies is explored, which enhances supply chain resilience through the use of real-time data and advanced analytics. Then, we highlighted the role of AI in transforming supply chains. By optimizing inventory management, improving demand forecasting, and facilitating data-driven decision-making, AI becomes a crucial tool for organizations striving to enhance their resilience in a rapidly changing market landscape. Finally, we discussed the democratization of AI through no-code and low-code platforms to lower barriers to adoption. These tools enable a wider range of users to implement AI solutions quickly and cost-effectively, fostering timely decision-making and enhancing operational agility. By combining these elements, organizations can efficiently manage the complexity of modern supply chains and capitalize on future opportunities.

Keywords: No-code · Low-code · Artificial Intelligence · Industry 4.0 · Supply Chain Resilience

JEL classification: L60 · L86 · M15 · O3

1 Introduction

In today's dynamic business management landscape, organizations encounter a variety of significant challenges. The complexities of globalization, along with current economic, environmental, and geopolitical dynamics, complicate supply chains and exacerbate disruptions [1]. Additionally, the recent COVID-19 pandemic has profoundly affected global production, distribution, and transportation networks, leading to severe repercussions across various sectors [2].

Y. Benadada et al. (Eds.): GOL 2026, LNNS 2034, pp. 89–96, 2027.
https://doi.org/10.1007/978-3-032-29677-1_9

To thrive in this competitive market, businesses must seize new opportunities and refine their operational processes. The interconnectedness fostered by globalization allows companies to expand into new, potentially lucrative markets. However, this interdependence also reveals critical vulnerabilities, compelling organizations to develop adaptive strategies and gain a profound understanding of local market dynamics to navigate the complexities of the contemporary business environment.

Effective supply chain management is essential for all stakeholders aspiring to implement a coherent business model that maximizes long-term performance throughout the entire chain. Achieving this goal requires the systematic and strategic alignment of logistics activities and manufacturing operations, both within organizations and in collaboration with external partners [3, 4]. In this context, data-driven technologies have become central to Industry 4.0 supply chains [5]. Additionally, the ability of Artificial Intelligence (AI) to learn from large datasets and constantly improve its decision-making skills offers considerable promise for supporting companies.

AI plays a crucial role in the digital transformation of businesses, greatly influencing human decision-making processes [6]. Nevertheless, companies face numerous challenges in implementing AI during its early stages [7]. In this regard, no-code and low-code AI tools facilitate rapid development by offering user-friendly drag-and-drop solutions, adaptable to various scenarios with customizable algorithms [8].

The No-Code AI (NCAI) machine learning models are designed to deliver rapid, cost-effective solutions, thereby improving overall business performance. Thus, the integration of advanced AI enables significant energy savings [9]. By leveraging these advanced technologies and strategies, organizations can not only mitigate the impact of current supply chain challenges, but also position themselves advantageously for future growth in an increasingly complex global marketplace.

2 Supply Chain Fundamentals

2.1 Supply Chain Disruption

Optimizing the supply chain has become a fundamental priority for modern businesses aiming to improve efficiency and reduce costs. However, many decision-makers remain relatively unaware of a critical phenomenon, known as the bullwhip effect. This effect, prevalent across various industries, refers to the tendency for order variations to increase as one moves up the supply chain on the demand side. Thus, minor fluctuations in consumer demand can lead to significant variations in production levels among suppliers at the other end of the chain.

The bullwhip effect was first identified by Forrester [10], who demonstrated that a 10% increase in retail demand could result in a 40% surge at the manufacturer level. His innovative simulation experiment, known as the "Beer Distribution Game," highlight the decision-making behaviors occurring within supply chains (Table 1). Further research in the early 1980s documented this amplification phenomenon between Procter & Gamble and its suppliers, coining the term "bullwhip effect" [11].

Subsequent analyses by scholars, including [12, 13], attributed this amplification to the irrational behavior of decision-makers. They observed that individuals often fail to consider the broader supply chain, harbor distrust toward the performance of other members, and base their decisions on distorted perceptions of demand. In contrast, [14] proposed alternative explanations suggesting that actors behave rationally within the context of the supply chain, complicating the discussion surrounding the bullwhip effect [15].

Table 1. Key dates in the history of the bullwhip effect.

Years	Authors	Research
1961	Forrester	Formulation of the demand amplification variability by the "industrial dynamics" approach Implementation of a simulation experience, the 'Beer Distribution Game'
1981	Schisgall	Documentation of the demand amplification between Procter & Gamble and its suppliers by calling it "Bullwhip effect"
1989	Sterman	Attributing amplification to "irrational behavior" of decision makers
1997	Lee et al	Proposition of four additional causes where it is assumed that the actors behave in a completely rational way

The fundamental principle underlying the "Beer Distribution Game" is to conduct simulations that reflect a real-time, multi-level system characterized by delayed demand, infinite capacity, processing, and shipping times (Fig. 1). Through careful adjustments in training and communication protocols, both practitioners and researchers can observe how each stage operates concerning its perception of demand. This includes assessing the impact of replenishment policies on inventory management and overall approaches to meeting customer needs.

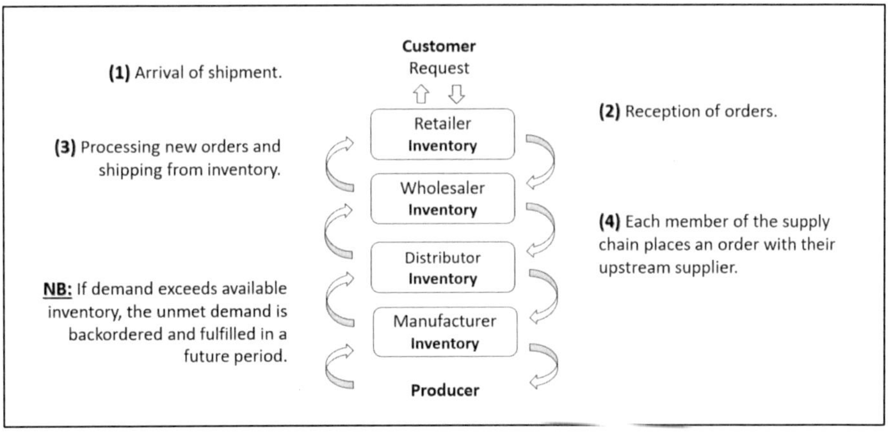

Fig. 1. Simulation experience highlighting decision-making behaviors within supply chains.

Therefore, as our understanding of the bullwhip effect evolves, it becomes increasingly important for decision-makers to adopt comprehensive strategies that consider behavioral dynamics throughout the supply chain. Implementing robust communication channels and fostering collaborative relationships among supply chain, mainly through the digital transformation offered by Industry 4.0, significantly reduce the amplitude of demand fluctuations, ultimately leading to more resilient and efficient supply chains.

2.2 Supply Chain Resilience

Supply chain resilience is defined as the ability of supply chains to persist, adapt, or change in response to various disruptions [16]. This resilience is increasingly crucial for organizations operating in an ever-changing global market marked by unpredictable events such as natural disasters, political instability, and economic fluctuations. To strengthen supply chain resilience, a collaborative approach among supply chain members is essential. This collaboration, as described in supply chain risk management, involves recognizing, assessing, and minimizing risks, thereby ensuring the smooth flow of goods, services, and information [17].

An effective strategy for strengthening supply chain resilience is to continuously adjust production and inventory levels to align them with customer demand. Increased demand variability often leads to inventory distortion, which can result in inventory level problems and significant revenue losses. Therefore, effective inventory management becomes a crucial factor for companies that want to maintain their operational efficiency [18]. As organizations strive to better cope with disruptions, researchers have proposed several new approaches to strengthen supply chain resilience [19].

The emergence of Industry 4.0 technologies has been pivotal in this regard [20]. These technologies, recognized for their profound impact on supply chain management, particularly in mitigating disruptions, have the potential to significantly improve resilience [21]. Industry 4.0 technologies promote both robustness and resilience by enabling end-to-end visibility within supply chains, which is essential for effectively managing disruptions [22].

Furthermore, these technologies facilitate collaboration among supply chain partners by improving information sharing and visibility throughout supply chains. This interconnectivity is particularly important for dealing with disruptions and enables rapid and informed responses, a necessity in today's dynamic business environment [23]. By leveraging innovative technologies, traditional industries can undergo significant transformations, thereby improving their productivity, resilience, and competitiveness [24].

3 Supply Chain Resilience Through Artificial Intelligence

3.1 Adoption of Artificial Intelligence

Artificial intelligence (AI) has emerged as a pivotal tool for strengthening supply chain resilience in today's complex global markets. This transformative technology contributes significantly to business performance, improves quality, and increases customer satisfaction [25]. In an era characterized by rapid technological advances, the adoption of AI is essential for organizations seeking to navigate the complex dynamics of supply chain disruptions.

Driving innovation is foundational in democratizing AI, thereby enabling businesses to mitigate inventory distortions effectively [26]. The integration of AI-based approaches equips firms with the tools necessary for analyzing factors that mitigate supply chain disruption management. These AI-driven methodologies not only enhance operational efficiency but also foster a proactive stance towards potential disruptions.

AI-based tools significantly improve planning and preparedness for disruptive events, which has become increasingly essential given the volatile nature of today's supply chain landscape [27]. With accurate AI-based forecasts, organizations can optimize inventory replenishment and production quantities, aligning these operations with consumer demand while ensuring financial targets are met. This capability is crucial for maintaining the delicate balance between supply and demand, particularly in uncertain market conditions.

Furthermore, AI is designed to facilitate intelligent decision-making by analyzing large data sets. This analytical capability enables AI systems to identify the most effective and efficient measures to take during disruptive events [28]. By providing more accurate forecasts of supply and demand, AI enables businesses to make informed decisions that ultimately strengthen their resilience.

The integration of AI within Industry 4.0 frameworks plays a vital role in enhancing supply chain resilience. By driving innovation, optimizing inventory management, and facilitating informed decision-making, AI serves as an essential instrument for organizations striving to thrive in a challenging and ever-evolving market environment. The comprehensive deployment of AI not only mitigate vulnerabilities in supply chains but also position companies to capitalize on future opportunities.

3.2 Democratization Through No-Code and Low-Code

The modern business landscape is constantly evolving, particularly due to disruptions in supply chains. Artificial intelligence (AI), with its inherent ability to analyze comprehensive data, is designed to facilitate intelligent decision-making by determining the most effective course of action to respond to these disruptions [28]. In particular, AI improves the accuracy of forecasts regarding fluctuations in supply and demand, enabling companies to adapt quickly to market conditions [28].

However, the process of developing AI models often faces significant obstacles, including financial constraints and complex technical requirements. Creating AI algorithms is a challenging task that typically requires specialized computer science knowledge, which is a barrier for many organizations. As a result, people who are not computer scientists often find themselves ill-equipped to effectively implement AI solutions. In addition, the process of onboarding staff members typically involves significant training costs, which delays the realization of innovative ideas [29].

The adoption of AI within organizations is recognized as a long and complex process [30]. Nevertheless, a paradigm shift towards human-centered approaches, particularly through No-Code Artificial Intelligence (NCAI), offers a viable solution. NCAI enables people without formal programming expertise to anticipate problems related to managing supply chain disruptions. This approach not only mitigates the high costs and technical limitations traditionally associated with AI development, but also accelerates the deployment of predictive analytics.

No-code platforms provide a comprehensive environment for creating applications without requiring any coding knowledge, thereby democratizing access to AI capabilities [31]. Similarly, low-code solutions can facilitate the generation of predictive models that are essential for addressing supply chain disruption management related issues. These models can be deployed to test various scenarios, optimize processes, and assess associated risks. As a result, the no-code low-code AI framework enables organizations to quickly conduct less complex experiments, ultimately allowing them to benefit from accelerated results [32].

In the study [31], the authors highlight the transformative role of NCAI in addressing inventory distortion problems, showcasing how this technology allows users to develop complex predictive models without any programming knowledge. By employing NCAI, researchers swiftly created a robust inventory forecasting model in under 30 s, significantly reducing the barrier to entry for businesses seeking to improve inventory management. This capability enables organizations to forecast customer needs based on historical sales data more effectively, ultimately minimizing overstock and out-of-stock situations.

Furthermore, integrating AI into decision-making can improve the cognitive performance of organizations by emphasizing human-centered knowledge. This approach represents transformative potential for AI, enabling companies to use AI-enhanced thinking in a more intuitive and adaptive way [33]. The adoption of no-code and low-code artificial intelligence offers a viable path for organizations seeking to mitigate the effects of business and logistical disruptions. By reducing barriers to AI implementation, these innovative solutions facilitate timely decision-making and strengthen operational resilience in an increasingly unpredictable business environment.

4 Conclusion

The democratization of artificial intelligence through no-code and low-code solutions greatly enhances supply chain performance by empowering employees to harness data-driven insights for informed decision-making. Key findings emphasize improved resilience and agility when integrating AI with Industry 4.0 technologies, underscoring the importance of aligning AI initiatives with supply chain strategies.

These solutions lower barriers to AI implementation, allowing a broader range of users to access advanced technologies. This inclusivity fosters innovation, as individuals from diverse backgrounds contribute to AI projects. Furthermore, organizations utilizing no-code and low-code AI can facilitate timely decision-making and strengthen operational resilience, critical factors in today's unpredictable business environment. The ability of AI to simulate complex scenarios enables proactive preparedness and strategic responses to potential disruptions.

Future research should focus on the impact of these tools on performance metrics, employee engagement, cross-industry applications, and cultural influences, providing deeper insights into the effective utilization of democratized AI for supply chain optimization. By embracing these technologies in the context of Industry 4.0, organizations can enhance their supply chains, ensuring robustness and agility amid ever-evolving market challenges.

References

1. Nikookar, E., Varsei, M., Wieland, A.: Gaining from disorder: making the case for antifragility in purchasing and supply chain management. J. Purch. Supply Manag. **27**(3), 100699 (2021). https://doi.org/10.1016/j.pursup.2021.100699
2. Pournader, M., Kach, A., Talluri, S.: A review of the existing and emerging topics in the supply chain risk management literature. Decis. Sci. **51**(4), 867–919 (2020)
3. Mentzer, J.T., et al.: Defining supply chain management. J. Bus. Logist. **22**, 1–25 (2001)
4. CSCMP: Supply Chain Management Terms and Glossary, CSCMP Media center (2013)
5. Batta, A., Gandhi, M., Kar, A.K., Loganayagam, N., Ilavarasan, V.: Diffusion of blockchain in logistics and transportation industry: an analysis through the synthesis of academic and trade literature. J. Sci. Technol. Policy Manag. **12**(3), 378–398 (2021). https://doi.org/10.1108/JSTPM-07-2020-0105
6. Duan, Y., Edwards, J.S., Dwivedi, Y.K.: Artificial intelligence for decision making in the era of big data—evolution, challenges and research agenda. Int. J. Inf. Manage. **48**, 63–71 (2019). https://doi.org/10.1016/j.ijinfomgt.2019.01.021
7. Chatterjee, S., Ghosh, S.K., Chaudhuri, R.: Knowledge management in improving business process: an interpretative framework for successful implementation of AI–CRM–KM system in organizations. Bus. Process. Manag. J. **26**(6), 1261–1281 (2020). https://doi.org/10.1108/BPMJ-05-2019-0183
8. Chang, Y.-H., Ko, C.-B.: A study on the design of low-code and no code platform for mobile application development. Int. J. Adv. Smart Conver. **6**(4), 50–55 (2017). https://doi.org/10.7236/IJASC.2017.6.4.7
9. Rizvi, A.T., Haleem, A., Bahl, S., Javaid, M.: Artificial Intelligence (AI) and its applications in Indian manufacturing: a review. Curr. Adv. Mech. Eng. **825** (2021). https://doi.org/10.1007/978-981-33-4795-3_76
10. Forrester, J.: Industrial Dynamics. MIT Press, Cambridge, MA (1961)
11. Schisgall, O.: In: Ferguson, J.G. (ed.) Eyes on Tomorrow: the Evolution of Procter & Gamble, 1st ed. Pub. Co. ; Distributed by Doubleday, Chicago , New York (1981)
12. Sterman, J.D.: Modeling managerial behavior: misperceptions of feedback in a dynamic decision-making experiment. Manage. Sci. **35**, 321–339 (1989)
13. Croson, R., Donohue, K.: Upstream versus downstream information and its impact on the bullwhip effect. Syst. Dyn. Rev. **21**, 249–260 (2005)
14. Lee, H.L., Padmanabhan, V., Whang, S.: Information distortion in a supply chain: the bullwhip effect. Manage. Sci. **43**, 546–558 (1997)
15. Wang, X., Disney, S.M.: The bullwhip effect: progress, trends and directions. Eur. J. Oper. Res. **250**(3), 691–701 (2016). https://doi.org/10.1016/j.ejor.2015.07.022
16. Nikookar, E., Gligor, D., Russo, I.: Supply chain resilience: when the recipe is more important than the ingredients for managing supply chain disruptions. Int. J. Prod. Econ. **272**, 109236 (2024). https://doi.org/10.1016/j.ijpe.2024.109236
17. Fiksel, J., Polyviou, M., Croxton, K.L., Pettit, T.J.: From risk to resilience: learning to deal with disruption. MIT Sloan Manag. Rev. **56**(2), 79–86 (2015)
18. Panigrahi, R.R.: 3 Inventory management practice and performances of manufacturing. Emerg. Trends Decis. Sci. Bus. Oper. **46** (2022)
19. Nikseresht, A., Shokouhyar, S., Tirkolaee, E.B., Nikookar, E., Shokoohyar, S.: An intelligent decision support system for warranty claims forecasting: merits of social media and quality function deployment. Technol. Forecast. Soc. Chang. **201**, 123268 (2024). https://doi.org/10.1016/j.techfore.2024.123268

20. Ismail, K., Nikookar, E., Pepper, M., Stevenson, M.: The implications of Industry 4.0 for managing supply chain disruption and enhancing supply chain resilience: a systematic literature review. Int. J. Prod. Res. **63**(19), 7278–7304 (2025). https://doi.org/10.1080/00207543.2025.2493948
21. Qader, G., Junaid, M., Abbas, Q., Mubarik, M.S.: Industry 4.0 enables supply chain resilience and supply chain performance. Technol. Forecast. Soc. Chang. **185**, 122026 (2022). https://doi.org/10.1016/j.techfore.2022.122026
22. Ghobakhloo, M., Iranmanesh, M., Foroughi, B., Tseng, M.L., Nikbin, D., Khanfar, A.A.: Industry 4.0 digital transformation and opportunities for supply chain resilience: a comprehensive review and a strategic roadmap. Prod. Plan. Control. **36**, 61–91 (2025)
23. Chowdhury, S., Rodriguez-Espindola, O., Dey, P., Budhwar, P.: Blockchain technology adoption for managing risks in operations and supply chain management: evidence from the UK. Ann. Oper. Res. **327**(1), 539–574 (2023). https://doi.org/10.1007/s10479-021-04487-1
24. Ralston, P., Blackhurst, J.: Industry 4.0 and resilience in the supply chain: a driver of capability enhancement or capability loss? Int. J. Prod. Res. **58**(16), 5006–5019 (2020). https://doi.org/10.1080/00207543.2020.1736724
25. Pournader, M., Ghaderi, H., Hassanzadegan, A., Fahimnia, B.: Artificial intelligence applications in supply chain management. Int. J. Prod. Econ. 108250 (2021)
26. Jauhar, S.K., Zolfagharinia, H., Amin, S.H.: A DEAANN-based analytical framework to assess and predict the efficiency of Canadian universities in a service supply chain context. Benchmarking: Int. J. (ahead-of-print) (2022). https://doi.org/10.1108/BIJ-08-2021-0458
27. Dey, P.K., Chowdhury, S., Abadie, A., Vann Yaroson, E., Sarkar, S.: Artificial intelligence-driven supply chain resilience in Vietnamese manufacturing small-and medium-sized enterprises. Int. J. Prod. Res. **62**, 1–40 (2023)
28. Modgil, S., Singh, R.K., Hannibal, C.: Artificial intelligence for supply chain resilience: learning from Covid-19. Int. J. Logist. Manag. **33**(4), 1246–1268 (2022)
29. Digital Transformation. 7 Digital Transformation Challenges & How to Face Them. https://www.walkme.com/blog/digitaltr-7-challenges. Accessed 12-09-2025
30. Enholm, I.M., Papagiannidis, E., Mikalef, P., Krogstie, J.: Artificial intelligence and business value: a literature review. Inf. Syst. Front. 1–26 (2021)
31. Jauhar, S.K., Jani, S.M., Kamble, S.S., Pratap, S., Belhadi, A., Gupta, S.: How to use no-code artificial intelligence to predict and minimize the inventory distortions for resilient supply chains. Int. J. Prod. Res. **62**(15), 5510–5534 (2023). https://doi.org/10.1080/00207543.2023.2166139
32. Garcia-Ortiz, J., Sanchez-Viteri, S.: Identification of the factors that influence university learning with low-code/no-code artificial intelligence techniques. Electronics **10**(10), 1192 (2021). https://doi.org/10.3390/electronics10101192
33. How, M.L., Chan, Y.J., Cheah, S.M., Khor, A.C., Say, E.M.P.: Artificial intelligence for social good in responsible global citizenship education: an inclusive democratized low-code approach. In: Proceedings of the 3rd World Conference on Teaching and Education, pp. 81–89 (2021)

Generative Artificial Intelligence, Development of (New) Green Products in the Fuzzy Front End and ESG of Supply Chains in the Creative Industry in a Global South Economy

Selma Oliveira[1](✉) and Jennyfer M. Ferreira[2]

[1] Fluminense Federal University, Niterói, Brazil
selmaregina@id.uff.br
[2] R. Des. Ellis Hermydio Figueira, 783 - Bloco A - Aterrado, Volta Redonda, Rio de Janeiro 27213-145, Brazil
jennyferf@id.uff.br

Abstract. Using a sample of companies from the creative industry in a Global South economy—in this case, Brazil—we uncover the contributions of generative artificial intelligence (GEN-AI) (in terms of benefits, novelty, creativity, feasibility, and information quality) to the green innovation process and the strengthening of environmental, social, and governance (ESG) performance, focusing on idea generation in the fuzzy front end of the green innovation process. Our study is original, fills a gap in the literature, and makes substantive contributions: (I) it shows the power of GEN-AI for the ideation process in the green new product development (NPD) process within the creative industry supply chain; (ii) it sheds light on creative industry managers who intend to progress with the integration of GEN-AI for green innovation and ESG reinforcement in the fuzzy front end of the innovation process; and finally, (iii) it contributes to achieving the Sustainable Development Goals (SDGs): 9—industry, innovation and infrastructure; 12—responsible consumption and production; and 17—partnerships and means of implementation.

Keywords: Generative Artificial Intelligence (GEN-AI) · Green and ESG Innovation · Creative Industry Supply Chain · Idea Generation · Fuzzy Front End

1 Introduction

Our study examines the contributions of generative artificial intelligence (GEN-AI) (in terms of benefits, novelty, creativity, feasibility, and information quality) to the fuzzy front end of the green innovation process and the strengthening of environmental, social, and governance (ESG) performance. We focus on idea generation. We shed light on the supply chains of the creative industry in a Global South economy—in this case, Brazil Understanding this approach is important for several reasons: ESG has assumed a critical role for socially responsible stakeholders [1]. At the same time, it is demonstrated that ESG reinforces the performance of global green innovation [2].

Y. Benadada et al. (Eds.): GOL 2026, LNNS 2034, pp. 97–110, 2027.
https://doi.org/10.1007/978-3-032-29677-1_10

Therefore, it is prevalent to consider ESG criteria and green innovation to bring benefits to companies and society [1]. In this context, supply chains consider green innovation to mitigate their environmental impacts [3] and gain a competitive advantage [4]. Similarly, studies suggest that creative industries are innovation-intensive and play a crucial role in sustainability, enhancing their practices and influencing broader social changes. The literature [5] defines green innovation as a set of activities that strengthen the development of new products and technologies with the aim of reducing environmental risks. Analogously, the development and implementation of green innovation are complex processes, especially in the phase of selecting the best idea to solve a problem [6]. At the same time, a rich body of theory reports that companies encounter difficulties in choosing methods or techniques to select the most suitable idea (in the fuzzy front end of the innovation process) to advance their innovations, which causes substantial impacts on innovation results [7]. GEN-AI can shape and encourage this process [8]. GEN-AI creates new content: texts, images, videos, etc. GEN-AI has the potential to create something new [9]. GEN-AI integrated into idea generation during the ideation phase of the innovation process can help with creativity, which is the central element of this phase.

Previous studies [9] have shown that GEN-AI reinforces idea generation for innovation. Similarly, previous studies have highlighted that AI significantly affects the creative design industry [10]. However, few studies have investigated how GEN-AI strengthens idea generation in the green innovation process (fuzzy front end) for ESG performance in the creative industry supply chain. We propose the following research question: How does GEN-AI strengthen idea generation for green innovation oriented towards ESG performance in the creative industry supply chain? Our study is original, fills a gap in the literature, and offers substantial contributions: (i) it shows the contributions of GEN-AI in idea generation in the fuzzy front end of the green innovation process oriented towards ESG performance; (ii) it sheds light on managers who intend to advance the diffusion of green innovation powered by GEN-AI; (iii) it expands the theoretical body on the use of GEN-AI in green innovation management in the creative industry supply chain; (iv) it contributes to the SDGs: 9—industry, innovation and infrastructure; 12—responsible consumption and production; and 17—partnerships for the goals. After the introduction, the next Sect. 2 presents the literature review; followed by the methodology (Sect. 3). Then, it addresses the results (Sect. 4). Finally (Sect. 5), we highlight the discussions and conclusions of the research.

2 Literature Review

Faced with major global challenges (climate crises, environmental degradation, etc.), supply chains are anchored in green innovation. Green innovation products are an organizational strategy that emphasizes the development and production of new products that consider aspects of technology and sustainability [11]. The relevance of implementing green innovation has been gaining prominence in the literature, especially among companies that wish to comply with the guidelines of environmental, social, and governance (ESG) issues [12, 13] in the context of the supply chain. Making an analogy with [14], "The green innovation process is multifaceted, and idea generation is a critical initial phase of development, seeking to bridge the gap between the problem domain and the desired solutions." Idea generation is a crucial phase and is the first step in the innovation process [15–18]. In other words, ideation is the "process of generating new concepts that satisfy a specific objective" [19, 20]. Generating ideas depends on creativity, divergent thinking, and at the same time is a collaborative process. It involves multiple orchestrators of (new) ideas [21]. Previous studies report several traditional ideation methods [21], for example, brainstorming, mind mapping, co-creation, etc. There is an understanding that the idea generation phases can be enhanced through collaboration with Generative AI (GEN-AI) [22]. In idea generation, AI increases the average creativity of ideas, substantively supports idea screening, etc. In short, AI has the potential to radically transform idea generation [23]. Previous research suggests that organizations can integrate GEN-AI tools (e.g., ChatGPT) into their ideation processes and generate multiple, more efficient, and diversified ideas in the initial ideation phases and subsequent refinement in later stages [22]. GEN-AI has the potential to reshape creativity and innovation. By enabling the rapid exploration of new ideas and facilitating divergent thinking, GEN-AI supports creative processes, from product design to predictive market simulations. At the same time, GEN-AI helps overcome cognitive biases and predict emerging trends, adding strategic value to organizations. Thus, GEN-AI is a catalyst for sustainability-oriented innovation in creative industries [24]. Finally, there are positive expectations regarding the impact of GEN-AI technology in strengthening the idea generation process. Few studies have investigated the incorporation of GEN-AI into the ideation process in the context of the fuzzy front end of innovation. This study takes this approach in the context of ESG-oriented green innovation in the creative industry supply chain.

3 Methodology

This research was developed in two phases: 1 - literature review; and 2 - Survey. Using the Web of Science, Science Direct, Scopus, IEEE Xplore databases, etc., we surveyed scientific articles and collected data on: generative artificial intelligence, considering its resources and contributions to integrating innovation development; specifically, green innovation in the context of the creative industry supply chain (not exclusively). We filtered the articles focusing on the ideation/idea generation process for innovation development. That is, this process involves the generation, development, and refinement of ideas for solving problems and/or exploring opportunities. Furthermore, this process involves different phases from problem identification to the evaluation and selection of ideas, converging towards the implementation of green innovation (in this study). The studies by [9, 22, 23, 25] were essential for the progress of this study. The second part of our research was the conducting of a survey. Using a Likert-scale questionnaire, data was collected from professionals in companies within the creative industry supply chain. The questionnaire was developed based on the theoretical framework and tested with two professionals with experience in green innovation and artificial intelligence. A few adjustments were recommended for the final refinement. The questionnaire is structured into 9 (nine) parts: general information, idea generation, market-based partnerships, performance in the fuzzy front end, attributes of idea generation, use of GEN-AI for the development of green innovation and reinforcement of ESG in the creative industry supply chain, contribution of GEN-AI to the development of new green products and ESG factors in the creative industry supply chain in terms of benefits, novelty, creativity, feasibility and quality of information (authenticity and added value), evaluation of ideas in the fuzzy front end using GEN-AI, and the impact of GEN-AI on the development of new green products and ESG. We used the following scales: 1—least important and 5—most important; and 1—Low importance, 2—Moderate/Medium importance; and 3—High importance. Using the LinkedIn professional network, we mapped and identified professionals to comprise the sample for this study. The questionnaires were submitted using the Google Forms platform. 50 questionnaires were sent and 10 were returned to comprise the sample for this study. The data was processed using descriptive statistical techniques. Figures 1, 2, 3, 4, and 5 show the profile of the respondents: qualifications, educational area, occupation, experience with green innovation development, experience with Artificial Intelligence, and the respondents' experience.

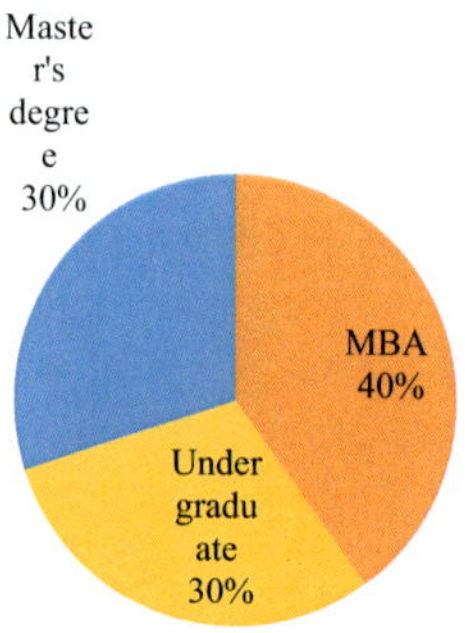

Fig. 1. Educational level

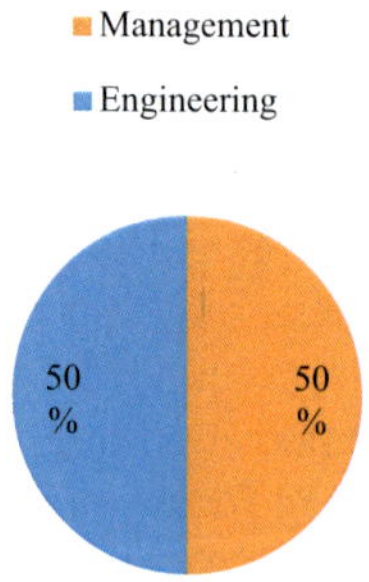

Fig. 2. Educational area

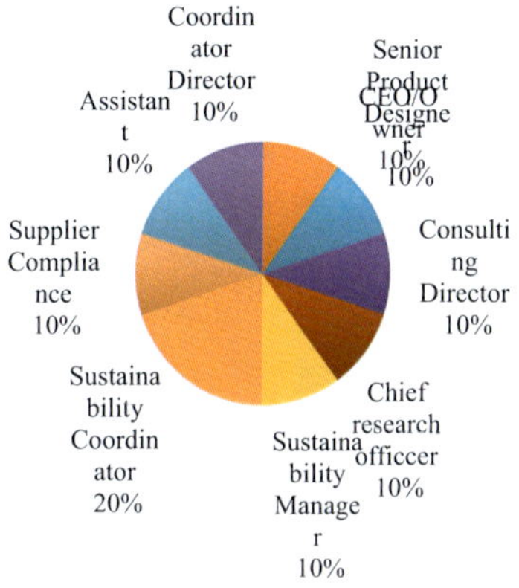

Fig. 3. Occupation

The results show that the respondents are Administrators and Engineers (50%), graduates (30%), master's degree holders (30%), and specialists (40%) (Figs. 1 and 2); and the majority hold positions related to sustainability (coordination and management) (30%) (Fig. 3). At the same time, the results show that respondents have substantial experience with the development of green innovation (50%) (Fig. 4) and moderate experience with artificial intelligence (60%) (Fig. 5).

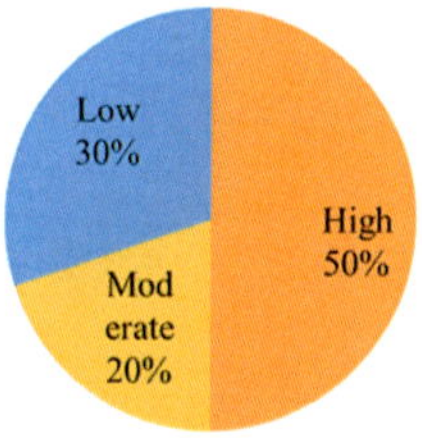

Fig. 4. Experience with the development of green innovation

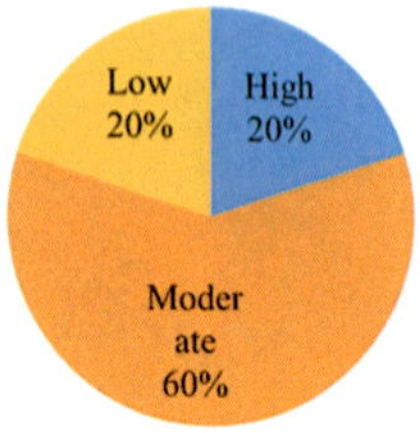

Fig. 5. Experience with artificial intelligence

4 Results

Using descriptive statistical techniques, this section presents the research results: systematic methods for generating ideas, partnerships for generating ideas, performance of the Fuzzy Front End of Green Innovation, attributes in idea generation, use of GEN-AI for the development of Green Innovation and strengthening of ESG, contribution of GEN-AI to the development of new green products and strengthening of ESG in the Supply Chain of the Creative Industry in terms of Benefits, Novelty/Creativity, Feasibility, and Information Quality (authenticity and added value). These results are detailed below. From the perspective of idea generation, the results suggest that there is no prevalence of one or another systematic method for generating ideas. On average (M = 2.8; M = 2.6; M = 2.7), the systematic methods for capturing and sharing ideas; recoding and capturing ideas from the external environment are adopted at a moderately low level (M = 2.7) (Fig. 6). At the same time, the most relevant partnerships for generating ideas are carried out with clients (24%) (Fig. 7).

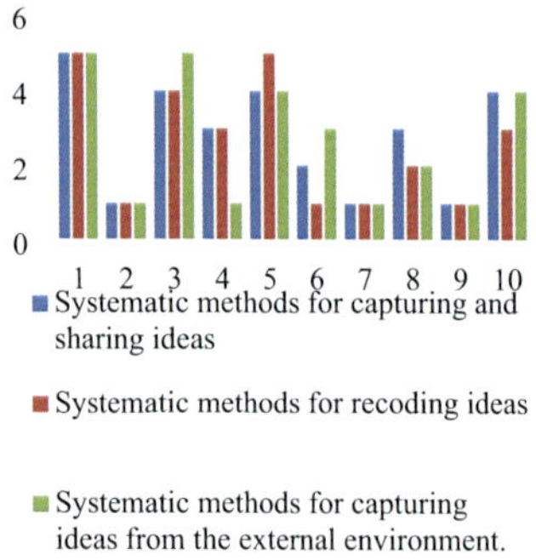

Fig. 6. Systematic methods for generating ideas

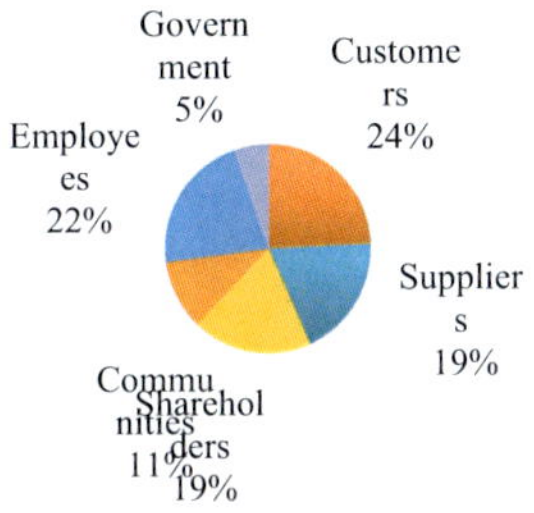

Fig. 7. Partnerships for generating ideas

The performance of companies in the fuzzy front end of green innovation shows moderate results (Fig. 8): satisfaction with the green innovation front-end in generating competitive advantage (M = 3.1); use of tools and methods to anticipate the next wave of green innovation (M = 3); green innovation front-end in meeting strategic objectives (M = 3); and feasibility of projects regarding technical requirements in the front-end (M = 3.4).

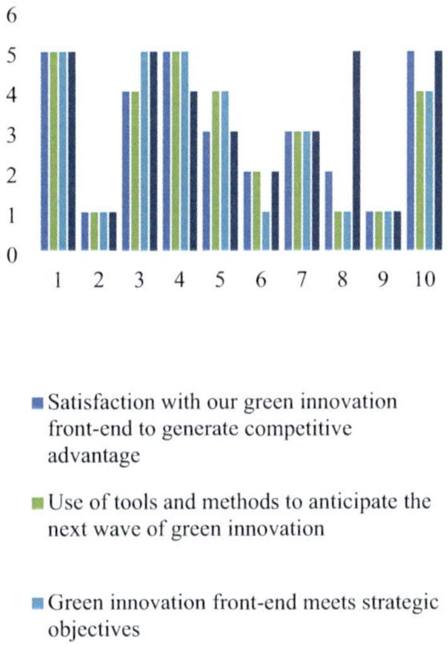

Fig. 8. Performance of the Green Innovation Fuzzy Front End

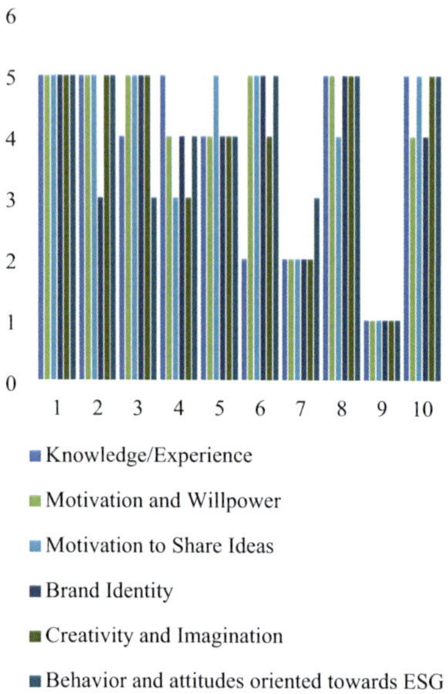

Fig. 9. Attributes in the ideation generation - in the ideation generation, our company takes into consideration: knowledge/experience, motivation to share ideas, creativity and imagination, motivation and willingness, ESG-oriented behavior and attitudes; brand identity.

On average (M = 3.12), the results show moderate performance in the front end of green innovation, with the following averages for performance: level of satisfaction in the front end of green innovation to generate competitive advantage (M = 3.1); use of tools and methods to anticipate the next wave of green innovation (M = 3); front end of green innovation meets strategic objectives (M = 3); and feasibility of projects regarding technical requirements in the front end (M = 3.4). On average (M = 3.9) (Fig. 9), in idea generation, companies take into account motivation and willingness (M = 4.0) (e.g., intrinsic motivation, entrepreneurial mindset of experienced empowerment, willingness to share ideas, problem-induced motivation, willingness to collaborate, excitement-driven motivation, and brand identity); brand identity (3.9); motivation to share ideas (4.0); knowledge and experience (M = 3.8) (e.g., knowledge and experience, need and solution knowledge, product knowledge, usage experience, and technical knowledge); creativity and skills (e.g., divergent thinking style, relevant creativity and imagination skills; intermediation centrality; age and cognitive ability) (M = 3.9); behavior and attitudes (e.g., financially attractive solution-oriented behavior, attention to others' ideas, reliability, credibility; early adopter mindset at the personal level of interaction, and relationship proximity) (M = 4).

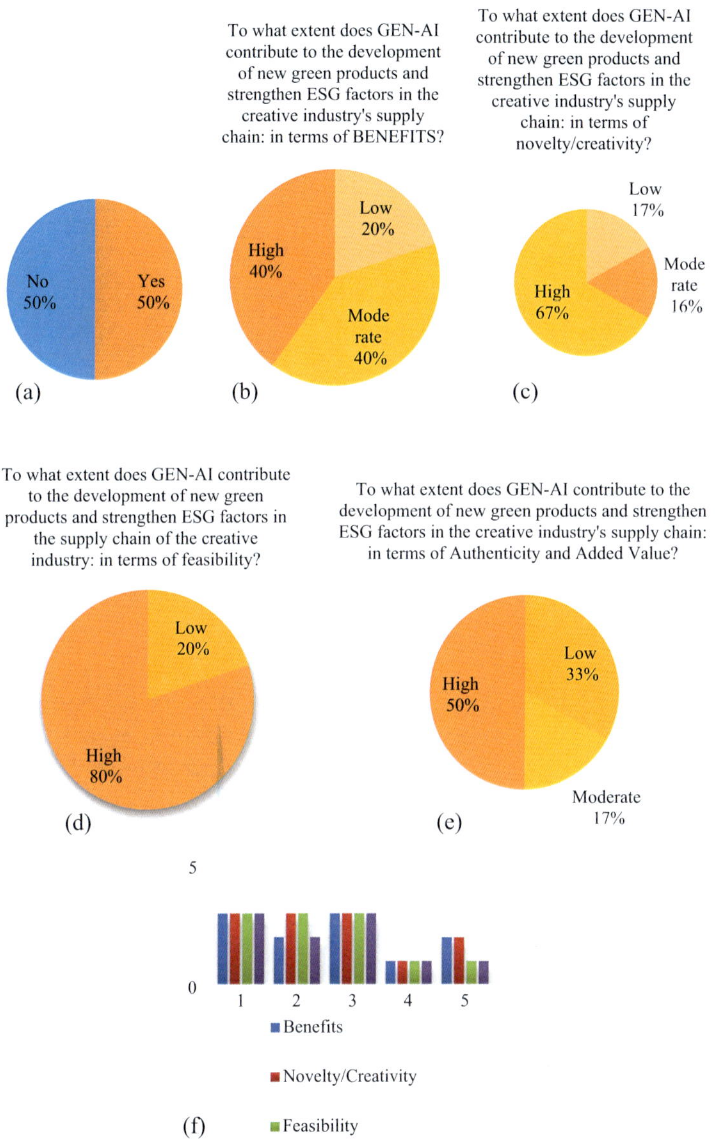

Fig. 10. (A): Use of GEN-AI for the development of Green Innovation and strengthening of ESG, (B): Benefits, (C): Novelties/Creativity, (D): Feasibility, (E): Authenticity/Added Value. (F): Contribution of GEN-AI to the development of new green products and strengthening ESG in the Creative Industry Supply Chain in terms of Benefits, Novelty/Creativity, Viability, and Information Quality (authenticity and added value)

The results (Fig. 10a) suggest that the companies (in this study) adopt GEN-AI for the development of new green products and moderately strengthen ESG factors in the creative industry's supply chain (50%). At the same time, GEN-AI contributes substantially (80%) in terms of feasibility (Fig. 10D); novelty/creativity (67%) (Fig. 10C); moderately high for generating benefits (80%) (10B); and also moderately high for authenticity/added value (67%) (Fig. 10E).

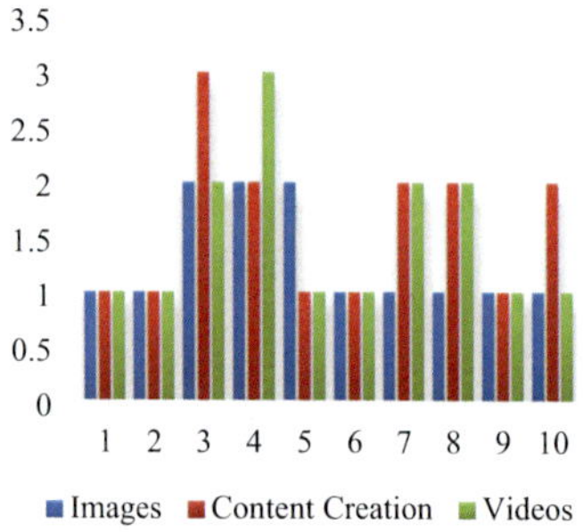

Fig. 11. Use of GEN-AI resources for the development of new green products and strengthening ESG in the creative supply chain—image generation, content creation, videos—scores: 1—low, 2—moderate; and 3—high.

In other words, the results (Fig. 10F) suggest that 50% of the companies analyzed adopt GEN-AI to strengthen the development of green innovation and reinforce ESG for the supply chains of the creative industry. On average, GEN-AI contributes mainly in terms of novelty (M = 2.4), creativity (M = 2.4), feasibility (M = 2.2), benefits (M = 2.2), and information quality (authenticity and added value) (M = 2.0). The results (Fig. 11) indicate that the companies (in this study) moderately use GEN-AI resources: Images (M = 1.3), Videos (M = 1.5), and Content Creation (M = 1.6) for the development of new green products and reinforcement of ESG in the Creative Supply Chain.

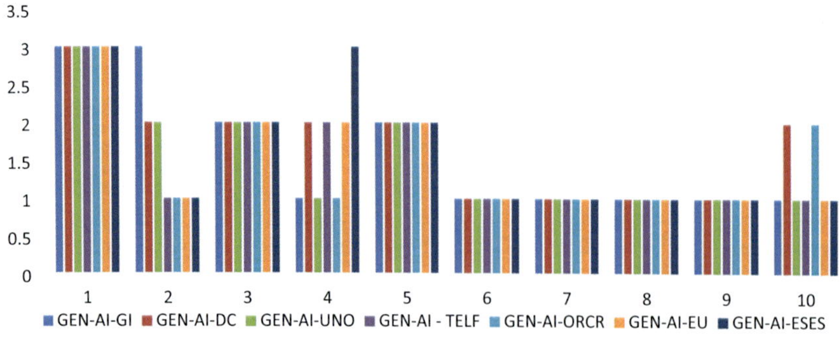

Fig. 12. Idea evaluation process for the development of (new) green products (fuzzy front end)—Scores: 1—Low; 2 – Moderate; 3—High

Legend

- GEN-AI-GI: In our company, we evaluate the idea generation process for green innovation (fuzzy front end) using GEN-AI.
- GEN-AI-DC: In our company, GEN-AI validates and refines design concepts, allowing for the improvement of initial ideas.
- GEN-AI-UNO: In our company, GEN-AI helps assess the extent to which each idea aligns with the project's design brief, objectives, and user needs.
- GEN-AI-TELF: In our company, GEN-AI evaluates the technical, economic, and logistical feasibility of implementing each idea within the given constraints.
- GEN-AI-ORCR: In our company, GEN-AI assesses the originality and creativity of each idea in terms of uniqueness, innovation, and potential for product differentiation in the market.
- GEN-AI-EU: In our company, GEN-AI evaluates the user experience and usability of each idea, ensuring that the final design meets the needs, preferences, and expectations of the target users.
- GEN-AI-ESES: In our company, GEN-AI evaluates the environmental, social, and economic sustainability of each idea, prioritizing solutions that minimize environmental impact and promote ethical practices.

Similarly, the evaluation of ideas for the development of (new) green products showed an average of 1.55 ($M = 1.55$). When evaluating the ideas for the development of new products (idea generation process, validation and refinement of design concepts to improve initial ideas, assessing the extent to which each idea aligns with the project's design brief, objectives and user needs, technical, economic and logistical feasibility of implementing each idea within the given constraints, originality and creativity of each idea, user experience and usability of each idea, and the environmental, social and economic sustainability of each idea, prioritizing solutions that minimize environmental impact and promote ethical practices), the results were as follows: GEN-AI – GI: 1.6; GEN-AI-DC: 1.7; GEN-AI-UNO: 1.5; GEN-AI-TELF: 1.5; GEN-AI-ORCR: 1.5; GEN-AI-EU: 1.5; GEN-AI-AI-ESES: 1.6. In other words, most companies develop their innovations using GEN-AI in a moderate way (Fig. 12).

5 Discussions and Conclusions

This study aims to explore the adoption of Generative Artificial Intelligence (GEN-AI) for the development of new green products in the fuzzy front end; and at the same time reinforce the environmental, social, and governance (ESG) issues of the creative industry's supply chains. Using a questionnaire developed based on theoretical foundations, we collected data from different professionals involved with experience in green innovation development and experience with AI. The sample portfolio indicated that most respondents have experience/knowledge with the development of green innovation (70%) and with artificial intelligence (80%). Using descriptive statistical techniques, the main findings are as follows: The companies (investigated) moderately adopt GEN-AI for the development of green innovations and reinforcement of ESG in the supply chains of the creative industry; there is a high prevalence of the use of systematic methods for idea generation; most partnerships for idea generation are carried out with clients; moderate performance in the front end of innovation, in idea generation, companies substantially consider: motivation and willingness; brand identity; and motivation to share ideas. At the same time, the results showed that GEN-AI contributes substantially to the development of green innovations in terms of feasibility, novelty/creativity, benefits, and authenticity/added value. Similarly, the results showed that companies moderately adopt simultaneously the GEN-AI resources: Images ($M = 1.3$), Videos ($M = 1.5$) and Content Creation ($M = 1.6$) for the development of new green products and reinforcement of ESG in the Creative Supply Chain. Finally, our findings highlight that the use of GEN-AI in the idea evaluation process for the development of green products is moderate ($M = 1.5$). Thus, this study proved that GEN-AI integrated into the innovation development process (ideation) in the fuzzy front end is worthwhile (in this sample). These results show congruence with the literature [22], which signals the potential contributions of GEN-AI to reinforce the ideation process in terms of benefits, novelty/creativity, information quality, and feasibility. Although [26] showed in his studies that collaborative fashion ideation with AI influences image thinking and creativity in a constantly changing media environment, image adoption is still used moderately in the companies in this study. Our study is original, fills a gap in the literature, and makes substantive contributions: (i) it shows the contributions of GEN-AI to the development process of new green products in the creative industry supply chain; (ii) it sheds light on managers in the creative industry who intend to progress with the integration of GEN-AI for green innovation development and ESG reinforcement, focusing on the fuzzy front end of the innovation process. That is, in the idea generation process (ideation); (iii) it expands the theoretical body on the use of artificial intelligence with ChatGPT to reinforce the green innovation and ESG process in the creative industry supply chain; and (iv) it contributes to achieving the Sustainable Development Goals (SDGs): 9—industry, innovation and infrastructure; 12—responsible consumption and production; and 17—partnerships and means of implementation. Although our study presents theoretical and practical implications, it is not without limitations. In this study, we considered a sample of 10 companies in the creative industry located in Brazil, which implies that the results obtained in this research cannot be generalized to other companies in Brazil and other countries. It would be interesting for future studies to consider samples of sub-national companies, and from emerging markets and the global South. Finally, we recommend the adoption of some theories, such as stakeholder theory, resource-based view theory, and signaling theory, to theoretically strengthen the research.

References

1. Oliveira, S., Trento, S.: Innovation intermediaries, spillover of green innovation, ESG performance in deep tech startups: does generative artificial intelligence matter? In: Moldovan, L., Gligor, A. (eds.) The 18th International Conference Interdisciplinarity in Engineering. Inter-Eng 2024. Lecture Notes in Networks and Systems, vol. 1250. Springer, Cham (2025)
2. Li, H., Zhang, X.: The impact of corporate ESG ratings on green innovation development. In: 2024 7th International Conference on Data Science and Information Technology (DSIT), Nanjing, China, pp. 1–11 (2024)

3. Wu, C., Shao, J., Lai, H.K., Shou, Y.: Radical or incremental? The effects of green innovation on the supply base stability of logistics service providers. In: IEEE Transactions on Engineering Management, vol. 72, pp. 1440–1453 (2025)
4. Lan, W., Xuepeng, Z., Qinlan, Y.: Study on the impact of green innovation on supply chain financing performance: the moderating role of digital application. In: 2023 6th International Conference on Artificial Intelligence and Big Data (ICAIBD), Chengdu, China, pp. 935–940 (2023)
5. Castellacci, F., Lie, C.M.: A taxonomy of green innovators: empirical evidence from South Korea. J. Clean. Prod. **143**(1), February 2017, 1036–1047 (2017)
6. Abadzhiev, A., Sukhov, A., Sihvonen, A., Johnson, M.: Managing the complexity of green innovation. Eur. J. Innov. Manag. **25**(6), 850–866 (2022)
7. Kim, S.J.: Generative artificial intelligence in collaborative ideation: educational insight from fashion students. IEEE Access **12**, 49261–49274 (2024)
8. Sedkaoui, S., Benaichouba, R.: Generative AI as a transformative force for innovation: a review of opportunities, applications and challenges. Eur. J. Innov. Manag. Vol. ahead-of-print No. ahead-of-print (2024)
9. Joosten, J., Bilgram, V., Hahn, A., e Totzek, D.: Comparing the ideation quality of humans with generative artificial intelligence. In: IEEE Eng. Manag. Rev. **52**(2), 153–164 (2024)
10. Clarencia, E., Tiranda, T.G., Achmad, S., Sutoyo, R.: The impact of artificial intelligence in the creative industries: design and editing, 2024 international seminar on application for technology of information and communication (iSemantic). Semarang, Indonesia **2024**, 440–444 (2024)
11. Levi-Bliech, M., Dahan, G.: The impact of green innovation products on an organization's social performance via green supply chain management. Green Technol. Sustain. Available online 16 September, 100273, In Press, Corrected Proof (2025)
12. Shan, H., Zhao, K., Liu, Y.: ESG performance and the persistence of green innovation: empirical evidence from Chinese manufacturing enterprises. Multinatl. Bus. Rev. **33**(2), 268–296 (2025)
13. Wu, L., Yi, X., Hu, K., Lyulyov, O. Pimonenko, T. The effect of ESG performance on corporate green innovation. Bus. Process. Manag. J. **31**(8), 8, 24–48 (2025)
14. Amabile, T.M.: A model of creativity and innovation in organizations. Res. Organ. Behav. **10**(2), 123–167 (1988)
15. Pescher, C., Tellis, G.J.: The role of artificial intelligence in the ideation process. J. Prod. Innov. Manag. **42**(5) (2025)
16. Eling, K., Herstatt, C.: Managing the front end of innovation—less fuzzy, yet still not fully understood. J. Prod. Innov. Manag. **34**(6), 864–874 (2017)
17. Kock, A., Heising, W., Gemuenden, H.G.: How ideation portfolio management influences front-end success. J. Prod. Innov. Manag. **32**(4), 539–555 (2014)
18. Utley, J., Klebahn, P.: Building an innovation pipeline. Stanf. Soc. Innov. Rev. (2023)
19. Freitas, J., Nave, G., Punton, S.: Ideation with generative AI—in Consumer: research and beyond. Published by Oxford University Press on behalf of Journal of Consumer Research, Inc. 52 (2025)
20. Koestler, A.: The Act of Creation. Macmillan (1964)
21. Biemans, W., Malshe, A.: How marketing and sales use digital tools for innovation ideation. Ind. Mark. Manage. **123**(November), 304–316 (2024)
22. Joosten, J., Bilgram, V., Hahn, A., Totzek, D.: Comparing the ideation quality of humans with generative artificial intelligence. IEEE Eng. Manage. Rev. **52**(2), 153–164 (2024)
23. Pescher, C., Tellis, G.J.: The role of artificial intelligence in the ideation process. J. Prod. Innov. Manag (2025)

24. Schilirò, D.: Generative artificial intelligence, creativity, and innovation. In: Crupi, A., Marinelli, L., Cacciatore, E. (Ed.) The Generative AI Impact (Emerald Studies in Sustainable Innovation Management), Emerald Publishing Limited, Leeds, pp. 199–211 (2025)
25. Bilgram, V., Laarmann, F.: Accelerating innovation with generative AI: AI-augmented digital prototyping and innovation methods. IEEE Eng. Manag. Rev. **51**(2), 18–25 (2023)

What you Have Is What You Get? Authentic Environmental, Social and Governance Information to Scale Green Innovations in Responsible Supply Chain in a Global South Economy

João Pedro Passos, Lucas Sampaio Motta, and Selma Oliveira(✉)

Fluminense Federal University, Volta Redonda, Rio de Janeiro, Brazil
{joaopassos,lucassampaiomotta,selmaregina}@id.uff.br

Abstract. While the potential benefits of environmental, social, and governance (ESG) information have been widely reported, the state-of-the-art literature rarely highlights authentic ESG information for scaling green innovations in responsible supply chains. Using integrated reporting from one company (case) from a Global South economy—namely, Brazil—this study examines the authenticity of ESG information for scaling green innovations in responsible supply chains under the lens of stakeholders and Generative Artificial Intelligence (GEN-AI). Our study fills a gap in the literature and makes significant contributions: (a) it provides an overview of the (prevalent) authentic ESG information for scaling green innovations in responsible supply chains under the lens of stakeholders and GEN-AI by ChatGPT; (b) it suggests that as companies strengthen responsible supply chains with authentic ESG information, green innovations are enhanced; (c) our study provides a timely warning to managers who intend to promote green innovations and strengthen the sustainability of supply chains enabled with authentic ESG information in a Global South economy; (d) furthermore, our study sheds light to advance the theoretical arguments on ESG for green innovation in sustainable supply chains; and lastly, (e) our study contributes to supply chains to improve the sustainable development goals (SDGs): (9) Industry, Innovation and Infrastructure—building resilient infrastructure, promoting inclusive and sustainable industrialization and fostering innovation; (12) responsible consumption and production—ensuring sustainable consumption and production patterns, etc.

Keywords: Generative Artificial Intelligence (GEN-AI) · Green and ESG Innovation · Creative Industry Supply Chain · Idea Generation · Fuzzy Front End

1 Introduction

It is generally accepted that environmental, social, and governance (ESG) disclosure improves corporate performance [1, 2] and, at the same time, supply chains worldwide [3]. Although the potential benefits of ESG and supply chain approaches have

Y. Benadada et al. (Eds.): GOL 2026, LNNS 2034, pp. 111–120, 2027.
https://doi.org/10.1007/978-3-032-29677-1_11

been widely reported by academics and practitioners, the state-of-the-art literature rarely highlights authentic ESG information for scaling green innovations in responsible supply chains. Using integrated reporting from one company from a Global South economy—namely, Brazil—this study examines the authenticity of ESG information for scaling green innovations in responsible supply chains under the lens of stakeholders and Generative Artificial Intelligence (GEN-AI). Understanding this relationship is a relevant priority for a number of reasons.

- Growing environmental awareness among individuals and societies has strengthened the demand for companies with a strong commitment to sustainability [4–7]. Investors believe that socially responsible companies bring the best returns [8]. At the same time, supply chain management (SCM) is under extra pressure to achieve profitability and sustainability simultaneously [9]. Climate change and recurrent natural disasters have substantially disrupted global supply chains [6], with significant implications for regional and international economic activities and social welfare [6].
- In this context, studies have shown that the adoption of green innovations led by responsible principles can lead to chains becoming successful, ensuring that they contribute positively to economic growth, environmental sustainability and social well-being [10]. The call for more green innovation stands out as a crucial approach, closely aligned with sustainable development goals. In short, as societies and global supply chains strive to achieve these commitments, those who lead in green innovation will play a prevalent role in bridging the gap.
- At the same time, there is a growing recognition that innovation must be integrated into supply chain operations in a way that is transparent, inclusive and aligned with global sustainability goals [10]. Drawing an analogy with the literature [11–13], responsible supply chain is dedicated to the integration of ethical standards, environmental sustainability, social well-being and economic viability.
- However, previous research suggests that the adoption of green innovation is strongly influenced and often depends on information and data related to [14, 15]. Therefore, the search for green innovation by companies is increasingly growing and challenging. One of the main challenges faced by managers is the availability of authentic ESG information disclosed at the company level. However, the adoption of ESG measures (for example, Brazil) is still not broad and comprehensive [16]. Although there are recent signs of greater openness to ESG disclosures, ESG disclosure rules lack clarity [17]; and this can generate errors and difficulties in sustainability practices and lead to a lack of transparency in reporting, damaging investor and consumer confidence. ESG information comes from different internal and external sources (internal and external reports, governments, media, etc.) [18]. One of the main challenges for companies is to obtain accurate and reliable ESG data and generate reports with authentic communication and disclosure frameworks [19]. In response to the gap between the demand for authentic ESG information by managers and the provision of such information by companies, there is an effort to disclose high-quality information on ESG issues [20]. Thus, there is an imperative need to uncover the level of quality of ESG information (level of authenticity) generated by companies to scale green innovations. To the best of our knowledge, there are no or almost no studies based on empirical data that relate the quality of (authentic) ESG information and green

innovation in the responsible supply chain. At the same time, there is no empirical evidence in the literature based on primary data from a Global South economy. In other words, the quality of ESG information disclosed at the company level is insufficient for informed decision-making. This study unravels the levels of quality (authenticity) of ESG information to scale green innovation in the responsible supply chain in a Global South economy.

Thus, this study has the following problem-question to be solved: What is the level of authenticity (quality) of the ESG information disclosed (by company) to scale green innovations in the responsible supply chain? To verify the quality (authenticity) of the ESG information disclosed in an Integrated Report of one company, two questionnaires were applied: the first was sent to a group of stakeholders (Managers and controllers). The second questionnaire was applied to generative artificial intelligence (GEN-AI) by ChatGPT. Studying company in the consumer goods sector located in Brazil can bring new insights to the debate on the adoption of ESG for green innovation in responsible supply chains. In summary, when considering authentic information to strengthen green innovation, the green orientation emerges as a strategic direction for companies in the supply chain, attracting more and more stakeholders due to its significant value for business viability and sustainable development.

This study fills a gap in the literature and makes significant contributions: (a) it provides an overview of the (prevalent) authentic ESG information for scaling green innovations in responsible supply chains under the lens of stakeholders and GEN-AI by ChatGPT; (b) it suggests that as companies strengthen responsible supply chains with authentic ESG information, green innovations are enhanced; (c) provides a timely warning to managers who intend to promote green innovations and strengthen the sustainability of supply chains enabled with authentic ESG information in a Global South economy; (d) furthermore, our study sheds light to advance the theoretical arguments on ESG for green innovation in sustainable supply chains; and lastly, (e) contributes to supply chains to improve the sustainable development goals (SDGs): (9) Industry, Innovation and Infrastructure - building resilient infrastructure, promoting inclusive and sustainable industrialization and fostering innovation; (12) responsible consumption and production—ensuring sustainable consumption and production patterns, etc. In summary, our findings therefore contribute to the discussion on the role of authentic ESG information for the diffusion of green innovations in socially responsible supply chains. Our study offers a timely warning to managers who intend to advance responsible supply chains enabled by authentic ESG information in the diffusion of green innovations. After this introduction, the remainder of the paper is structured in five additional sections. Section 2 presents the theoretical concepts. Section 3 describes the research methodology used. Section 4 reports the results. Finally, Sect. 5 highlights the discussions and conclusions and suggestions for future research.

2 Literature Review and Hypothesis Development

Previous research suggests that a responsible supply chain is an important factor in attracting investors, as it signals a company's commitment to sustainability, reduces risks, improves reputation, and strengthens the company's efficiency and profitability

[21]. At the same time, studies indicate that the supply chain should be a key area for organizations pursuing a sustainability agenda [22]. Previous studies [14] suggest that companies have been oriented towards green innovation, which is considered a key component to achieve the dual goals of environmental protection and economic growth [23]. Green innovation has been suggested by researchers as a strategic tool to maintain sustainable development while gaining competitive advantages [9 citing 24]. Green innovation is critical to achieving environmental protection-oriented goals while emphasizing economic growth [23, 15].

Green innovation is innovative solutions that seek to reduce environmental impact and promote sustainability, for example, developing products with lower environmental impacts, recycled materials, products with greater durability and lower energy consumption, etc. At the same time, it calls for the implementation of more efficient production processes, with less consumption of natural resources, less waste generation and lower pollutant emissions, etc. Previous research suggests that companies that actively disclose ESG information can promote corporate green innovation. In short, the disclosure of ESG information not only serves the diverse interests of stakeholders but also motivates companies to proactively engage in green innovation. Previous studies have shown that ESG information positively influences the generation of green innovations in companies [15].

Although some studies have identified the positive link between ESG information and green innovations, the level of authenticity (quality) of ESG information to strengthen green innovation in the supply chain is still inconclusive. Arguments in the literature highlight that green innovation require more information and knowledge due to environmental factors such as resource and energy conservation, waste and pollution prevention, recycling and reuse of waste, etc. [1, 25]. In this context of responsible supply chain, the quality of ESG information is crucial to generate green innovations. At the same time, the lack of clarity, consistency, and reliability of data can generate risks and losses for companies and investors. Therefore, offering authentic (reliable and transparent) ESG information is still a challenge [22].

Drawing an analogy with [26], the quality of ESG information can be defined as the suitability of information to be useful to its users (consumers, investors, etc.). In this same understanding, authentic ESG information is related to the state of being real, genuine, complete, transparent, true, and consistent with internal values and actions [27]. By analogy with previous research [28, 29], the quality of ESG information depends on how it is judged by the user (be it an individual or a group) and at the same time on the context in which it is considered, since it is only through the perception and observation made by the user that one can, in fact, assess the quality of ESG information. However, previous research [30] highlights that the quality and clarity of companies' ESG information varies due to the lack of consistent disclosure standards and rules. Some studies [31–33] suggest that disagreements in the quality of ESG information may also arise from intentional decisions by companies to strategically disguise or distort ESG information [34].

We found a wide range of criteria to assess the quality of information (availability, authenticity, integrity, etc.). However, this work will follow the classification carried out by [26], indicating that it is the study with the largest number of citations, and thus considered a classification with credentials accepted by the academic community.

These authors suggest four categories of criteria to classify the quality of information: intrinsic (accuracy, credibility, objectivity and reputation), contextual (added value, relevance, timeliness, completeness, adequate quantity), representational (interpretability, easy understanding, consistent representation, concise representation) and accessibility (accessibility and secure access).

Drawing an analogy with the literature, the quality (authenticity) of ESG information needs to be accurate, transparent, precise, comparable, relevant, and transparent to ensure that stakeholders can trust and use this data effectively. Companies disclose ESG information in various forms, for example, ESG reports, integrated reports, etc. In other words, an Integrated Report combines financial and non-financial information, highlighting how a company manages its ESG factors and generates value for its stakeholders [35]. The ESG report, in turn, focuses on communicating specific data about the company's environmental, social, and governance impacts [17]. Both reports facilitate communication with investors and other stakeholders, showing how the company manages its ESG factors and their impacts. Existing studies [34] indicate that the inaccuracy of ESG information shapes shareholders' responses to ESG news [36]. We are facing the opacity of ESG information.

3 Research Methodology

3.1 Measures and Scales

For data collection in this study, a questionnaire was developed based on the research structure and theoretical framework presented in Sect. 2. In order to verify inconsistencies such as redundancies, application time, degree of complexity, etc., a questionnaire was tested with three professionals from the controllership area of multinational companies based in Brazil. The experts did not report significant difficulties when answering the questionnaire and considered it adequate. Only a few terms were summarized to simplify and facilitate understanding in the final refinement. The construct quality of ESG information to scale green innovations was measured using a scale (1 = low; 2 = medium and 3 = high). In this study, we adopted some metrics classified by Wang and Strong [26] to measure the quality of ESG information: credibility, added value, relevance and completeness. The questionnaire directed to ChatGPT also considered the constructs classified by [26] to assess the sentiment perceived by ChatGPT in relation to the quality of ESG information: credibility, added value, relevance and completeness. Thus, we adopted the same scores for the interview with ChatGPT: 1—low; 2—medium; and 3—high. However, the (4) questions directed to ChatGPT were all open-ended.

3.2 Sample and Data Collection Procedures

To assess the quality of ESG information (authenticity), we used the Integrated Report of a consumer goods company operating in Brazil—year 2022. Thus, the quality of ESG information (2022) was assessed by consulting stakeholders—in this case, managers and controllers of multinational companies and through ChatGPT. The data collection procedures are described below: Data collection: Professionals – The professional social network LinkedIn was used to map professionals to answer the questionnaires. Managers and controllers of multinational companies headquartered in Brazil were mapped and identified. We sought to identify professionals with relationships and experience with strategic management, innovation and sustainability. Thus, 23 questionnaires were submitted and 5 questionnaires were returned answered. Of this total, 1 was eliminated, leaving 4 complete questionnaires to make up the sample for this study. Data Collection: GEN-AI/ChatGPT – Data collection using ChatGPT was conducted in three rounds of questions (interview) by the researchers of this research. The responses (from the three rounds) showed similar results. The data were analyzed using descriptive statistics techniques. The data were analyzed using descriptive statistics techniques.

4 Results

Using descriptive statistics techniques, Figs. 1 and 2 show the results of the sentiment averages of generative artificial intelligence and stakeholders in relation to the quality of ESG information disclosed in the 2022 IR of the company analyzed in this research.

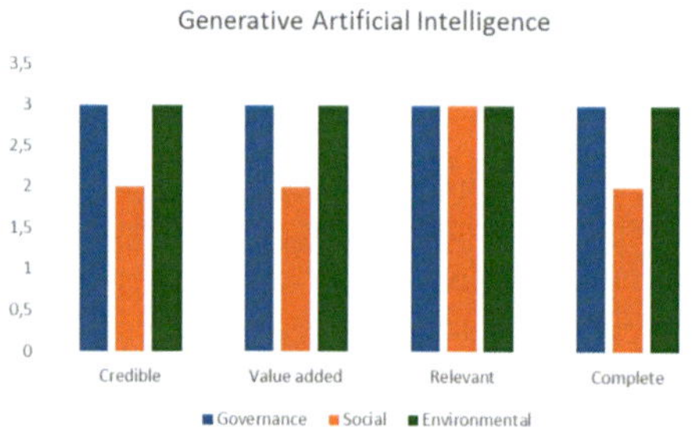

Fig. 1. Result of GEN-AI sentiment measurements on ESG information disclosed

Fig. 2. Result of average stakeholder sentiment regarding disclosed ESG information

On average, GEN-AI perceives the ESG information disclosed by the sample company as highly relevant (M = 3). At the same time, social information is perceived as moderately credible, valuable and complete (M = 2). In other words, on average, our findings show that environmental and governance information exhibits high credibility, value-added, relevance, and completeness. At the same time, social information indicates moderate credibility, value-added, and completeness. However, all information is highly relevant. In the opinion of stakeholders, the ESG information disclosed is highly relevant (M = 3). At the same time, environmental information has high added value and is complete (M = 3). Likewise, social information has high added value. On the other hand, ESG information has medium credibility (M = 2.33) (Fig. 3).

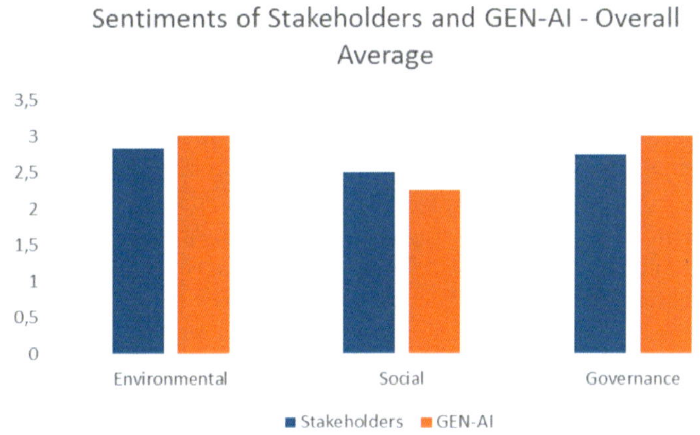

Fig. 3. Comparative result of the averages of stakeholder sentiment and GEN-AI on the quality of ESG information disclosed

The results suggest that GEN-AI has a better perception of the quality of ESG information (M = 2.75) when compared to stakeholder sentiment (M = 2.69).

5 Discussions and Conclusions

Using integrated reporting (IR) from a company (case) in a Global South economy—namely, Brazil—this study examines the authenticity of ESG information to scale green innovations in responsible supply chains under the lens of stakeholders and GEN-AI. We use a case study of a food company in Brazil (named Company A). Our study sheds light on an economy in the Global South. At the same time, our research generates insights to contribute to the current debate around ESG implementation worldwide. The research results suggest that ESG information presents high authenticity, especially from the perspective of GEN-AI and with emphasis on environmental and governance information. The criterion relevance of information received the best evaluation under the lens of stakeholders and GEN-AI. However, our findings reveal that other ESG information criteria are relevant to contribute to scaling green innovations.

Therefore, our study fills a gap in the literature and makes significant contributions: (a) it provides an overview of the authentic (prevalent) ESG information to scale green innovations in responsible supply chains under the lens of stakeholders and GEN-AI by ChatGPT; (b) it suggests that as companies strengthen responsible supply chains with authentic ESG information, green innovations are enhanced; (c) our study provides a timely warning to managers who intend to promote green innovations and strengthen the sustainability of supply chains enabled with authentic ESG information in a Global South economy; (d) further, our study sheds light to advance the theoretical arguments on ESG for green innovation in sustainable supply chains; and lastly, (e) our study contributes to supply chains to enhance the sustainable development goals (SDGs): (9) Industry, Innovation and Infrastructure—building resilient infrastructure, promoting inclusive and sustainable industrialization and fostering innovation; (12) responsible consumption and production—ensuring sustainable consumption and production patterns, etc.

The findings of our research may be useful for academics, entrepreneurs, and end users who face the challenges of adopting ESG information to scale green innovations in economies in the Global South that share characteristics with Brazil. The authors of this paper acknowledge that this study has some limitations that should be addressed. In our research framework, some limitations arose due to the sample size of this study. Our sample consisted of only one company. This study represents a first effort to understand the relevance of authentic ESG information to scale green innovations in the responsible supply chain. Therefore, the results of this research should be considered pertinent to the sample of this research and should not be extrapolated beyond that limit. Furthermore, the methodology used in this study was based on a cross-sectional study, in which data were collected at a specific point in time. The sample was also limited in terms of size. Thus, looking ahead, we understand that future studies could expand the sample and consider other sectors. Our research points to opportunities for future research on the adoption of ESG information to diffuse green innovations in companies in a sustainable supply chain. Furthermore, different theories could add new insights to the debate on the adoption of ESG for green innovation in companies from the Global South.

Acknowledgement. Thanks to the Carlos Chagas Filho Foundation for Research Support of the State of Rio de Janeiro—FAPERJ.

References

1. Chen, Z., Xie, G.: ESG disclosure and financial performance: moderating role of ESG investors. Int. Rev. Financ. Anal. **83**, 102291 (2022). https://doi.org/10.1016/j.irfa.2022.102291
2. Bamel, N., Khatri, K., Bamel, U., Kumar, S.: ESG disclosure and firm performance in global south economy: does industry profile and board independence moderate the relationship. Int. Rev. Econ. Fin. **100**, 104093 (2025). https://doi.org/10.1016/j.iref.2025.104093
3. Truant, E., Borlatto, E., Crocco, E., Sahore, N.: Environmental, social and governance issues in supply chains. A systematic review for strategic performance. J. Clean. Prod. **434**, 140024 (2024). https://doi.org/10.1016/j.jclepro.2023.140024
4. Hart, S.L., Milstein, M.B.: Creating sustainable value. Acad. Manag. Perspect. **17**(2), 56–67 (2003). https://doi.org/10.5465/ame.2003.10025194
5. Garavaglia, S., Van Landuyt, B.W., White, B.J., Irwin, J.: The ESG stopping effect: do investor reactions differ across the lifespan of ESG initiatives? Acc. Organ. Soc. **113**, 101441 (2024). https://doi.org/10.1016/j.aos.2023.101441

6. Kamal, M.M., Song, M., Mangla, S.K., Gupta, N., Kumar, V., Tjahjono, B.: Green innovation: disruptive technologies for carbon neutrality and sustainable supply chains. In: Editorial, IEEE-TEM's—IEEE transactions on engineering management. https://www.ieee-tems.org/call-for-papers-green-innovation-disruptive-technologies-for-carbon-neutrality-and-sustainable-supply-chains/. Access in Jan 2025 (2024)
7. Laine, M. (2024). Accounting, reporting and verification of impact: implications for sustainability: a commentary on Adams et al. (2024) "Styles of verification and the pursuit of organisational repair: the case of social impact". Acc. Organ. Soc., 113, 101540. https://doi.org/10.1016/j.aos.2023.101540
8. Tan, W., Liu, Y., Dong, Q., Chen, X.H.: From values to value: exploring the dual impact of national spirit on corporate ESG performance. J. Account. Lit. (2025). (ahead of print)). https://doi.org/10.1108/JAL-08-2024-0222
9. Qu, K., Liu, Z.: Green innovations, supply chain integration and green information system: a model of moderation. J. Clean. Prod. **339**, 130557 (2022). https://doi.org/10.1016/j.jclepro.2022.130557
10. Fan, Y., Jonsson, P., Shou, Y., Zhou, H.: Navigating the new era: responsible technological transformation in global manufacturing operations. Int. J. Prod. Econ. (2025) https://www.sciencedirect.com/special-issue/316570/navigating-the-new-era-responsible-technological-transformation-in-global-manufacturing-operations
11. Choi, T.M., Kumar, S., Yue, X., Chan, H.L.: Disruptive technologies and operations management in the industry 4.0 era and beyond. Prod. Oper. Manag. **31**(1), 9–31 (2022). https://doi.org/10.1111/poms.13622
12. Konstantakis, K.N., Michaelides, P.G., Xidonas, P., Prelorentzos, A.G.N., Samitas, A.: Responsible artificial intelligence for measuring efficiency: a neural production specification. Ann. Oper. Res. 1–27 (2024). https://doi.org/10.1007/s10479-024-05929-2
13. Mhlanga, D.: Responsible Industry 4.0: A Framework for Human-Centered Artificial Intelligence, 298p. Taylor & Francis, Routledge (2023). ISBN 9781032493534
14. Li, J., Lian, G., Xu, A.: How do ESG affect the spillover of green innovation among peer firms? Mechanism discussion and performance study. J. Bus. Res. **158**, 113648 (2023). https://doi.org/10.1016/j.jbusres.2023.113648
15. Ruan, L., Yang, L., Dong, K.: Corporate green innovation: the influence of ESG information disclosure. J. Innov. Knowl. **9**(4), 100628 (2024). https://doi.org/10.1016/j.jik.2024.100628
16. Stott, M.: Reputation fears propel surge of ESG investment in Brazil—special report Brazil: sustainable business. Financial Times. Available in: https://www.ft.com/content/7ac277f2-b0e7-4cd8-b60a-bdb03c7092f4. Access in May 2025 (2021)
17. Tsang, A., Frost, T., Cao, H.: Environmental, social, and governance (ESG) disclosure: a literature review. Br. Account. Rev. **55**(1), 101149 (2023). https://doi.org/10.1016/j.bar.2022.101149
18. Asif, M., Searcy, C., Castka, P.: ESG and industry 5.0: the role of technologies in enhancing ESG disclosure. Technol. Forecast. Soc. Chang. **195**, 122806 (2023). https://doi.org/10.1016/j.techfore.2023.122806
19. Hutton, A.P., Marcus, A.J., Tehranian, H.: Opaque financial reports, R2, and crash risk. J. Financ. Econ. **94**(1), 67–86 (2009). https://doi.org/10.1016/j.jfineco.2008.10.003
20. Zhang, C., Zhang, S., Zhang, Y., Yang, Y., Lan, K.: Does green finance policy contribute to esg disclosure of listed companies? A quasi-natural experiment from China. SAGE Open. **14**(1), 21582440241233376 (2024). https://doi.org/10.1177/21582440241233376
21. Gunasekaran, A., Lai, K-H, Cheng, E.: Responsive supply chain: a competitive strategy in a networked economy. J. Clean. Prod. **278**(1), 123521 (2021). https://doi.org/10.1016/j.omega.2006.12.002

22. IBM: Crie uma cadeia de suprimentos mais sustentável e resiliente. Relatório. https://www.ibm.com/br-pt/resources/business-operations/supply-chain-sustainability. Access in April 2025 (2024)
23. Takalo, S.K., Tooranloo, H.S., Parizi, Z.S.: Green innovation: a systematic literature review. J. Clean. Prod. **279**, 122474 (2021). https://doi.org/10.1016/j.jclepro.2020.122474
24. Liu, M., Liu, L., Feng, A.: The impact of green innovation on corporate performance: an analysis based on substantive and strategic green innovations. Sustainability. (2024)
25. Li, G., Wang, X., Wu, J.: How scientific researchers form green innovation behavior: an empirical analysis of China's enterprises. Technol. Soc. **56**, 134–146 (2019). https://doi.org/10.1016/j.techsoc.2018.09.012
26. Wang, R.Y., Strong, D.M.: Beyond accuracy: what data quality means to data consumers. J. Manag. Inf. Syst. **12**(4), 5–33 (1996). https://doi.org/10.1080/07421222.1996.11518099
27. Kannegiesser, M., Günther, H.O., Gylfason, Ó.: Sustainable development of global supply chains—part 2: investigation of the European automotive industry. Flex. Serv. Manuf. J. **26**, 48–68 (2014). https://doi.org/10.1007/s10696-013-9177-4
28. Repo, A.J.: The dual approach to the value of information: an appraisal of use and exchange values. Inf. Process. Manag. **22**(5), 373–383 (1986). https://doi.org/10.1016/0306-4573(86)90072-5
29. Strong, D.M., Lee, Y.W., Wang, R.Y.: Data quality in context. Commun. ACM. **40**(5), 103–110 (1997). https://doi.org/10.1145/253769.253804
30. Berg, F., Kölbel, J.F., Rigobon, R.: Aggregate confusion: the divergence of ESG ratings. Rev. Finan. **26**(6), 1315–1344 (2022). https://doi.org/10.1093/rof/rfac033
31. Bond, P., Zeng, Y.: Silence is safest: information disclosure when the audience's preferences are uncertain. J. Financ. Econ. **145**(1), 178–193 (2022). https://doi.org/10.1016/j.jfineco.2021.08.012
32. Pope, S., Peillex, J., El Ouadghiri, I., Gomes, M.: Floodlight or spotlight? Public attention and the selective disclosure of environmental information. J. Manag. Stud. **61**(4), 1230–1265 (2024). https://doi.org/10.1111/joms.12920
33. Ryou, J.W., Tsang, A., Wang, K.T.: Product market competition and voluntary corporate social responsibility disclosures. Contemp. Account. Res. **39**(2), 1215–1259 (2022). https://doi.org/10.1111/1911-3846.12748
34. He, H., Chen, Y., Guo, R., He, L., Wan, H.: Cost of vagueness: stakeholders' responses to firms' ESG information. J. Bus. Ethics. **1–24** (2025). https://doi.org/10.1007/s10551-025-06001-0
35. IFRS Foundation: Annual Report: Better Information for Better Decisions. Available in: https://www.ifrs.org/content/dam/ifrs/about-us/funding/2022/ifrs-foundation-annual-report-2022.pdf. Access in Jan 2025 (2022)
36. Serafeim, G., Yoon, A.: Stock price reactions to ESG news: the role of ESG ratings and disagreement. Rev. Acc. Stud. **28**(3), 1500–1530 (2023). https://doi.org/10.1007/s11142-022-09675-3

AI-Driven Trade Facilitation and Smart Transport Systems for Resilient Supply Chains

Advanced AI and Optimization Techniques for Managing Smart Water Distribution Networks

Radia Hajji[1(✉)] and Fatima El Khoukhi[2]

[1] IA Laboratory, FS Meknes, Moulay Ismail University of Meknes, Meknes, Morocco
ra.hajji@edu.umi.ac.ma
[2] IA Laboratory, FLSH Meknes, Moulay Ismail University of Meknes, Meknes, Morocco
f.elkhoukhi@flsh.umi.ac.ma

Abstract. The water utilities use technology advancements to improve the efficiency of Water Distribution Networks (WDNs) in response to increasing demands on the world's water resources. Smart Water Distribution Networks (SWDNs) are the result of the digitization of WDNs, which is made possible by high-frequency sensors and sophisticated data analytics. The broad adoption of these technologies is still limited, despite their promise to increase operational effectiveness, lower water losses, and promote sustainable management. Artificial Intelligence (AI) and optimization methods have become advantageous instruments in recent years to help manage SWDNs in a sustainable and knowledgeable manner. AI techniques have demonstrated great promise in water quality modeling, demand forecasting, infrastructure failure detection, and system performance optimization. Simultaneously, advanced optimization strategies have been successfully applied to optimize operations in water distribution networks, wastewater systems, irrigation scheduling, and reservoir management. This paper presents a comprehensive literature review synthesizing finding from 30 scientific articles published between 2019 and 2025, including 10 review papers sourced from ScienceDirect, MDPI (Multidisciplinary Digital Publishing Institute), and the International Water Association (IWA). The paper highlights the current state-of-the-art in AI enhanced and optimization driven approaches for WRM. Special focus is given to their integration in domains such as SWDNs, water supply and wastewater systems, reservoir operation and irrigation systems.

Keywords: Smart water distribution networks · Artificial Intelligence Technologies · Optimization techniques · Mathematical programming · Real time monitoring

1 Introduction

Water distribution networks (WDNs) are crucial infrastructures for ensuring regular water delivery, but their management is challenging due to old assets, urban demand, and environmental pressures. Around 126 billion m3 of water is lost annually due to leaks and bursts, causing economic damages and additional financial costs [1–6]. Currently

Y. Benadada et al. (Eds.): GOL 2026, LNNS 2034, pp. 123–135, 2027.
https://doi.org/10.1007/978-3-032-29677-1_12

more than ever, effective water resources management (WRM) necessitates a combination of strong strategy, forecasting, and methods for minimizing risk. The stochastic nature of water inflows, fluctuating needs, environmental flow restrictions, and the long-term effects of climate change make these operations intrinsically complicated. Water availability is still essential for maintaining ecosystems, industry, hydropower, and agriculture beyond municipal water supplies. Against these barriers, Integrated WRM has surfaced as a multidisciplinary framework that balances technical, environmental, and social factors to manage water consumption across various industries [1–4, 7].

The WDNs is the essential infrastructure in charge of ensuring the supply of drinking water to consumers, is at the operational center of WRM. However, because of things like infrastructure degradation, demand shifts, and energy limitations, operational complexity is rising. Optimization approaches have long been used to improve pump scheduling [8, 9], reduce leakages [10, 11], and improve network design [12, 13]. To enhance decision-making in all facets of WRM, the Smart Water Network Management (SWNM) paradigm has surfaced in recent years. It combines digital sensing, data-driven analytics, and intelligent control technologies. Artificial intelligence (AI) and hybrid AI approaches, which combine data-based learning with conventional optimization and hydraulic models, are primarily responsible for this change. In the meantime, mathematical optimization offers a strong and interpretable framework for water system management. Although linear, nonlinear, and evolutionary algorithms are well-established [3, 4]. In this way, the foundation of contemporary smart water management is the integration of intelligent modeling and optimization, where historical knowledge and current data come together to enable effective, robust, and sustainable system operation.

The paper describes the primary mathematical models, optimization strategies, and AI algorithms applied to SWNM based on an examination of 32 scientific publications published between 2019 and 2025, including 10 review papers picked up from ScienceDirect, MDPI (Multidisciplinary Digital Publishing Institute), and the International Water Association (IWA). It also identifies current research gaps and suggests prospective pathways for creating intelligent, climate-resilient, and hybrid solutions.

The structure of the present paper is as follows: The introduction is given in Sect. 1. The methodology of search it is presented in Sect. 2. A comprehensive review of the literature and a comparison of optimization and AI applications in WRM are given in Sect. 3. Emerging trends and insights into the integration of various techniques are covered in Sect. 4, and important innovations and suggestions for furthering SWNM are presented in Sect. 5.

2 Methodology

A systematic literature review was conducted following the guidelines of [31] to investigate recent AI-driven, optimization-based, and hybrid approaches applied to Smart Water Distribution Networks (SWDNs). This methodology ensures a transparent, reproducible, and comprehensive assessment of the relevant literature.

The literature search was performed using major scientific databases, including Scopus, Web of Science, IEEE Xplore, and ScienceDirect, targeting peer-reviewed journal and conference papers published between 2019 and 2025. Discipline specific keywords related to SWDNs, optimization techniques, AI, digital twins, and IoT were combined to retrieve relevant publications.

All identified studies were screened at the title and abstract levels using predefined inclusion and exclusion criteria. Only publications focusing on water distribution systems and employing optimization, AI, or hybrid methods were retained, while purely conceptual studies or works lacking methodological detail were excluded. The selected papers were then reviewed in full text, and key information regarding application domain, methodological approach, and reported advantages and limitations was extracted for critical and comparative analysis.

To reduce the risk of missing relevant studies, a snowballing technique based on the reference lists of the selected papers [32] was applied. This process resulted in a final sample of 30 representative studies, which form the basis of this review.

3 Literature Review

SWDNs are the digital evolution of the standard water distribution systems, combining sensing, communication, and computational intelligence technologies, this enables for real-time monitoring, adaptive control, and predictive maintenance. SWDNs give water utilities the ability to optimize energy use, increase reliability, and guarantee equitable distribution by integrating Internet of Things (IoT) devices, Supervision Control and Data Acquisition (SCADA) systems, and sophisticated decision support algorithms. Recent research has shown that the most effective smart water networks combine optimization frameworks, AI, and hybrid approaches, providing a comprehensive foundation for intelligent, data driven management under uncertainty [6].

Optimization based methods remain fundamental for SWDNs planning and operation. Multi-objective optimization (MOO) and mixed integer linear programming (MILP) formulations such as MILPNet and piecewise linear MILP have been employed to model hydraulic behaviors, device operations, and energy constraints with high precision, supporting optimal pump scheduling and water distribution system operation [9, 20, 21]. Evolutionary algorithms and swarm intelligence methods, including the genetic algorithm (GA), particle swarm optimization (PSO), marine predators algorithm (MPA), and modified Rao II algorithm, have demonstrated strong capability in handling non-linear and non-convex optimization problems, especially for SWDNs design, pump scheduling, and reservoir management where conflicting objectives must be balanced [8, 13, 22–25]. In these studies, MOO frameworks have been used to balance cost, reliability, and equity in network design, while Pareto based solution sets are employed to support multi criteria decision making for selecting optimal solutions [22]. Furthermore, graph theory based optimization has proven effective for optimizing valve placement, enhancing the resilience and seismic reliability of smart networks by integrating topological indicators and minimizing hydraulic simulation complexity [26]. Although their resilience, optimization techniques have several challenges, including high computing costs, the need for professional calibration, and a propensity to overlook data-driven uncertainty or socio-hydrological interactions, particularly in the face of severe climate uncertainty. [1, 2]. These limitations have spurred increasing adoption of AI techniques to enhance automation and predictive performance in smart systems.

AI has become a critical enabler of data driven decision making in smart SWDNs by learning from large volumes of operational and environmental data. ML and DL methods such as support vector machines (SVM), random forest (RF), artificial neural networks (ANN), deep neural networks (DNN), long short term memory (LSTM), gated recurrent units (GRU), and convolutional neural networks (CNN) have been applied to demand forecasting, water quality prediction, rainfall estimation, and leak detection, thereby supporting near real time operational control [10, 13, 27]. For instance, a hybrid model combining feature selection and a single hidden layer feed forward neural network optimized by differential evolution (DE) has been used to forecast water demand with improved accuracy by integrating meteorological and socioeconomic drivers [7]. Similarly, DNN have enhanced dissolved oxygen prediction in lakes [12], while real time ML frameworks have improved the detection of leaks and bursts in complex SWDNs [10]. Also, [6] applies a CNN with advanced metering infrastructure (AMI) data and a digital twin to detect leaks in eleven SWDNs. Results show that network type and pipe characteristics (diameter and length) significantly affect detection accuracy. However, their performance is dependent on factors such as sensor calibration, data quality, and model interpretability, which can restrict adoption in contexts with limited resources or data [19]. These difficulties have spurred the development of hybrid models, which blend the structured optimization of conventional mathematical programming with the predictive capability of AI.

Hybrid frameworks are capable of forecasting important hydrological variables and feeding them straight into real-time control systems by combining prediction model with optimization algorithms. For instant, decision-making under climatic uncertainty is supported by adaptive neuro fuzzy inference systems (ANFIS) combined with meta-heuristic algorithms such as simulated annealing (SA), GA, and artificial bee colony (ABC) that have improved rainfall prediction accuracy [16]. In order to enable proactive water planning in climate-sensitive areas, several research have investigated statistical ARIMA models combined with ANN (SARIMA ANN) for long-term rainfall forecasting [17]. When tank storage dynamics are taken into consideration, hybrid models that combine knowledge based GA with hydraulic simulators have demonstrated significant energy savings (average 15%, maximum 25%) as well as decreases in operating costs and CO_2 emissions in the context of energy optimization [8]. Further demonstrating the versatility of hybrid intelligence in handling intricate, multifaceted problems, improved ant optimization (IAO) in conjunction with capsule neural network (CapsNNs) architectures has been applied to the prediction of drought and water stress [28]. The literature shows that modern SWDNs are shaped by the convergence of optimization, AI, and hybrid intelligence. Optimization establishes the mathematical foundation for resource allocation, pump control, and network design, while AI contributes to pattern recognition, anomaly detection, and forecasting. Hybrid frameworks combine both to enable real-time adaptability, self-learning, and resilient operations.

Table 1 summarizes key studies addressing water resource management through these methods, outlining their models, objectives, algorithms, and applications to reveal prevailing trends, research gaps, and opportunities for integrating intelligent frameworks in future studies.

Table 1 gives a comparative overview of recent studies tackling SWDNs using various methodological approaches, including AI, Hybrid AI, and MOO techniques. To gain a better understanding of these trends and their consequences for SWDNs, the next section presents a detailed study and explanation of the variations among methodologies and applications.

Table 1. Comparative table of studies on WRM optimization

References	Objective	Method type	Adapted algorithms	Domain/Application
[6], 2025	Evaluate the effectiveness of a CNN model for leak detection in smart water distribution systems using AMI data and digital twins	AI	CNN	Leak detection and leak size estimation
[16], 2024	Forecast monthly rainfall using hybrid SARIMA ANN models for long term planning in regions with climate variability	AI	SARIMA ANN vs ANN	Rainfall prediction
[17], 2024	Predict daily rainfall using hybrid ANFIS models with metaheuristic optimization and data preprocessing	Hybrid AI	ANFIS ABC, ANFIS GA, ANFIS SA	Rainfall prediction
[10], 2024	Predict leaks in pipe networks by linking predicted probabilities to actual leakage rates for prioritizing maintenance	AI	ANN	Leak detection
[28], 2024	Predict monthly water stress and drought intensity using a capsule neural network combined with an optimization algorithm	Hybrid AI	CapsNNs + IAO algorithm	Water stress prediction

(*continued*)

Table 1. (*continued*)

References	Objective	Method type	Adapted algorithms	Domain/Application
[9], 2024	Use piece-wise linear MILP approximations to solve optimal pump scheduling issues with variable speed pumps	MOO	MILP with piecewise linear approximation	Pump scheduling
[21], 2024	Develop a dynamic MILP framework to simulate and optimize WDS operations under hydraulic and energy constraints	MOO	MILPNet	WDS operations
[8], 2024	Determine optimal pump operation strategies that balance energy efficiency and water quality in urban systems	MOO	PSO, Non-dominated Sorting Genetic Algorithm II (NSGA-II)	Pump scheduling
[24], 2024	Reduce construction effort of large scale WDS while maintaining near global optimal design performance	MOO	Modified Rao II algorithm	WDS design
[26], 2024	Optimize isolation valve placement using graph theory to enhance reliability and seismic resilience	MOO	Graph theory indicators	WDS resilience / Valve optimization
[23], 2024	Design water distribution networks by balancing cost, reliability, and equity through MOO	MOO	NSGA-II	WDS design

(*continued*)

Table 1. (*continued*)

References	Objective	Method type	Adapted algorithms	Domain/Application
[22], 2024	Manage reservoir levels under flood risk using a dynamic multi-objective model with risk and runoff integration	MOO	Dynamic Multi-objective Multi Strategy Co evolution (DMMC)	Reservoir management
[29], 2023	Optimize water allocation to enhance aquifer recharge, reduce deficits, and maximize agricultural returns	MOO	PSO, GA, MPA	Agricultural water management
[20], 2022	Assess economically viable and environmentally sustainable water supply options in arid regions	MOO	MILP	WDS design
[30], 2021	Use SVM models with changes to the input to make monthly rainfall and pan evaporation predictions more accurate	AI	SVM	Rainfall prediction
[15], 2021	Accurately predict groundwater levels using a hybrid model with decomposition, input selection, and uncertainty analysis	Hybrid AI	VBELM	Groundwater level prediction
[12], 2021	Predict dissolved oxygen in lake water using DNN to support real time decision making and water quality monitoring	AI	DNN	Water quality prediction / Lakes

(*continued*)

Table 1. (*continued*)

References	Objective	Method type	Adapted algorithms	Domain/Application
[27], 2020	Enhance leak detection using statistical analysis of acoustic emission signals with data improvement techniques for small sample sizes	AI	Statistical signal analysis + data enhancement	Leak detection
[11], 2019	Forecast water demand using AI with optimized feature selection including climatic and socioeconomic factors	AI	Single hidden layer feed forward neural network + Differential Evolution	Water demand forecasting
[25], 2019	Optimize daily pump scheduling to minimize energy cost, emissions, and operational constraints while meeting demand	MOO	GA	Pump scheduling

4 Analysis and Discussion

4.1 Trends and Comparative Insights on AI and Optimization Techniques in WRM Studies

The comparative analysis of studies published between 2019 and 2025 (see Figs. 1 and 2), based on the classification presented in Table 1, highlights the evolving adoption of computational methods in WRM and SWNM. This temporal distribution reflects a clear methodological transition driven by increasing system complexity, data availability, and operational requirements.

In the early phase of the review period, traditional machine learning models such as ANNs and SVMs were predominantly applied to demand forecasting and anomaly detection tasks due to their relatively low data requirements and straightforward implementation [12, 30]. While effective in controlled or data-limited settings, these models often exhibit limited generalization capabilities when exposed to non-stationary demand patterns or real-time operational variability.

As sensing infrastructures expanded through SCADA systems and Advanced Metering Infrastructure, researchers progressively adopted more advanced approaches, including DL architectures and hybrid AI–optimization frameworks, to manage high-dimensional and streaming data [10, 16]. External pressures such as rapid urbanization, climate variability, and the need for resilient infrastructure further accelerated this shift, exposing the limitations of static or rule-based decision-support methods. From 2023 onward, the increased use of CNNs, CapsNNs, and other deep architectures is evident in applications such as rainfall prediction, drought monitoring, and leak detection [6, 16, 27, 28]. These methods demonstrate superior predictive accuracy and feature extraction capabilities.

In parallel, multi-objective optimization techniques, including NSGA-II, PSO, and MILP, remain widely used to address trade-offs between energy consumption, operational cost, and system reliability [9, 23]. Their main strength lies in their ability to explicitly incorporate physical constraints and deliver interpretable Pareto-optimal solutions, making them attractive for real-world operational planning [1, 4]. Nonetheless, their computational burden and reduced adaptability to real-time conditions limit their standalone applicability in highly dynamic environments.

Overall, the observed trends indicate a gradual convergence between AI-driven prediction models and optimization-based decision frameworks, setting the foundation for more adaptive and intelligent water management systems.

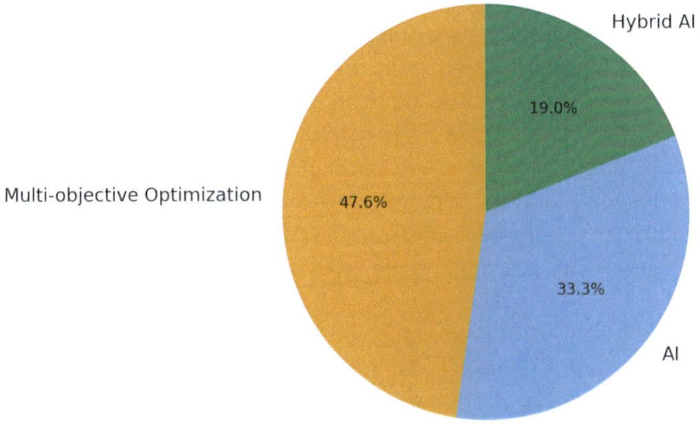

Fig. 1. Rate use of each method type in WRM studies

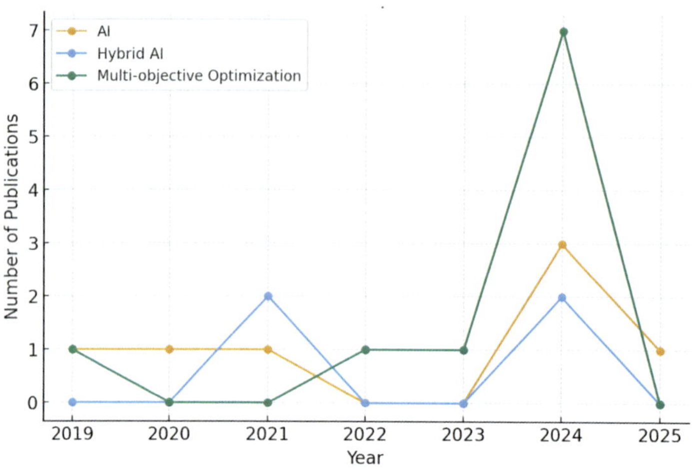

Fig. 2. Evolution use of method types over time (2019–2025)

4.2 Comparative Applications of Optimization, AI, and Hybrid AI Techniques in SWDNs

Because AI and optimization techniques have been applied across a broad range of SWNMs applications, leveraging their complementary strengths in prediction, control, and decision support. Key application domains include leak detection and size estimation [10, 14], rainfall and water demand forecasting [16, 17, 30], pump and reservoir operation scheduling [21, 22], network design and resilience enhancement [8], as well as drought and groundwater prediction [15, 28, 29].

Optimization based approaches are particularly effective for structured decision-making problems such as pump scheduling and network design, where system constraints and objectives can be clearly defined. However, their performance may degrade under uncertain or rapidly changing conditions, especially when real-time responsiveness is required.

AI-based methods excel in pattern recognition and forecasting tasks, enabling early detection of anomalies and improved short-term prediction accuracy. Despite these advantages, their effectiveness strongly depends on data quality and availability, and their black-box nature can limit interpretability and acceptance by water utilities.

Hybrid AI–optimization approaches have gained increasing attention in recent years, particularly in applications that require the integration of prediction and control. Examples include the coupling of deep learning models with optimization frameworks for drought management and leak prioritization [6], as well as the integration of ANFIS with metaheuristic algorithms for rainfall prediction [16]. These hybrid systems enhance accuracy, adaptability, and uncertainty handling, making them well suited for complex, data-rich operational environments.

As real-time monitoring data become more comprehensive, future applications are expected to expand toward predictive maintenance and adaptive control. This evolution reflects a transition toward integrated, data-driven management systems capable of autonomously adjusting to changing operational and environmental conditions, thereby improving system robustness, energy efficiency, and sustainability.

5 Conclusion and Future Directions

Considering Based on the analysis of more than 30 studies published between 2019 and 2025, this review provides a comprehensive synthesis of recent computational methodologies applied to SWNM. The integration of real-time data, IoT sensors, digital twin technologies, and AI-driven analytics has facilitated a paradigm shift from reactive to predictive and proactive water management strategies. This transition has contributed to measurable improvements in system robustness, energy efficiency, resource conservation, and operational performance.

The diversity of applications, ranging from MOO for network design to AI based leak detection and drought monitoring, highlights the flexibility and effectiveness of these advanced computational approaches. However, the analysis also reveals that no single methodology is universally optimal; instead, the suitability of each approach depends on the specific application context, data availability, and operational constraints.

Looking ahead, emerging trends such as generative AI, reinforcement learning, and adaptive optimization are expected to further enhance system autonomy, scalability, and resilience. The continued integration of cloud computing, IoT sensor networks, cybersecurity frameworks, and digital twins will support real-time decision-making while mitigating systemic risks. Future research should increasingly focus on balancing trade-offs between cost-effectiveness, reliability, equity, and environmental sustainability, thereby aligning short-term operational objectives with long-term strategic planning. Finally, ensuring accessibility to advanced AI technologies and prioritizing workforce training will be essential for the equitable and effective deployment of smart water management solutions, particularly in resource-constrained regions.

References

1. Sinha, G.K., Purwar, A.K.: Optimization models in water resources management and security: a critical review. Int. J. Math., Eng. Manag. Sci. **9**, 129–146 (2024)
2. Mala-Jetmarova, H., Sultanova, N., Savic, D.: Lost in optimisation of water distribution systems? A literature review of system design. Water **10** (2018)
3. Kumar, V., Yadav, S.M.: A state-of-the-Art review of heuristic and metaheuristic optimization techniques for the management of water resources. Water Supply **22** (2022)
4. Janga Reddy, M., Nagesh Kumar, D.: Evolutionary algorithms, swarm intelligence methods, and their applications in water resources engineering: a state-of-the-art review. H2Open J. **3**(1), 135–188 (2020)
5. Shirajuddin, T.M., Muhammad, N.S., Abdullah, J.: Optimisation problems in water distribution systems using non-dominated sorting genetic Algorithm II: an overview. Ain Shams Eng. J. **14** (2023)
6. Jun, S., Jung, D.: Exploration of deep learning leak detection model across multiple smart water distribution systems: detectable leak sizes with AMI meters. Water Res. **29** (2025)
7. Krishnan, S.R., et al.: Smart water resource management using artificial intelligence—a review. Sustainability **14**, 13384 (2022)

8. Brentan, B., Mota, F., Menapace, A., Zanfei, A., Meirelles, G.: Optimizing pump operations in water distribution networks: balancing energy efficiency, water quality and operational constraints. J. Water Process. Eng. **63** (2024)
9. Janus, T., Ulanicki, B., Diao, K.: Optimal scheduling of variable speed pumps with mixed integer linear programming. Water Supply **24**(7), 2409–2426 (2024)
10. Joseph, K., et al.: Leak and burst detection in water distribution network using logic- and machine learning-based approaches. Water **16**(14), 1935 (2024)
11. Sörensen, J., Nilsson, E., Nilsson, D., Gröndahl, E., Rehn, D., Giertz, T.: Evaluation of ANN model for pipe status assessment in drinking water management. Water Supply **24**, 1985–1998 (2024)
12. Karamoutsou, L., Psilovikos, A.: Deep learning in water resources management: τhe case study of Kastoria Lake in Greece. Water **13**(23), 3364 (2021)
13. Negm, A., Ma, X., Aggidis, G.: Deep reinforcement learning challenges and opportunities for urban water systems. Water Res. **253**, 121145 (2024)
14. Kamyab, H., et al.: The latest innovative avenues for the utilization of artificial Intelligence and big data analytics in water resource management. Results Eng. **20**, 101566 (2023)
15. Wu, M., Feng, Q., Wen, X., Yin, Z., Yang, L., Sheng, D.: Deterministic analysis and uncertainty analysis of ensemble forecasting model based on variational mode decomposition for estimation of monthly groundwater level. Water **13**(2), 139 (2021)
16. Latif, S.D., Mohammed, D.O., Jaafar, A.: Developing an innovative machine learning model for rainfall prediction in a semi-arid region. J. Hydroinformatics **26**(4), 904–914 (2024)
17. Pham, B.T., Bui, K-T.T., Prakash, I., Ly, H.-B.: Hybrid artificial intelligence models based on adaptive neuro fuzzy inference system and metaheuristic optimization algorithms for prediction of daily rainfall. Phys. Chem. Earth, Parts A/B/C **134**, 103563 (2024)
18. Zaresefat, M., Derakhshani, R.: Revolutionizing groundwater management with hybrid AI models: a practical review. Water **15**(9),1750 (2023)
19. Oğuz, A., Ertuğrul, Ö.F.: A survey on applications of machine learning algorithms in water quality assessment and water supply and management. Water Supply **23**(2), 895–922 (2023)
20. Kizhisseri, M.I., Mohamed, M.M., Hamouda, M.A.: A mixed-integer optimization model for water sector planning and policy making in arid regions. Water Resour. Ind. **28**, 100193 (2022)
21. Thomas, M., Sela, L.: A mixed-integer linear programming framework for optimization of water network operations problems. Water Resour. Res. **60**(2), e2023WR034526 (2024)
22. Zhang, L., Feng, Z., Yao, X., Niu, W., Wang, Y., Mo, L.: A multi-objective operation optimization method for dynamic control of reservoir water level in evolving flood season environments. J. Hydrol. **643**, 131940 (2024)
23. Ramani, K., Umamahesh, N. V.: Multi-criteria decision-making methods for optimal design of intermittent water distribution network. Water Supply **24**(4), 1176–1195 (2024)
24. Gangwani, L., Palod, N., Dongre, S., Gupta, R.: Optimal pipe-sizing design of water distribution networks using modified Rao-II algorithm. J. Water Clim. Chang. **15**(8), 3775–3791 (2024)
25. Luna, T., Ribau, J., Figueiredo, D., Alves, R.: Improving energy efficiency in water supply systems with pump scheduling optimization. J. Clean. Prod. **213**, 342–356 (2019)
26. Liu, J., et al.: Optimization of isolation valve placement to improve reliability and resilience of water distribution systems. Reliab. Eng. Syst. Saf. **252**, 110423 (2024)
27. Luong, T.T.N., Kim, J.-M.: The enhancement of leak detection performance for water pipelines through the renovation of training data. Sensors **20**(9), 2542 (2020)
28. Wang, Y., Razmjooy, S.: Prediction of drought hydrological and water scarcity based on optimal artificial intelligence by developing a metaheuristic optimization algorithm. Phys. Chem. Earth, Parts A/B/C **135**, 103669 (2024)

29. Jain, S., Ramesh, D., Trivedi, M.C., Edla, D.R.: Evaluation of metaheuristic optimization algorithms for optimal allocation of surface water and groundwater resources for crop production. Agric. Water Manag. **279** (2023)
30. Meidute-Kavaliauskiene, I., Jabehdar, M.A., Davidavičienė, V., Ghorbani, M.A., Sammen, S.Sh.: A simple way to increase the prediction accuracy of hydrological processes using an artificial intelligence model. Sustainability **13**(14), 7752 (2021)
31. Catherine, P., Jason, B.: The benefits of publishing systematic quantitative literature reviews for PhD candidates and other early-career researchers. High. Educ. Res. Dev. **33** (2013)
32. Mahin, N., Hamideh, G., Fereshteh, G.: Snowball sampling: a purposeful method of sampling in qualitative research. Strides Dev. Med. Educ. **10** (2017)

Enhancing Traceability and Compliance in Air Transport of Dangerous Goods Through Distributed Ledger Technology: Policy Perspectives from Morocco

Fayssal moukafi(✉), Hiba Belgaid, and Amine Dafir

Laboratory for Research on the New Economy and Development (LARNED), Hassan II University of Casablanca, 0250 Casablanca, Morocco
fayssal.moukafi-etu@etu.univh2c.ma

Abstract. Traceability, transparency, and regulatory compliance are paramount in the air transport of hazardous goods (DG), yet these requirements are becoming more difficult to meet due to the rising complexity of global supply chains. The potential of Distributed Ledger Technology (DLT) to improve compliance and traceability during the air transport of hazardous materials in Morocco is explored in this study. This research examines the air transport industry in Morocco through the lens of the macro environmental variables impacting DLT adoption using the PESTEL framework. International standards published by ONDA, ICAO, and IATA, as well as official institutional papers and industry reports, formed the basis of the qualitative triangulation that formed the basis of the study. Results show a supportive institutional environment, with the help of ONDA's Digital Strategy 2030, a changing regulatory landscape, and increasing air freight activity. Problems with interoperability, data governance, cybersecurity, and lack of digital skills are still major issues. Safe and verifiable DG data management may be possible with the help of DLT integrated with AI-based and Internet of Things technologies, according to the article. In order to help Morocco with its digital revolution in air logistics, we have some policy suggestions.

Keywords: Distributed Ledger Technology (DLT) · Dangerous Goods · Air Transport

1 Introduction

One of the components of critical infrastructure systems and of the essentials foundations of the international trade economy is Air freight.

The air transport infrastructure in international trade depends on a complex socio-technical system aviation. When it's interrupted, the consequences are immediate and severe strategically, socially, and economically [1].

The sheer value of air cargo illustrates this perfectly. Despite making up just 2–3% of total freight tonnage, it carries nearly 35% of the value in global trade [2]. We saw

Y. Benadada et al. (Eds.): GOL 2026, LNNS 2034, pp. 136–145, 2027.
https://doi.org/10.1007/978-3-032-29677-1_13

clear proof of its importance in the post-pandemic recovery, where its role in stabilizing supply chains helped traffic return to 94% of 2019 levels by 2023 [3].

Aviation's complexity is also its biggest weakness, creating very real problems for operational reliability, safety, and compliance. You see this most clearly with the transport of dangerous goods (DGs) things like chemical or radioactive materials. There are strict rules for this, like the ICAO and IATA regulations, but actually getting people to follow them is another matter. The problem comes down to basics: databases that don't connect, paperwork done by hand, and a total lack of communication between airlines, customs, and regulators. This is what causes the constant information delays and errors. As Imeri et al. [4] found, it's incredibly difficult to maintain reliable data across the DG supply chain when your core systems are vulnerable to tampering and inconsistency.

Distributed Ledger Technology (DLT), or blockchain, presents a direct model for solving these failures. The technology's foundation is a shared ledger where all participants validate data together, ensuring it remains immutable and transparent without any central party in control. IATA explored this for aviation in a 2018 White Paper, identifying concrete uses: digital certificates for aircraft parts, live cargo tracking, and automated compliance checks with smart contracts. Such applications would improve efficiency and data integrity, while also helping environmental goals by removing paper. However, in their reports also acknowledged the real-world obstacles. Industry-wide adoption is still held back by governance complexities, technical scaling limits, and the simple lack of a harmonized legal framework.

Focusing specifically on dangerous goods, Imeri et al. [4] proposed a model using a blockchain digital certificate. This certificate records all stages of the transport loading, storage, environmental state on a distributed ledger that cannot be changed. The proposed system is also compatible with IoT sensors. Temperature, vibration, and pressure are constantly monitored through these sensors. Once a problem is detected, it automatically triggers an alert. A full end-to-end visibility is provided by putting these technologies together. It also improves compliance and makes accountability clear between consignors, carriers, and authorities.

For Morocco, these global developments are critically important. The country's air transport sector is a foundation of its economic growth and a central piece of its logistics strategy. While the government is pursuing modernization through the ONDA Digital Strategy 2030 and Loi n° 40–13, a significant gap remains in practice. The system for tracking dangerous goods is still fragmented and disconnected between agencies. This failure is rooted in three core problems: a lack of integration between digital systems, the absence of any regulatory framework for distributed records, and the need for pilot projects to test blockchain's real-world viability [5]. To address these gaps, the present study investigates how DLT can enhance traceability and regulatory compliance in Morocco's air transport of dangerous goods. Using a PESTEL analytical framework, it identifies the political, economic, social, technological, environmental, and legal factors shaping DLT adoption. The goal is to provide a roadmap for a secure, transparent, and internationally compliant digital air-cargo ecosystem in Morocco.

2 Literature Review and Theoretical Framework

2.1 Distributed Ledger Technology and Traceability in Aviation

Blockchain, securely manages data in a decentralized network using cryptographically linked blocks. This is how the system ensures data integrity and immutability [6, 7].In the world of aviation logistics, these features have huge potential. They can directly improve interoperability and accountability in areas like cargo handling, maintenance, and auditing. Vevera and Georgescu [1], argue that blockchain should be seen as part of the Critical Air Transport Infrastructure (CATI), where data you can trust is essential. DLT is effective due to its tamper-resistant consensus mechanism. We see this in things like smart contracts for compliance and provenance tracking to verify parts. This is the same vision IATA pushed in its 2018 White Paper [3]. We also have empirical studies from Kouhizadeh et al. [8] and Poleshkina [9] that confirm blockchain cuts documentation time and reduces fraud. But the technology still faces real-world hurdles. As Stoic's study shows, many struggles such as with governance, scalability, and integration with legacy systems continue to be the primary barriers to adoption [10].

2.2 Distributed Ledger Technology and Traceability in Aviation

Transporting dangerous goods is a high-risk logistics activity, especially by air. These materials, from flammable liquids to medical waste, need rigorous oversight to prevent serious harm to people and the environment [4]. We have detailed rules from ICAO and IATA. But in practice, our information systems are failing. They are fragmented, centralized, and still depend on manual work that creates delays and compliance risks.

The model from Imeri et al. [4] is a direct response to this. It uses a blockchain certificate with IoT devices to log real-time data such as temperature, vibration, location on a ledger that cannot be altered. Smart contracts check the rules automatically. What this does is enforce transparency and accountability. This idea fits perfectly within Design Science Research [11], treating blockchain as a new tool both technical and organizational to build trust across the aviation network.

2.3 Blockchain and Critical Transport Infrastructure (CATI)

Building on the systemic approach, Vevera and Georgescu [1] conceptualize CATI as a complex ecosystem of "technical elements, facilities, distributed machinery, communication links, and organizations" whose disruption would cause severe socio-economic damage (p. 422). Within such infrastructures, blockchain functions as both a technological safeguard and a strategic enabler of resilience. By decentralizing data control and ensuring immutability, blockchain reduces the single points of failure that characterize traditional centralized systems.

The authors illustrate blockchain's potential in CATI through applications such as:

- Maintenance record authentication for aircraft parts.
- Digital identity management to control access to critical facilities.
- Automated settlement systems for aviation procurement, with smart contracts facilitating transparent, auditable transactions.

These use cases reflect a shift toward security conscious digitalization, where emerging technologies support both operational continuity and national defense resilience [12, 13]. Vevera and Georgescu [1] also warn against the uncritical adoption of blockchain across all aviation processes, noting that the technology's benefits must be balanced with careful governance and cybersecurity oversight. Their proposed approach favors a hybrid implementation model, in which credible authorities such as IATA or national regulators manage permissioned blockchain networks, ensuring both scalability and institutional trust. This is consistent with IATA's Travel Grid concept, a modular ecosystem where verified participants share standardized data via APIs [3].

3 Methodology: PESTEL Analytical Framework

This study employs a qualitative and exploratory methodology, drawing upon a system-based analysis framework applied to aviation digitalization, as established by Jurgelāne-Kaldava et al. [14]. This sector is conceptualized as a socio-technical system in which technologies such as DLT, regulatory frameworks, and institutional actors engage interactively, rather than as a solely technical infrastructure.

Within this framework, this technology is analyzed not merely as an information and communication instrument but also as a potential governance tool capable of augmenting institutional trust and ensuring regulatory compliance in the transportation of hazardous materials. To analyze the macro-environmental factors impacting DLT adoption in an emergent country such as Morocco, this analysis of PESTEL was employed, as it facilitates a systematic examination of political, economic, social, technological, environmental, and legal aspects within highly regulated industries such as aviation. The analysis is based on qualitative data triangulation, integrating official documents and statistics from national agencies (ONDA, DGAC), international regulatory standards and reports published by ICAO and IATA, as well as secondary documentary sources pertaining to airport operations and customs procedures. Documentary evidence was methodically coded in accordance with the six PESTEL dimensions. Triangulation with extant scholarly literature was employed to improve the reliability and consistency of the results.

4 Results and Discussion

4.1 Application of the PESTEL Model to the Moroccan Context

4.1.1 Political Factors

Morocco's political commitment to modernizing its aviation sector is clear and well-documented. This is not just ambition; it is anchored in law, specifically the law n° 40–13 [15]. The national airports office, ONDA, has also built these goals into its Plan Stratégique et Transformation Digitale 2030 [16]. This push has tangible results. National organizations that have international standards such as the Ministry of Transport and Logistics, ONDA, and even the DGAC are actively enhancing with IATA on global standards. The sector's growth is undeniable: the Ministry of Transport (2024) confirmed a 22 percent jump in air freight by August 2024 [17].

However, despite all this enhancement, the coordination between the key ministries that oversee this mode of transport, customs, and digital governance remains fragile. The absence of institutional synergy is the single biggest problem. Overcoming it is the essential next step to actually build the DLT-enabled ecosystem needed to link regulators, carriers, and freight forwarders.

4.1.2 Economic Factors

There has been consistent expansion in Morocco's air freight market in recent years. In the second quarter of 2024 alone, aviation cargo traffic grew by roughly 24%, reaching 46,209 tons, according to the Moroccan Airports Authority [18]. Regulatory compliance systems are under increasing financial and operational strain as a result of this increase, especially in the transportation of hazardous materials. Reducing transaction costs, enhancing information sharing, and boosting coordination among logistics operators are three ways in which Distributed Ledger Technology (DLT) might improve efficiency from an economic standpoint. Vevera and Georgescu [1] pointed out that solutions built on the blockchain can help build a more transparent and robust aviation infrastructure. On the other hand, public-private partnerships and focused innovation financing are necessary to overcome the high investment costs that small and medium-sized operators face. This will help to guarantee that digital adoption is inclusive.

4.1.3 Social Factors

The success of any aviation innovation in Morocco will depend on digital maturity and public trust in its organizations. While there is clear improvement on the digital front as highlighted in the UN E-Government Survey 2024 [19, 20] this positive trend is undermined by a persistent technical skills gap. This gap remains a major barrier, creating a patchy and uneven capacity across different sectors to actually adopt and use new technologies effectively.

As Jurgelāne-Kaldava et al. [14] emphasize, human capital and inter-organizational learning are vital for successful transformation. Introducing specialized training on blockchain governance, cybersecurity, and DG handling through ONDA and DGAC programs would reinforce DLT adoption and compliance capabilities.

4.1.4 Technological Factors

Morocco already possesses advanced digital infrastructure. The PortNet single-window platform and the customs BADR/RIAD systems enable paperless trade and comply with ISO/IEC 27001:2022 standards [21] customs circular requires electronic submission of all supporting documents (Administration des Douanes et Impôts Indirects [21]. These systems can host permissioned blockchain nodes for tamper-proof DG certificates. However, a major barrier to automation persists: the lack of interoperability between the systems used by ONDA, customs, and ground handlers. The model proposed by Imeri et al. [4], which integrates DLT with IoT sensors, is designed to solve this by enabling continuous compliance monitoring and rapid anomaly detection.

4.1.5 Environmental Factors

Air-freight expansion has multiplied documentation flows, increasing paper use and emissions. ONDA [18] data highlight how digital documentation supports Morocco's *Green Logistics 2025* goals. Blockchain-enabled DG certificates can significantly reduce paper dependency and transport between agencies. Coupling DLT with IoT-based monitoring further strengthens proactive risk management and aligns with the *United Nations Sustainable Development Goals 9* (Industry and Innovation) and *12* (Responsible Consumption and Production).

4.1.6 Legal Factors

Legally, electronic data governance and digital documentation are already under Morocco's purview. The National Data Protection Commission (CNDP) is tasked with overseeing digital data processing according to Law No. 09-08 on personal data protection [24], while Law No. 53-05 on electronic data interchange ensures the legal validity of certifying authority and electronic signatures [23]. Secure digital governance is laid forth by these legislative instruments taken as a whole. Worldwide acceptance of DLT-based documentation for hazardous goods transport depends on conformity with international aviation standards, especially ICAO Annex 18, the ICAO Technical Instructions for the Safe Transport of Dangerous Goods by Air (Doc 9284, 2025–2026 edition) [25], and the IATA Dangerous Goods Regulations (66th edition, 2025) [26] (Table 1).

Table 1. Condensed PESTEL factors influencing DLT adoption in Morocco's air transport of dangerous goods.

PESTEL	Key driver / constraint	Evidence (statistics)	Implication for DG traceability and compliance
Political	Strong aviation legal base	Law 40-13 (BO 7410, 2025)	Institutional mandate for ICAO-aligned digital governance
Economic	Rapid air-freight growth	46,209 t H1-2024 (+24%) [18]	Rising compliance workload → need for automated traceability
Social	Expanding e-government capacity	EGDI rank 90/193 (2024) [19, 20]	Digital readiness exists but skills gap persists
Technological	Mature digital platforms	PortNet ISO/IEC 27001; BADR [21, 22]	Secure dematerialization enables DLT interoperability
Environmental	Shift toward paperless logistics	Freight + 24% H1-2024 [18]	Reduced paper use and improved auditability

(*continued*)

Table 1. (*continued*)

PESTEL	Key driver / constraint	Evidence (statistics)	Implication for DG traceability and compliance
Legal / Safety	ICAO/IATA DG framework	ICAO Doc 9284 (2025–26); IATA DGR (66th ed.) [25, 26]	Legal base exists; explicit DLT recognition still required

Source Author's elaboration based on ONDA (2024), UN DESA (2024), ICAO Doc 9284 (2025–2026), and IATA DGR (66th ed., 2025)

4.2 Discussion

Distributed ledger technology (DLT) may find widespread use in air transport logistics in Morocco, according to a PESTEL research that looked at the country from many angles. In the highly regulated and risk-sensitive arena of dangerous goods transit, however, these enabling characteristics do not necessarily translate into effective implementation. The results show that there is a disconnect between plans and preparations for implementation. From an institutional and political perspective, national digital strategies and aviation policies demonstrate a clear commitment to digital transformation. Yet, as shown in studies on critical transport infrastructure, the absence of coordinated governance mechanisms may limit the operational impact of advanced digital technologies such as DLT [1]. There is a pressing demand for secure, auditable, and real-time traceability solutions due to the increased exposure to compliance and safety concerns connected with dangerous commodities caused by the fast rise of air freight activities. Technologically, existing platforms such as PortNet and BADR provide a solid foundation for interoperability and data exchange. Nevertheless, fragmented data governance and uneven digital capabilities particularly among small and medium-sized operators remain significant barriers, echoing challenges identified in other air cargo digitalization initiatives [14]. From a legal standpoint, while Morocco's framework on electronic signatures and data protection is broadly aligned with international requirements, the lack of explicit legal recognition for DLT-based documentation continues to constrain large-scale adoption. Overall, DLT can enhance compliance and risk management for dangerous goods only if supported by regulatory clarification, coordinated inter-agency governance, and targeted capacity-building in line with ICAO and IATA standards [25, 26].Based on these findings, a phased and coordinated roadmap from regulatory alignment to pilot implementation, interoperability, and international coordination emerges as a pragmatic way to translate Morocco's favorable structural conditions into effective DLT adoption for dangerous goods air transport (Table 2).

Table 2. Phased roadmap for DLT adoption in the air transport of dangerous goods in Morocco.

Phase	Main objective	Key stakeholders	Key actions / values
Phase 1: Regulatory alignment	Establish a clear legal and institutional basis for DLT use in DG air transport	DGAC, ONDA, Customs (ADII), CNDP	Legal clarity; regulatory alignment with ICAO/IATA standards; institutional coordination
Phase 2: Pilot implementation	Test DLT solutions in a controlled operational environment	ONDA, airlines, freight forwarders, ground handlers	Traceability; transparency; operational learning; risk control
Phase 3: Interoperability and scaling	Ensure system compatibility and wider adoption	ONDA, Customs, IT providers, SMEs	Interoperability; inclusiveness of SMEs; data governance; scalability
Phase 4: International alignment	Achieve international recognition and cross-border compatibility	ICAO, IATA, partner airports, international carriers	International compliance; mutual recognition; trust; global integration

Source Author's elaboration based on PESTEL analysis and international regulatory frameworks

5 Conclusion

In this study, the PESTEL framework was used to analyze how Distributed Ledger Technology (DLT) may improve traceability and regulatory compliance during the air transport of hazardous products in Morocco. In spite of rising air freight activity and supportive national digital plans, the results show that there are still obstacles to overcome in the areas of interoperability, data governance, cybersecurity, and digital skills. This study shows that DLT, in conjunction with IoT and AI-based systems, may increase the dependability of dangerous commodities data management and make it easier to comply with international aviation requirements. This has important policy implications. This study addes to the existing art of literature a contextualized analysis of the use of this technology within the regulated aviation sector of a developing country. To assess the operational advantages of DLT-based solutions in aviation logistics, future research could build upon this study by undertaking empirical field surveys, stakeholder interviews, or experimental project evaluations.

References

1. Vevera, A.-V., Georgescu, A.: The impact of blockchain on critical transport infrastructure—the case of aviation. Land Forces Acad. Rev. **29**(3), 421–431 (2024)
2. Turner, Y.: The importance of air cargo to the world economy. Air Cargo Week (2021)
3. International Air Transport Association (IATA): Blockchain in Aviation: exploring the fundamentals, use cases and industry initiatives. International Air Transport Association (2018)
4. Imeri, A., Feltus, C., Agoulmine, N., Khadraoui, D.: The Blockchain-Based Digital Certificate for the Transport of Dangerous Goods. Springer, Cham (2022)
5. Peoples, J.: The Economics of International Airline Transport (2014)
6. Dinu, A., Vasile, P.C.: Shaping the educational landscape: the rise and potential of verifiable credentials in higher education. In: Proceedings of the 18th International Conference on Virtual Learning, pp. 123–132 (2023)
7. Baboi, M.: Security of consensus mechanisms in blockchain. Rom.Ian Cyber Secur. J. **5**(2), 45–53 (2023)
8. Kouhizadeh, M., Saberi, S., Sarkis, J.: Blockchain technology and the sustainable supply chain: Theoretically exploring adoption barriers. Int. J. Prod. Econ. **231**(C) (2021). https://doi.org/10.1016/j.ijpe.2020.107831
9. Poleshkina, I.: Blockchain in air cargo: challenges of a new world. MATEC Web Conf. **341**, 00021 (2021)
10. Stoica, G.M.: Account abstraction on Ethereum. Rom. Cyber Secur. J. **5**(2), 37–44 (2023)
11. Bichler, M.: Design science in information systems research. MIS Q. **48**(2), 133–135 (2006). https://doi.org/10.1007/s11576-006-0028-8
12. Bucovetchi, O., Georgescu, A., Badea, D., Stanciu, R.D.: Agent-based modeling—support for emphasizing the air transport infrastructure dependence on space systems. Sustainability **11**(19), 5331 (2019)
13. Jungwirth, R., et al.: Fortifying defence: strengthening critical energy infrastructure against hybrid threats. European Defence Agency (2023)
14. Jurgelāne-Kaldava, I., Effenberger, W.V., Batenko, A., Čižiūnienė, K.: Digitalization of air cargo supply chains: a case study of Latvia. Systems **13**(6), 468 (2025). https://doi.org/10.3390/systems13060468
15. Secrétariat Général du Gouvernement (SGG): Loi n 40–13 portant Code de l'Aviation Civile—Bulletin Officiel du Royaume du Maroc (2017)
16. Office National des Aéroports (ONDA): Stratégie digitale 2030 du secteur aérien marocain. ONDA, Casablanca (2023)
17. Ministère du Transport et de la Logistique: Le trafic aérien et le fret en progression à fin août 2024 (2024)
18. Office National des Aéroports (ONDA): Communiqué de presse—Trafic 1er semestre 2024 (2024)
19. United Nations Department of Economic and Social Affairs (UN DESA): E-Government Survey 2024—Data Center. United Nations, New York (2024)
20. Hespress English: Morocco ranks 90th in the 2024 UN e-Government Index (2025)
21. PortNet S.A.: PortNet ISO/IEC 27001:2022 Renewal and Vision (2025) https://www.portnet.ma/web/portail-federateur/b/portnet-s-a-certifiee-iso-27001-2022. Last accessed 26 Oct 2025
22. Administration des Douanes et Impôts Indirects (ADII): BADR—Modification des catégories de documents à joindre à la déclaration. EcoActu (2024)
23. Direction Générale de la Sécurité des Systèmes d'Information (DGSSI): Loi n 53-05 relative à l'échange électronique de données juridiques (n.d.)

24. Commission Nationale de Contrôle de la Protection des Données à Caractère Personnel (CNDP): Loi 09-08 relative à la protection des personnes physiques à l'égard du traitement des données à caractère personnel (n.d.)
25. International Civil Aviation Organization (ICAO). Technical Instructions for the Safe Transport of Dangerous Goods by Air (Doc 9284), 2025–2026 Edition. Montreal: ICAO (2024)
26. International Air Transport Association (IATA): Dangerous Goods Regulations (DGR), 66th edn. IATA, Montreal (2025)

Towards Improved Decision Tree Interpretability via Large Language Models: A Case Study in Motorcycle Accident Severity

Ilyas Modni(✉), Maissae Haddouchi, Zoubida Chorfi, and Abdelaziz Berrado

AMIPS Research Team, EMI, Mohammed V University in Rabat, Rabat, Morocco
{i.modni,maissaehaddouchi}@research.emi.ac.ma, {Chorfi,Berrado}@emi.ac.ma

Abstract. While decision trees are fundamentally interpretable, their complex rule sets can become a barrier to human understanding, particularly in high-stakes fields like transportation safety. This paper proposes a hybrid LLM-assisted approach for translating the complex rule sets of a decision tree into clear and actionable insights. Our workflow involves training a pruned CART decision tree on a dataset of Moroccan motorcycle accidents, extracting its predictive rules and passing them along with relevant interpretive context to a Large Language Model (LLM) via a structured prompt. The LLM successfully synthesizes this information to produce a global interpretation of risk factors and targeted actionable safety recommendations. This work demonstrates how LLMs can serve as analytical partners, translating the quantitative logic of a traditional model into narrative explanations that are accessible and meaningful for domain experts and policymakers in transportation safety.

Keywords: Large Language Models · Interpretability · Decision Trees · Explainable AI (XAI) · Prompt Engineering · Road Safety

1 Introduction

Decision trees are a cornerstone of machine learning, widely recognized for their transparency. Often called "white-box" models, they allow us to see the exact logic behind their predictions, which is crucial in fields where understanding the "why" is just as important as the "what" [11,14]. However, as models grow in complexity and predictive power, their decision processes become increasingly difficult to interpret. As a decision tree grows highly branched to capture subtle data patterns, understanding its decision process becomes increasingly difficult [3,5]. This is a significant hurdle in sensitive areas like road safety, where the insights from a model must be clear and actionable to be of any real-world use [10,16]. The challenge of interpreting complex decision trees becomes particularly apparent when analyzing the severity of motorcycle accidents, a common task for such models [8]. A model might be very accurate in predicting how

Y. Benadada et al. (Eds.): GOL 2026, LNNS 2034, pp. 146–155, 2027.
https://doi.org/10.1007/978-3-032-29677-1_14

severe an accident will be, but if it produces a long list of complicated rules, it is of little help to safety experts and policymakers. To bridge this gap between a statistically sound model and a human-understandable explanation, we propose a new, hybrid approach. Our method combines the logical structure of decision trees with the advanced language modeling capabilities of Large Language Models (LLMs) to translate complex rules into clear, meaningful narratives, reflecting a growing trend in AI research [7,18].

This paper presents a four-step workflow designed to enhance the interpretability of decision tree models. Our process begins by training a decision tree on a dataset of motorcycle accident severity and extracting its learned predictive rules. These rules, along with important contextual information, are then provided to a Large Language Model, which in the final step generates easy-to-read explanations that highlight key risk factors and offer practical safety insights. Through this approach, we aim to make the valuable insights from decision tree models more accessible and impactful for improving road safety. The following sections of this paper will detail our methodology, present a case study, and discuss the implications of this work for the future of interpretable AI.

2 Related Work

The idea of making artificial intelligence easier to understand is not new, but the way we approach it is changing. Simple decision trees are inherently interpretable, but as they grow deeper and more complex, understanding their rules becomes increasingly challenging. The combination of decision trees with Large Language Models (LLMs) is a promising approach to help summarize and contextualize these complex rules, providing richer explanations for domain experts.

Recent studies have begun to explore how LLMs can act as analytical partners for machine learning models. Some of the most innovative work has focused on integrating LLMs directly into the model-building process itself. For example, GPT-HTree uses LLMs to generate human-readable descriptions for hierarchically-clustered data [13], while the GPTree framework leverages LLMs to dynamically generate questions and tree splits, often with expert feedback [18]. This is not just limited to decision trees; similar work has been done in complex fields like cybersecurity, where LLMs help explain how an intrusion detection system identifies threats [17].

The core idea behind this research is that explanations generated by LLMs can make a real difference in how well people understand and trust AI systems. A key study confirmed this, showing that when people were given LLM-generated explanations for a model's decisions, their comprehension improved significantly [20]. Beyond just explaining existing models, some researchers are using LLMs to improve the models themselves. Systems like DeLTa [19] and Aug-Tree [15] use LLMs to refine and enhance the rule sets of decision trees, making them more robust and insightful.

In contrast, our work focuses on the practical challenge of the post-hoc interpretation of a trained Tree-based model. Our contribution is twofold: first, we propose a workflow where the LLM is provided with a comprehensive case file including the model's complete rule set and its branch impurity statistics to ground its narrative synthesis. Second, we apply and validate this workflow in the high-stakes domain of road accident analysis, demonstrating its value in translating a model's complex logic into actionable, policy-oriented recommendations for experts in the field.

3 Methodology: An LLM-Assisted Approach to Decision Tree Interpretation

This section presents our proposed hybrid approach for interpreting decision trees. As shown in Fig. 1, the workflow translates complex model rules into clear and actionable insights by combining the predictive logic of decision trees with the narrative capabilities of a large language model.

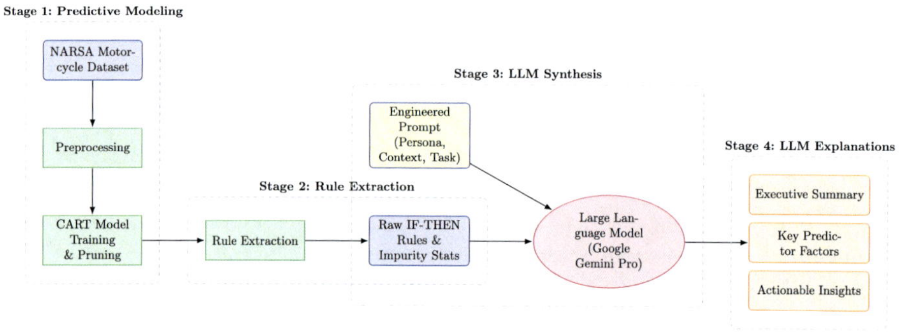

Fig. 1. The proposed four-stage hybrid interpretation approach.

3.1 Step 1: Predictive Model Training and Evaluation

The initial step is to build a robust predictive model. We use a Decision Tree Classifier based on the CART (Classification and Regression Trees) algorithm to predict motorcycle accident severity [1].

Data and Preprocessing. The model was trained on a dataset from Morocco's National Road Safety Agency (NARSA). The binary target variable indicates accident severity: Fatal (1) or Non-Fatal (0). Relevant input features related to human, environmental, and vehicle factors were selected. Categorical features were one-hot encoded, and the data was split into a training set (70%) and a testing set (30%) using stratified sampling to preserve class proportions.

Handling Class Imbalance. The dataset is highly imbalanced. To address this, we integrated the `RandomUnderSampler` technique into our training pipeline [9]. This balances the class distribution by randomly removing instances from the majority class (Non-fatal), improving the model's sensitivity to the critical minority class.

Model Optimization via Pruning. To prevent overfitting and find the optimal tree complexity, we employed cost-complexity pruning [1]. We first grew a maximal tree on the full training data to obtain a range of candidate `ccp_alpha` values, where `ccp_alpha` is the complexity parameter that controls the trade-off between tree size and training error. Then, using a 5-fold stratified cross-validation pipeline, we performed a `GridSearchCV` to identify the optimal `ccp_alpha` that maximizes mean accuracy [12]. This process yields a final, pruned tree that generalizes better to unseen data.

3.2 Step 2: Extracting Decision Rules

Once the final pruned decision tree is trained, its internal logic is exported into a structured text format. Using standard libraries, every path from the tree's root to a terminal leaf is translated into an IF-THEN rule representing a specific combination of conditions. These rules, along with branch impurity metrics, form the textual basis for the next stage.

3.3 Step 3: Instructing the LLM via Prompt Engineering

This step is the core of our hybrid approach, where we communicate with our chosen Large Language Model, Google's Gemini Pro (gemini-2.5-pro) [6], through a carefully constructed set of instructions known as Prompt Engineering. Rather than asking a simple question, we assign the LLM a comprehensive task, providing it with all the necessary context to perform a deep analysis. Our prompt is structured around three key components:

- **Assigning a Persona:** We give the LLM a specific role to play: "You are an AI researcher specializing in road safety analytics." This ensures the tone, focus, and depth of the output are aligned with our research goals.
- **Providing the "Case File" (The Context):** We provide all the necessary information for the analysis. This includes a full description of the dataset features, a feature label mapping that acts as a glossary for encoded names, the complete list of extracted decision tree rules, and their associated branch impurity.
- **Defining the Mission (The Task):** We give the LLM a clear and structured set of objectives: "Provide a thorough interpretation, including key predictive factors, the model's logic, and practical safety recommendations derived from the rules."

This entire process relies on In-Context Learning (ICL) [2,4]. The LLM is not retrained on our dataset but uses its vast pre-existing knowledge to reason over the specific information provided within the prompt.

3.4 Step 4: Generating Explanations and Insights

In the final step, the LLM processes the prompt to generate a global interpretation of the model. This structured output is designed to be a multi-faceted analytical report, comprising several key components:

- **An Executive Summary:** A high-level narrative that synthesizes the model's overall findings and identifies the most critical risk factors at a glance.
- **A Breakdown of Key Predictive Factors:** A detailed analysis of the most influential features, explaining their hierarchy (e.g., primary vs. secondary) and their specific roles in the model's logic.
- **A Translation of Decision Rules:** The conversion of the most significant logical paths into practical, human-readable IF-THEN risk scenarios (e.g., highest-risk, lowest-risk).
- **Actionable Safety Recommendations:** A set of concrete recommendations for safety interventions, derived directly from the risk patterns identified by the model.

This multi-component approach transforms the decision tree from a simple predictive tool into a rich decision-support system, providing insights in a format ready for domain experts and policymakers.

4 Experimental Results

This section begins with a quantitative analysis of the pruned CART model, covering both its predictive performance and its structural complexity. This complexity provides the direct motivation for the final part of our results: the detailed qualitative insights generated by the LLM.

4.1 Predictive Model Performance

The credibility of the interpretation hinges on the model's validity. The pruned CART classifier was evaluated on the held-out test set. Performance metrics are in Table 1.

Table 1. Classification report for the pruned CART model on the test set.

Class	Precision	Recall	F1-Score	Support
Non-fatal	0.96	0.67	0.79	17919
Fatal	0.15	0.66	0.25	1621

The model achieves an overall accuracy of 0.67 and an AUC of 0.71, indicating a moderately effective ability to discriminate between fatal and non-fatal accidents. For the critical Fatal class, the model achieves a strong Recall of 0.66, meaning it correctly identifies 66% of all actual fatal accidents. This high sensitivity is vital for a safety application. However, it comes at the cost of very low Precision (0.15), indicating that many non-fatal accidents are incorrectly flagged as fatal (false positives). This trade-off is common in imbalanced classification problems. While not perfect, the model is a reasonable basis for demonstrating our interpretation workflow.

4.2 Structural Complexity of the Pruned Tree

Before presenting the LLM's narrative interpretation, it is important to quantify the structural complexity of the final pruned CART model. This complexity underscores the analytical challenge that motivates our LLM-assisted approach. The final model had a significant depth of 15, comprising a total of 157 nodes. This structure resulted in 79 terminal leaves, each corresponding to a distinct predictive rule. On average, a path from the root to a leaf involved 9.46 distinct features, making the rules too long and numerous for straightforward manual interpretation. This level of complexity is precisely the kind of output that our workflow is designed to synthesize and make accessible.

4.3 LLM-Generated Global Interpretation

The LLM provided a comprehensive interpretation of the decision tree's logic, which is summarized below.
Executive Summary. The decision tree model identifies a clear hierarchy of risk factors, with helmet usage as the most critical determinant. Fatal accidents are strongly associated with the absence of a helmet, especially during high-risk time periods. Even when a helmet is worn, collisions involving other vehicles, driver demographics, and trip purpose create complex risk profiles, which the model successfully untangles.
Key Predictive Factors. The model prioritizes features that offer the most significant information gain.

- **Primary Factor (Behavioral):** Helmet Usage (`POR_CAS`) is the root node, making it the most dominant predictor. Not wearing a helmet immediately places an accident in a high-risk category.
- **Secondary Factors:** These become critical after the initial split. They include:
  - **Temporal:** Hour of Accident (`HEU_ACC`) consistently appears as a key differentiator, likely corresponding to daytime vs. nighttime.
  - **Situational:** Vehicles Involved (`VEH_IMPLQ`) and Number of Pedestrians (`NBR_PIET`) are significant risk amplifiers.
  - **Socio-Demographic:** Insurance Status (`ASS_map_Non assuré`), Driver Profession (`COD_PRF_CON`), and Trip Purpose (`COD_TRA`) are used as proxies for higher-risk behavior.

Model Logic and Feature Interactions. The tree recursively splits data to reduce impurity, with the first split on helmet use providing the greatest information gain. The structure highlights key feature interactions, showing that risk is often compounding. For instance, a helmeted rider involved in a multi-vehicle collision, who is uninsured and riding at night, faces a high probability of a fatal outcome. *Insights from Decision Rules.* Translating paths into IF-THEN rules reveals practical scenarios.

- **Highest-Risk Scenario:** IF a rider is not wearing a helmet AND the accident occurs during a high-risk time period, THEN the model predicts FATAL.
- **Compounded Risk Scenario:** IF a rider wears a helmet BUT is in a multi-vehicle collision AND is uninsured AND the trip is for leisure AND occurs at a high-risk time, THEN the model predicts FATAL.
- **Lowest-Risk Scenario:** IF a rider wears a helmet AND the accident is not a severe collision type AND occurs during a low-risk time, THEN the model predicts NON-FATAL.

Practical Safety Recommendations. Based on the model's logic, the LLM generated the following actionable recommendations:

- **Helmet Law Enforcement:** The overwhelming importance of `POR_CAS` suggests that strict enforcement of mandatory helmet laws is the single most effective intervention.
- **Time-Specific Awareness Campaigns:** Public safety messages should warn riders of the elevated risks of riding at night or during other high-risk periods identified by the model.
- **Defensive Riding Training:** Since collisions with other vehicles are a major risk factor even for helmeted riders, training should focus on defensive techniques to avoid multi-vehicle accidents.
- **Targeted Outreach:** The model's use of features like profession and insurance status suggests risk is not evenly distributed. Outreach programs could be targeted towards specific at-risk demographic groups.

5 Discussion

Our findings show that combining a classic decision tree with a modern LLM offers a promising approach to generating actionable insights than the model alone. In this section, we discuss the value of the LLM's analysis, the validation steps, and the current limitations of this approach.

5.1 Qualitative Validation of the LLM's Analysis

A key finding from our work is the LLM's ability to move beyond a simple list of rules to articulate a plausible narrative about risk. Instead of merely identifying helmet usage and time of day as important factors, it framed their interaction as "compounding risk", illustrating how multiple variables combine to create dangerous scenarios. This reasoning resembles that of a human expert more closely than standard model outputs. To ensure the credibility of these insights, we conducted a qualitative consistency check, focusing on descriptive consistency rather than quantitative alignment.

First, we compared the primary factors identified by the LLM, such as helmet usage, against the structure of our trained CART model. Helmet usage was confirmed as both the root node and the top-ranked feature by Gini importance. Secondary factors reported by the LLM also corresponded to the top-ranked features in the tree model.

Regarding the IFTHEN rules generated by the LLM, the"Highest-Risk Scenario" was correctly provided. The "Compounded Risk Scenario" and the "Lowest-Risk Scenario" were generally consistent, but we identified a few issues. In one of the conditions,(`ASS_map_Non_ASSURE <= 0.5`) actually corresponds to an insured rider, not an uninsured one as stated by the LLM. This suggests that the model misinterpreted the condition, likely because it does not numerically process inequality signs. This observation indicates that an LLM may misrepresent conditions if feature mappings are undocumented, leading to inaccurate rule explanations. In addition, in the"Compounded Risk Scenario", the LLM selected only some conditions from the tree extracted rule while ignoring others probably because the original rule was too long to be easily understood by a general user. However, the rationale behind this selection remains unclear. Finally, after reviewing the Practical Safety Recommendations, we found that they were logical and aligned with the need to address the identified risk factors and scenarios.

Based on these checks, we found a good qualitative consistency between the LLM's narrative and the key quantitative outputs of the model. However, we acknowledge that this manual validation, while effective in this context, is subjective and not easily scalable. This highlights an important limitation and a promising direction for future work.

5.2 Limitations and Future Directions

While our approach is promising, it is important to address its limitations, which in turn point to critical directions for future research.

The first limitation is the nature of the LLM's inferential reasoning itself. The generated narrative, while plausible, is not provably faithful and may be influenced by linguistic or contextual biases rather than being a pure reflection of the model's statistical structure. To mitigate this, a key area for future work is to investigate how to better ground the LLM's narrative. This includes assessing how its responses change and potentially improve in faithfulness when the prompt is enriched with more detailed quantitative information, such as rule support, confidence values, and feature importance scores. Second, and relatedly, is the broader challenge of formally verifying an explanation's faithfulness. Our qualitative check was a necessary but subjective step. The field currently lacks standardized, automated metrics for this task. Therefore, a critical direction for future research is the development of such quantitative metrics to move from plausible narratives to provably faithful explanations.

Finally, our study used a single LLM (Google Gemini Pro). Different models may interpret the same rules differently due to their unique architectures. Future work could compare various LLMs to evaluate the consistency and diversity of their analytical capabilities, helping to establish the robustness of the LLM-assisted approach.

6 Conclusion

This paper addressed the persistent challenge of making decision tree models not just statistically sound, but also truly understandable for human experts. We introduced and successfully demonstrated a hybrid, LLM-assisted approach that translates the complex rules of a CART model into a coherent and actionable insights. Our case study on motorcycle accident severity showed that this workflow is a promising approach at extracting and articulating nuanced insights, such as the concept of compounding risk, that are often buried in a model's logic.

The primary contribution of our work is a four-stage workflow that bridges the gap between model output and real-world application. We demonstrated that this approach moves beyond the numerical outputs of traditional XAI tools by producing narrative explanations that help contextualize model outputs. Furthermore, we grounded our work in scientific practice by performing manual validation of the LLM's faithfulness and by transparently reporting the predictive performance of the underlying CART model.

While acknowledging that the quality of any interpretation depends on the predictive performance of the base model, our work establishes a valuable proof of concept. The proposed workflow represents a significant step towards making machine learning outputs more accessible, trustworthy, and impactful for the domain experts. Future research should develop formal metrics for faithfulness, evaluate how explanations improve with richer quantitative input, and explore the interpretive capabilities of different models, further advancing the frontier of explainable AI.

Acknowledgements. This work was supported by the Ministry of Transport and Logistics, NARSA and CNRST.

References

1. Breiman, L., Friedman, J., Olshen, R.A., Stone, C.J.: Classification and Regression Trees. Chapman and Hall/CRC, New York (2017). 10.1201/9781315139470
2. Brown, T., Mann, B., Ryder, N., Subbiah, M., Kaplan, J.D., Dhariwal, P., Amodei, D.: Language models are few-shot learners. Adv. Neural. Inf. Process. Syst. **33**, 1877–1901 (2020)
3. Cano, A., Zafra, A., Ventura, S.: An interpretable classification rule mining algorithm. Inf. Sci. **240**, 1–20 (2013). https://doi.org/10.1016/j.ins.2013.03.038

4. Dong, Q., Li, L., Dai, D., Zheng, C., Ma, J., Li, R., Xia, H., Xu, J., Wu, Z., Liu, T., Chang, B., Sun, X., Li, L., Sui, Z.: A Survey on In-context Learning (2024). 10.48550/arXiv.2301.00234
5. Freitas, A.A.: Comprehensible classification models: a position paper. SIGKDD Explor. Newsl. **15**(1), 1–10 (2014). https://doi.org/10.1145/2594473.2594475
6. Comanici, G., et al.: Gemini 2.5: Pushing the Frontier with Advanced Reasoning, Multimodality, Long Context, and Next Generation Agentic Capabilities (2025). 10.48550/arXiv.2507.06261
7. Hao, S., Gu, Y., Ma, H., Hong, J.J., Wang, Z., Wang, D.Z., Hu, Z.: Reasoning with Language Model is Planning with World Model (2023). 10.48550/arXiv.2305.14992
8. Iranitalab, A., Khattak, A.: Comparison of four statistical and machine learning methods for crash severity prediction. Accid. Anal. Prev. **108**, 27–36 (2017). https://doi.org/10.1016/j.aap.2017.08.008
9. Lemaître, G., Nogueira, F., Aridas, C.K.: Imbalanced-learn: a python toolbox to tackle the curse of imbalanced datasets in machine learning. J. Mach. Learn. Res. **18**(17), 1–5 (2017)
10. Li, L., Shrestha, S., Hu, G.: Analysis of road traffic fatal accidents using data mining techniques. In: 2017 IEEE 15th International Conference on Software Engineering Research, Management and Applications (SERA). pp. 363–370 (2017). 10.1109/SERA.2017.7965753
11. Murdoch, W.J., Singh, C., Kumbier, K., Abbasi-Asl, R., Yu, B.: Definitions, methods, and applications in interpretable machine learning. Proc. Natl. Acad. Sci. **116**(44), 22071–22080 (2019). https://doi.org/10.1073/pnas.1900654116
12. Pedregosa, F., Varoquaux, G., Gramfort, A., Michel, V., Thirion, B., Grisel, O., Blondel, M., Prettenhofer, P., Weiss, R., Dubourg, V., Vanderplas, J., Passos, A., Cournapeau, D., Brucher, M., Perrot, M., Duchesnay, É.: Scikit-learn: machine learning in Python. J. Mach. Learn. Res. **12**(85), 2825–2830 (2011)
13. Pei, T., Alican, F., Yin, A.O., Ihlamur, Y.: GPT-HTree: A Decision Tree Framework Integrating Hierarchical Clustering and Large Language Models for Explainable Classification (2025). 10.48550/arXiv.2501.13743
14. Rudin, C.: Stop explaining black box machine learning models for high stakes decisions and use interpretable models instead. Nat. Mach. Intell. **1**(5), 206–215 (2019). https://doi.org/10.1038/s42256-019-0048-x
15. Singh, C., Askari, A., Caruana, R., Gao, J.: Augmenting interpretable models with large language models during training. Nat. Commun. **14**(1), 7913 (2023). https://doi.org/10.1038/s41467-023-43713-1
16. Sohail, A., Cheema, M.A., Ali, M.E., Toosi, A.N., Rakha, H.A.: Data-driven approaches for road safety: a comprehensive systematic literature review. Saf. Sci. **158**, 105949 (2023). https://doi.org/10.1016/j.ssci.2022.105949
17. Wu, D., Peng, Z., Chen, M., Liu, Y.: Transforming network intrusion detection using large language models. In: 2025 IEEE 22nd Consumer Communications and Networking Conference (CCNC). pp. 1–9 (2025). 10.1109/CCNC54725.2025.10976105
18. Xiong, S., Ihlamur, Y., Alican, F., Yin, A.O.: GPTree: Towards Explainable Decision-Making via LLM-powered Decision Trees (2024). 10.48550/arXiv.2411.08257
19. Ye, H., Li, J., Zhao, H., Guo, D., Chang, Y.: LLM Meeting Decision Trees on Tabular Data (2025). 10.48550/arXiv.2505.17918
20. Ziems, N., Liu, G., Flanagan, J., Jiang, M.: Explaining Tree Model Decisions in Natural Language for Network Intrusion Detection (2023). 10.48550/arXiv.2310.19658

Towards a Circular Supply Chain 5.0: A Conceptual Framework Integrating Circular Economy, Intertwined Supply Networks, and Industry 5.0

Sanae Tissir[1(✉)], Abdelkabir Charkaoui[1], Khadija Echefaj[1], and Anass Cherrafi[2]

[1] Faculty of Sciences and Technique, Hassan First University of Settat, Settat, Morocco
sanaetissir10@gmail.com, {abdelkabir.charkaoui, k.echefaj}@uhp.ac.ma

[2] EST-Safi, Cadi Ayyad University, Marrakech, Morocco
a.cherrafi@uca.ac.ma

Abstract. Industry 5.0 represents a significant transformation in production systems, emphasizing the central role of humans, sustainability, and resilience in supply chain design and operations. Circular economy principles and Circular Intertwined Supply Networks (CISN) are being recognized as strategic solutions to address environmental, economic, and social challenges. However, existing literature lacks a unified theoretical framework that integrates these key dimensions. This article introduces a comprehensive conceptual framework for Circular Supply Chain 5.0, focusing on circular flows, actor interconnections, smart technologies, and collaborative governance with a human-centered approach. The methodology involves a critical analysis of interdisciplinary literature, systematic examination of interactions, and development of a conceptual model with a schematic representation. This theoretical framework establishes a solid foundation for future research on the drivers, barriers, and prerequisites for developing sustainable and resilient supply chains aligned with the principles of Industry 5.0.

Keywords: Industry 5.0 · Circular economy · CISN · Sustainability · Resilience · Human Centricity

1 Introduction

Current global supply chains are in the midst of significant transformations, driven by environmental pressures, socio-economic crises, and rapid technological advancements. Factors such as climate change, resource scarcity, and increasing sustainability regulations are compelling industries to reassess their traditional production and logistics models. Recent crises like the COVID-19 pandemic and geopolitical tensions have also highlighted the vulnerabilities of globalized supply chains, emphasizing the need for increased resilience [1]. Furthermore, the evolution towards Industry 4.0 and now Industry 5.0 is revolutionizing operational strategies by combining advanced technologies with

Y. Benadada et al. (Eds.): GOL 2026, LNNS 2034, pp. 156–166, 2027.
https://doi.org/10.1007/978-3-032-29677-1_15

human-centered governance to achieve competitiveness, resilience, and sustainability simultaneously [2]. Despite the emergence of the circular economy (CE) as a promising approach to waste reduction and resource regeneration, its application in supply chains has primarily focused on end-of-life strategies such as recycling, remanufacturing, and reverse logistics, lacking systemic integration or design-oriented approaches. Circular intertwined supply networks (CISN) have shown potential for agility and disruption-resistance but fall short on aligning with sustainability and regenerative objectives [3]. While the concept of Industry 5.0 emphasizes a human-centric and sustainability-driven perspective in industrial systems, practical frameworks for supply chain management are still limited. These fragmented approaches underscore the need for an integrated framework that combines circularity, resilience, sustainability, and human-centered governance in supply chain design and operations.

To address these gaps, this research focuses on developing an integrated conceptual model of supply chain management that incorporates circular economy principles, resilient intertwined networks, and the human-centric vision of Industry 5.0 to establish sustainable and regenerative supply chains. The primary objectives of this article are to critically analyze existing academic and institutional contributions from 2010 to 2024 to construct a robust theoretical foundation and to propose an integrative conceptual framework titled Integrated Circular Supply Chain 5.0 (ICSC-5.0) that offers both theoretical insights and practical value. The structure of this article is organized as follows: the first section introduces the theoretical underpinnings and key concepts of the study (Industry 5.0, circular economy, intertwined networks), followed by the methodology section outlining a systematic review of 90 academic and institutional publications. The third section presents a descriptive analysis of the selected literature, while the fourth section elaborates and illustrates the ICSC-5.0 conceptual model. A critical discussion of the model's contributions, implications, and limitations is provided in the fifth section, with the conclusion highlighting theoretical and managerial contributions and proposing directions for future empirical research.

2 Literature Review

2.1 Industry 5.0 and Its Fundamental Principles

Industry 5.0 represents a significant shift in industrial systems, moving away from the efficiency and technology-focused approach of Industry 4.0. This new paradigm places a strong emphasis on human involvement in technological innovation, promoting collaboration between advanced technologies and human intelligence [4]. It is built on three core principles: (i) Human centricity, where industrial transformations are guided by decision-making that prioritizes inclusiveness and social well-being. (ii) Proactive sustainability, promoting environmentally conscious practices that go beyond zero-waste to create net-positive impacts. (iii) Collaborative technologies, such as AI, robotics, IoT, and blockchain, which work alongside human labor to create value in production networks that are both resilient and ethical. These principles position Industry 5.0 as a fertile ground for the reimagining of supply chains that embrace circularity, resilience, and human values simultaneously. Several studies have analyzed the evolution and emerging research trends of Industry 5.0, highlighting its growing importance for future industrial systems [8].

2.2 Foundational Concepts in Circular and Networked Supply Chains

The circular economy (CE) reshapes supply chains by promoting regenerative design, modularity, repairability, and closed-loop processes to extend product lifespans. In supply chains, CE translates into actions such as reverse logistics, industrial symbiosis, and the use of digital tools like digital twins to improve lifecycle efficiency. Its success, however, relies on integration within complex and interconnected supply networks (CISNs), which stress connectivity, multi-tier coordination, and rapid adaptability to disruptions. Unlike linear models, CISNs adopt a systemic view where firms act as linked nodes in a resilient ecosystem, shifting CE from fragmented end-of-life strategies to proactive, design-driven circularity embedded in networked operations.

2.3 Limits of Existing Models

Despite ongoing advancements in supply chain models, there are significant limitations that persist. Traditional linear supply chains continue to dominate, prioritizing efficiency and cost reduction over factors like resilience, adaptability, and sustainability [5]. Although Supply Chain 4.0 frameworks have incorporated technologies like IoT, big data, and automation, they often focus primarily on technical aspects and overlook governance, ethics, and socio-environmental impacts Circular economy models show promise but are often limited to waste management and recycling, neglecting upstream product design and network resilience.Additionally, many existing frameworks lack mechanisms for collaborative governance, stakeholder engagement, and the integration of human values, all of which are crucial for achieving sustainable transformation [6]. Addressing these limitations requires an integrated framework that combines circular economy principles, interconnected supply chain networks, and a human-centric approach from Industry 5.0 to create a coherent and operational model.

3 Methodology

This study employs a qualitative and conceptual approach to develop an integrated circular supply chain model that is interconnected, resilient, sustainable, and human-centric, in accordance with the core principles of Industry 5.0. The methodological goals are as follows:

- Establish a strong theoretical foundation by identifying, categorizing, and critically evaluating relevant academic literature.
- Introduce an integrative conceptual framework (ICSC-5.0) that addresses the gaps present in current research, specifically the fragmented treatment of circular economy, interconnected networks, and human-centric governance.

The methodology adheres to the guidelines outlined by Tranfield et al. (2003) for systematic literature reviews, combining thorough documentary search coverage with discerning inclusion and exclusion criteria. This method ensures both scientific rigor and conceptual alignment with the study's objectives.

3.1 PRISMA Screening and Selection Process

The literature selection process followed the PRISMA guidelines and comprised four stages: identification, screening, eligibility, and inclusion. An initial set of 247 records was retrieved from major scientific databases. After duplicate removal, 198 publications were screened based on titles and abstracts, leading to the exclusion of 42 studies outside the scope of circular supply chains, Industry 5.0, or network resilience. Full-text assessment of 156 publications resulted in the exclusion of 66 studies due to limited theoretical relevance, yielding a final sample of 90 publications for qualitative synthesis and conceptual development (see Fig. 1).

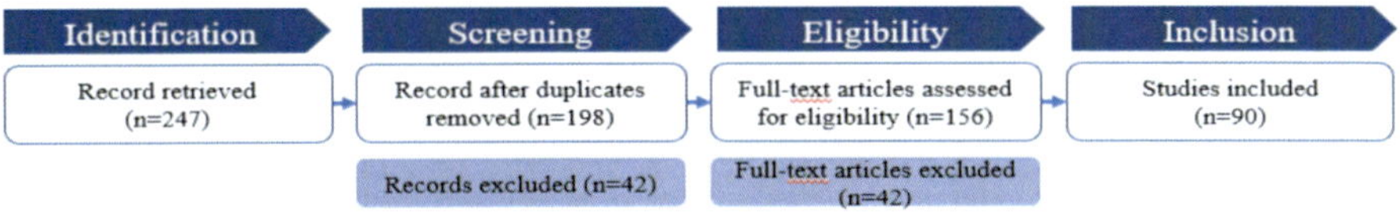

Fig. 1. PRISMA process

3.2 Exclusion and Inclusion Criteria

In order to maintain scientific rigor, specific selection criteria were applied as outlined in the table below (Table 1).

Table 1. Publication selection criteria

Criterion	Inclusion	Exclusion
Timeframe	2010–2024	Before 2010 (except essential historical references)
Scientific quality	Q1 or Q2 journals (Scimago, WoS)	Non-indexed or low-impact journals
Document type	Peer-reviewed articles, book chapters, institutional reports	Notes, editorials, unpublished communications
Theme	CE, I5.0, CISN, resilience, sustainability, circular governance	Out of scope
Language	English, French	Other languages

The selected publications were analyzed using a structured coding framework to support systematic synthesis. Each study was examined across five analytical dimensions: circularity mechanisms, supply network resilience, technological enablers, governance and human-centric aspects, and value creation outcomes. This coding process enabled consistent cross-study comparison and facilitated the identification of dominant theoretical patterns and research gaps, which directly informed the development of the ICSC-5.0 framework.

4 Descriptive Analysis

This descriptive analysis examines 90 selected articles from several perspectives, including their evolution over time, geographical distribution, publishing journals, and the themes addressed. Rather than relying on a simple narrative synthesis, it highlights the main contributions of these studies, particularly regarding circular economy implementation, supply chain resilience, the role of digital technologies, governance models, and value creation mechanisms. This stage is essential in a systematic literature review, as it enables the identification of dominant research trends, theoretical convergences, and structural gaps that form the foundation of the ICSC-5.0 framework (Fig. 2).

- **Temporal Distribution of Publications (2010–2024)**

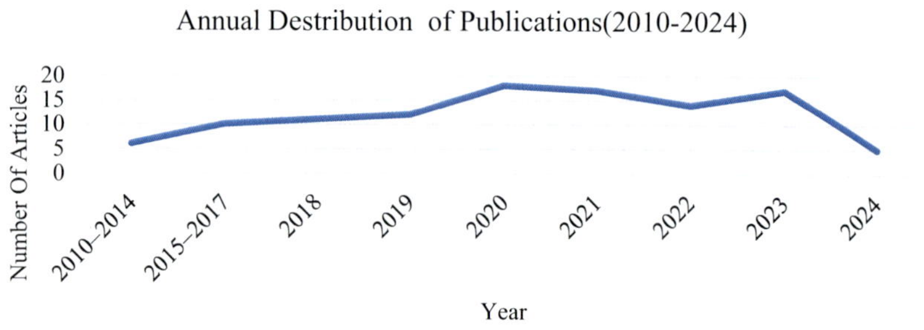

Fig. 2. Annual distribution of publications (2010–2024)

The analysis shows a steady rise in academic interest in integrating Circular Economy (CE), Industry 5.0 (I5.0), and Intertwined Supply Networks (CISN) from 2010 to 2024. Between 2010 and 2015, studies were mostly exploratory, focusing on end-of-life CE strategies or early resilience concepts [7].

From 2016 to 2019, publications doubled, marking the first efforts toward integration.

After 2020, research activity surged to over 15 papers per year, driven by COVID-19 disruptions and the demand for resilient, sustainable systems.

By 2023–2024, more than 20 papers were published annually, reflecting the maturity of Industry 5.0 discussions on human-centric and technology-enabled sustainability.

Thus, the 2010–2024 period captures both the conceptual emergence and the rapid convergence of sustainability-focused supply chain research (Fig. 3).

- **Geographical Distribution of Research**

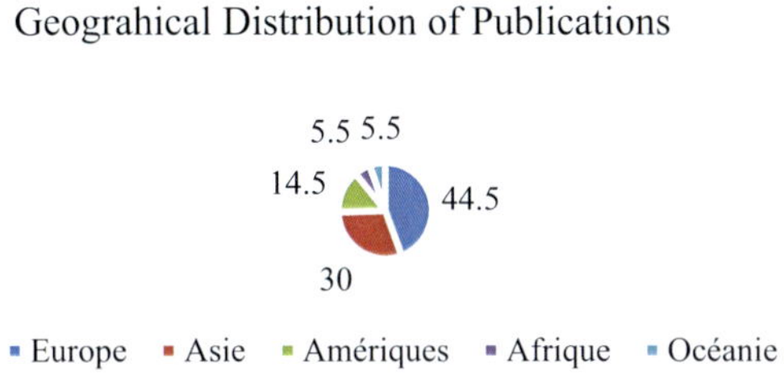

Fig. 3. Geographical distribution of research

The geographical analysis shows that Europe dominates the research landscape with 42% of publications, followed by Asia (31%) and North America (18%), while Africa, Oceania, and Latin America together account for the remaining 9%. Europe's leadership stems from strong policy support, notably the EU Circular Economy Action Plan, which has encouraged both academia and industry to adopt circular economy (CE) principles. Countries such as the UK, Germany, and the Netherlands have produced key studies on closed-loop design, digital twins, and CE governance models. In Asia, China and India focus on scaling CE practices in manufacturing sectors, while Japan leads research on human-technology interaction aligned with Industry 5.0 principles. In North America, especially the US and Canada, research emphasizes technological enablers. Overall, this geographic diversity highlights how different socio-economic and regulatory contexts shape CE adoption, reinforcing the need for flexible, region-specific implementations of the Integrated Circular Supply Chain 5.0 (ICSC-5.0) framework.

- **Conceptual Framework**

The Integrated Circular Supply Chain 5.0 (ICSC-5.0) proposes a unified theoretical framework that integrates the Circular Economy, interconnected supply networks, and Industry 5.0 into a single operational system linking design, technology, and governance. Its five pillars are derived from dominant theoretical streams in the literature and reflect core functional dimensions repeatedly identified across these research domains.

Unlike traditional linear or Supply Chain 4.0 models, the ICSC-5.0 functions as a dynamic and interconnected system. Through closed-loop flows of materials, value, and information, its interactive pillars; design, coordination, execution, governance, and value creation, collectively enhance adaptability, performance, and sustainability across the supply network.

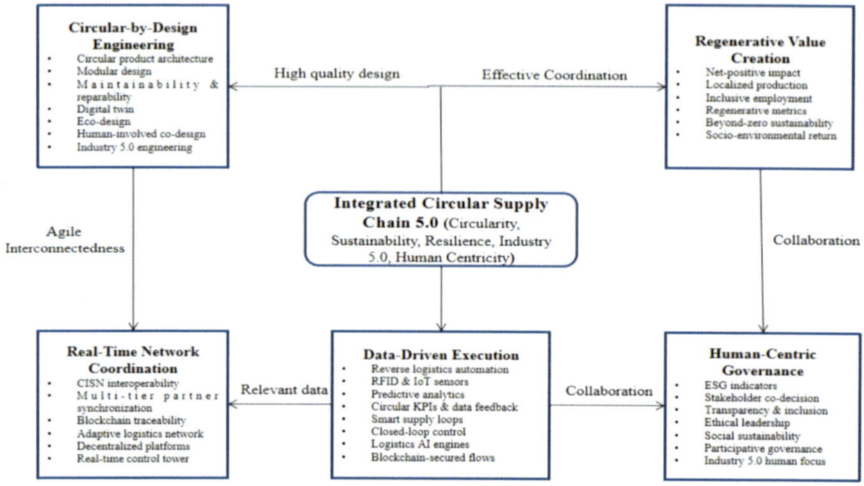

Fig. 4. Conceptual framework

- **Circular-by-Design Engineering**

This function forms the foundation of regeneration by embedding circularity from the earliest design stages. It promotes modular design, maintainability, and repairability, using tools like digital twins and AI simulations to anticipate life-cycle impacts. Collaboration with stakeholders ensures sustainability requirements are integrated from the start [9]. Unlike traditional CE models that emphasize end-of-life processes, ICSC-5.0 shifts focus upstream, making design a strategic driver of circularity and sustainability.

- **Real-Time Network Coordination**

This function enhances adaptive resilience in multi-actor, multi-tier supply chains through Circular Interconnected Supply Networks (CISNs). It integrates interoperable platforms, shared data infrastructures, blockchain-based traceability, and dynamic logistics reconfiguration. The COVID-19 crisis exposed the fragility of linear supply chains. ICSC-5.0 [10] counters this by adopting flexible architectures that adjust flows in response to disruptions, supporting agility and performance in volatile environments.

- **Data-Driven Execution**

This function ensures the closure of circular loops through digitalization and automation. It integrates IoT, RFID, and predictive analytics for real-time tracking, maintenance, and optimization, with blockchain-secured flows for trust and traceability. Data becomes the unifying element across the model, enabling real-time KPIs to monitor and adjust material flows dynamically [11]. This improves efficiency, supports decision-making, and strengthens feedback between design, governance, and execution.

- **Human-CentricGovernance**

This function serves as the ethical and social backbone of the model. It promotes stakeholder collaboration, embeds ESG criteria, and fosters transparency, inclusivity, and ethical leadership. Unlike many CE models lacking strong governance, ICSC-5.0 ensures legitimacy and accountability by centering human needs and collective decision-making. This human-centric approach transforms sustainability from a compliance goal into a shared organizational value [12].

- **Regenerative Value Creation**

This function represents the ultimate goal of ICSC-5.0: achieving net-positive environmental and social impact. It includes local production, inclusive employment, ecosystem restoration, and regenerative metrics to assess benefits. Inspired by the "beyond-zero" paradigm, it moves beyond harm reduction to actively restore natural and social systems. The interdependence between all five functions design, coordination, execution, governance, and regeneration creates feedback loops that sustain continuous improvement.

- **Systemic Integration**

ICSC-5.0 functions as a complex adaptive system capable of self-regulation and learning. Each function feeds into and reinforces the others data informs governance, coordination supports design, and regeneration drives system evolution. This synergistic structure ensures that the framework remains adaptive, scalable, and applicable across diverse industrial contexts.

In comparison with existing supply chain and circular economy frameworks, the ICSC-5.0 model adopts a more integrative and systemic perspective. While prior models typically emphasize isolated dimensions such as digitalization, circular resource flows, or resilience capabilities, the proposed framework simultaneously integrates circular-by-design principles, real-time network coordination, data-driven execution, and human-centric governance within a regenerative logic. This holistic configuration allows the ICSC-5.0 framework to address both operational efficiency and long-term sustainability challenges, thereby extending and complementing existing theoretical approaches.

5 Discussion

The Integrated Circular Supply Chain 5.0 (ICSC-5.0) presents a unified systemic framework that integrates circular economy principles, supply chain resilience, and Industry 5.0 within a single operational logic. It emphasizes a proactive, design-driven approach, embedding circularity from the early design stage while strengthening resilience through adaptive and collaborative Intertwined Supply Networks. These networks coordinate flows, actors, and decisions in real time, allowing rapid responses to disruptions. The experience of global crises such as COVID-19 has shown that firms capable of dynamic reconfiguration recover faster and sustain performance under volatility. The ICSC-5.0 operationalizes this adaptability by using digital coordination platforms that integrate information sharing, decision-making, and synchronization across multiple supply chain tiers. This systemic connectivity ensures that resilience becomes an inherent feature of circular supply chain operations rather than an emergency response mechanism. The model also leverages data-driven technologies to translate circular and resilient principles into daily operations. IoT and RFID support traceability and product tracking, blockchain reinforces trust and transparency across partners, and AI enables predictive analytics for disruption anticipation and recovery planning. Together, these tools create an intelligent network that learns from its own data flows. Continuous feedback loops between design, coordination, and governance allow the system to evolve and optimize performance over time, embedding learning and adaptation at its core. At the governance level, the ICSC-5.0 incorporates Industry 5.0 values human-centricity, sustainability, and resilience into managerial and institutional practices. Governance in this framework is participatory and ethical, built on transparency, stakeholder engagement, and ESG-based evaluation. Instead of viewing governance as a top-down control mechanism, the model treats it as a catalyst for collaboration and legitimacy. It positions inclusivity and accountability at the center of decision-making, ensuring that social and environmental dimensions are integrated alongside economic objectives. By merging structural capabilities (design, operations, coordination) with informational capabilities (data, intelligence, governance), the ICSC-5.0 views sustainability and resilience as interconnected outcomes. This integration produces regenerative effects environmental recovery, social value creation, and economic stability that continuously feed back into system improvement. The framework thus represents a shift from static sustainability models toward a dynamic, learning-oriented ecosystem. Conceptually, the ICSC-5.0 advances three main contributions: it establishes a functional integration logic that connects the principles of circularity, resilience, and Industry 5.0; it introduces a theory of interdependencies linking design, data, and governance; and it translates Industry 5.0 values into operational practices. For practitioners, it provides a strategic roadmap for embedding circularity throughout R&D, operations, coordination, and governance structures. The ICSC-5.0 framework offers a new perspective by integrating Industry 5.0 principles into a systemic view of circular supply chains. Although conceptual at this stage, it provides a solid foundation for future empirical research. Each of its five pillars can be translated into measurable indicators, paving the way for quantitative validation, for example through structural equation modeling. Qualitative approaches such as Delphi studies, expert panels, or in-depth case analyses can further complement the assessment by testing the robustness and adaptability of the framework across different industrial contexts.

6 Conclusion

This study introduces the Integrated Circular Supply Chain 5.0 (ICSC-5.0), a conceptual framework designed to address fragmentation in the literature on circular economy, interconnected and resilient supply networks, and Industry 5.0. Based on a systematic review of scholarly publications published between 2010 and 2024, the study establishes a robust theoretical foundation and ensures conceptual saturation. The analysis highlights recurring limitations in existing research. Circular economy studies mainly emphasize end-of-life strategies, often overlooking design-oriented and regenerative approaches.

Research on interconnected supply networks prioritizes agility and resilience but remains weakly aligned with sustainability objectives. Likewise, Industry 5.0 literature promotes human-centric and sustainability principles with limited operationalization in supply chain contexts. The ICSC-5.0 framework addresses these gaps by integrating human, technological, and ecological dimensions within a unified supply chain architecture. The framework is structured around five interdependent pillars: Circular-by-Design Engineering, Real-Time Network Coordination, Data-Driven Execution, Human-Centric Governance, and Regenerative Value Creation. Together, these pillars enable supply chains to evolve beyond linear, efficiency-driven models toward adaptive, transparent, inclusive, and regenerative systems. This study contributes theoretically by bridging previously disconnected research streams, methodologically by demonstrating the value of cross-disciplinary systematic reviews, and practically by offering a reference model for managers and policymakers navigating the digital and ecological transitions. However, the framework remains conceptual and requires empirical validation through case studies, simulations, or surveys, with applicability potentially varying across sectors and regions. Future research should focus on operationalizing the framework, developing integrated performance indicators, and examining the role of digital technologies in regenerative supply chain design. Overall, the ICSC-5.0 provides a robust foundation for advancing sustainable, resilient, and human-centered supply chains in line with Industry 5.0 principles.

References

1. Chowdhury, P., Paul, S.K., Kaisar, S., Moktadir, M.A.: COVID-19 pandemic related supply chain studies: a systematic review. Transp. Res. Part E Logist. Transp. Rev. **148**, 102271 (2021). https://doi.org/10.1016/j.tre.2021.102271
2. Adel, A.: Future of industry 5.0 in society: human-centric solutions, challenges and prospective research areas. J. Cloud Comput. **11**(1), 40 (2022). https://doi.org/10.1186/s13677-022-00314-5
3. Echefaj, K., Charkaoui, A., Cherrafi, A., Garza-Reyes, J.A., Khan, S.A.R., Chaouni Benabdellah, A.: Sustainable and resilient supplier selection in the context of circular economy: an ontology-based model. Manag. Environ. Qual. Int. J. **34**(5), 1461–1489 (2023). https://doi.org/10.1108/MEQ-02-2023-0037
4. Agrawal, S., Agrawal, R., Kumar, A., Luthra, S., Garza-Reyes, J.A.: Can industry 5.0 technologies overcome supply chain disruptions? A perspective study on pandemics, war, and climate change issues. Published online 15 Sept 2023. [En ligne]. Disponible sur: https://doi.org/10.1007/s12063-023-00410-y
5. Agrawal, S., Singh, R.K., Murtaza, Q.: A literature review and perspectives in reverse logistics. Resour. Conserv. Recycl. **97**, 76–92 (2015). https://doi.org/10.1016/j.resconrec.2015.02.009
6. Echefaj, K., Charkaoui, A., Cherrafi, A., Ivanov, D.: Design of resilient and viable sourcing strategies in intertwined circular supply networks. Ann. Oper. Res. **337**(1), 459–498 (2024). https://doi.org/10.1007/s10479-024-05873-1
7. Ivanov, D.: Collaborative emergency adaptation for ripple effect mitigation in intertwined supply networks. Ann. Oper. Res. (2023). https://doi.org/10.1007/s10479-023-05408-0
8. Akundi, A., Euresti, D., Luna, S., Ankobiah, W., Lopes, A., Edinbarough, I.: State of industry 5.0 analysis and identification of current research trends. Appl. Syst. Innov. **5**(1), 27 (2022). https://doi.org/10.3390/asi5010027

9. Hernandez Marquina, M.V., Le Dain, M.-A., Joly, I., Zwolinski, P.: Exploring determinants of collaboration in circular supply chains: a social exchange theory perspective. Sustain. Prod. Consum. **50**, 1–19 (2024). https://doi.org/10.1016/j.spc.2024.07.017
10. Liu, H., Han, Y., Zhu, A.: Modeling supply chain viability and adaptation against underload cascading failure during the COVID-19 pandemic. Nonlinear Dyn. **110**(3), 2931–2947 (2022). https://doi.org/10.1007/s11071-022-07741-8
11. Bag, S., Wood, L.C., Xu, L., Dhamija, P., Kayikci, Y.: Big data analytics as an operational excellence approach to enhance sustainable supply chain performance. Resour. Conserv. Recycl. **153**, 104559 (2020). https://doi.org/10.1016/j.resconrec.2019.104559
12. Bauer, M.C., Grosse, E.H.: Enhancing human-centricity for strategic alignment: the value of coaching in digital transformation. IFAC-PapersOnLine **58**(19), 551–556 (2024). https://doi.org/10.1016/j.ifacol.2024.09.270

Intelligent Multi-agent Systems and Metaheuristics for Logistics Management

A Multi-agent Decision Support System for Integrated Product and Downstream Supply Chain Design Optimization

Hafsa Bouidmar(✉), Oulfa Labbi, and Abdeslam Ahmadi

LMIMP Laboratory, ENSAM, Moulay Ismail University, Meknes, Morocco
h.bouidmar@edu.umi.ac.ma

Abstract. This paper proposes a Multi-Agent Decision Support System (MADSS) for the integrated optimization of product design and downstream supply chain configuration. Departing from the conventional sequential decision-making processes, the proposed framework enables simultaneous coordination between design and logistics functions through autonomous agents representing product design, logistics, and reverse operations. These agents collaborate through structured communication and negotiation protocols to achieve consistent and optimal decisions. A bi-objective Mixed Integer Linear Programming (MILP) model, embedded within the Downstream Optimization Agent, optimizes facility selection, product allocation, and recovery strategies, minimizing both operational costs and environmental impacts, according to the Pareto optimality Principle. The hybrid architecture combines decentralized agent negotiation with centralized optimization to enhance system adaptability and robustness. The negotiation process of the proposed MADSS is illustrated by an UML sequence diagram. The proposed MADSS provides a flexible and scalable framework for integrated product–supply chain decision-making, contributing to sustainable and resilient supply chain management.

Keywords: Product-supply chain integration · Downstream Supply Chain · Multi-Agent Decision Support System (MADSS) · Mixed Integer Linear Programming (MILP) optimization · sustainability

1 Introduction

The increasing vulnerability of global supply chains, accentuated by the complexity of networks and the occurrence of unforeseen events, calls into question the effectiveness of traditional management methods. This highlights the urgency of adopting more proactive resilience strategies [1]. Yet, product design and supply chain organization are often approached separately, leading to structural inefficiencies, leading to higher costs, limited responsiveness, and complications when incorporating subsequent changes [2].

Thus, simultaneous integration of product design and logistics planning appears to be a promising strategy for improving overall performance [3]. Within this context, the paper focuses particularly on the downstream supply chain and proposes a multi-agent

Y. Benadada et al. (Eds.): GOL 2026, LNNS 2034, pp. 169–178, 2027.
https://doi.org/10.1007/978-3-032-29677-1_16

decision-making framework to jointly optimize product design and the corresponding downstream supply chain. The proposed multi-agent decision support system (MADSS) integrates a bi-objective mixed integer linear programming (MILP) model to guide joined product and downstream supply chain optimization. This system guides interactions between agents and ensures optimality on a global scale. Each agent represents a key player in the supply chain and dynamically evaluates different configurations through automated negotiations, ensuring consistent coordination between product design and downstream logistics.

The main contributions of this work lie in the development of formal MAS architecture linking design to downstream supply chain planning exclusively. It integrates an optimization agent based on a MILP model for the distribution network design and implements automated communication and negotiation protocols to achieve optimal trade-offs according to the Pareto principle.

The article is structured as follows: Sect. 2 reviews the literature, Sect. 3 presents the proposed MAS architecture, Sect. 4 details the integrated MILP model, Sect. 5 details system 'an interaction through negotiation agent within MADSS and Sect. 6 concludes by proposing future research perspectives.

2 Literature Review

2.1 Integrated Product and Supply Chain Design

Integrated product and supply chain design has recently gained increasing attention as it allows for better coordination of strategic and operational decisions throughout the product lifecycle [3]. Unlike traditional sequential methods, this integrated approach seeks to simultaneously optimize the product architecture and supply chain configuration, while considering cost, sustainability, and resilience constraints [4, 19].

Aguila and ElMaraghy [5] developed a mixed integer linear programming (MILP) model to jointly design the product architecture and supply chain, incorporating a natural catastrophe risk index to identify resilient configurations. Moreover, Baud-Lavigne et al. [6] addressed the joint design of a product family and its supply network, considering the substitution of components in the bill of materials (BOM). Their MILP model has demonstrated that simultaneous optimization generates significant cost savings compared to separate planning, while improving flexibility in the event of fluctuations in demand.

Overall, these contributions highlight that joint product-supply chain optimization, supported by analytical models such as MILP, allows for better alignment of technical design and logistics performance. However, it is important to note that integrated optimization often relies on centralized models, which can be rigid and inflexible in uncertain environments. This constraint highlights the importance of decentralized approaches, such as multi-agent systems (MAS), which promote local decision-making while ensuring global coordination, thereby improving flexibility and responsiveness.

2.2 Multi-agent Systems for Supply Chain Optimization

In parallel, multi-agent systems (MAS) have emerged as promising enablers for intelligent coordination in supply chain management, due to their distributed, adaptive, and collaborative nature [7]. MASs provide a cooperative environment where agents can interact, coordinate, and communicate to achieve common goals [8].

Giannakis and Louis [9] proposed a MAS architecture combined with Big Data analytics to improve supply chain agility through responsiveness, flexibility, and speed. Their system incorporates semantic technologies and introduces a "big data agent" to capture, process, and disseminate information in real-time, showing how MAS can overcome the trade-off between agility and complexity. Djennas et al. [10] combined MAS with fuzzy logic and genetic algorithms to manage distribution in dynamic supply chains. Their model emphasizes flexibility as a critical survival factor and demonstrates how decentralized agents improve coordination and adaptability under uncertainty. More recently, Plinere et al. [11] developed modular MAS architecture with specialized agents (sales, inventory management), relying on ontologies to improve semantic interoperability. Their results show improved delivery times, optimized inventory management and greater customer satisfaction despite uncertain demand.

As far as the use of MAS for product design is concerned, Ballouki et al. [20] demonstrated that the effectiveness of MAS in integrated product and supply chain design has been proven through hierarchical agent-based architecture. Moreover, MAS has been used to manage complexity and uncertainty in product design. Li et al. [21] designed complex products, by modeling how risks propagate between interacting agents. In addition, the work of Chiu and Tsai [22] highlights the importance of MAS in supporting adaptive, user-centered design strategies, especially in the context of product-service systems.

These studies confirm that MAS can effectively manage the distributed and evolving nature of supply chains, especially in downstream operations such as distribution, risk management, and coordination with customers. By integrating data processing, behavioral simulation, and learning capabilities, MASs pave the way for harmonized optimization and collaborative decision-making in complex supply networks [12].

3 Proposed Multi-agent System Framework

3.1 Description of the MAS Architecture

The Multi-Agent System (MAS) architecture we propose is designed as a collaborative, distributed framework, where autonomous agents play a critical role in both product design and the downstream supply chain. Each agent brings their own expertise with unique decision-making skills, covering areas like product engineering, distribution management, and logistics coordination. These agents interact through a shared communication protocol and a central coordination layer, ensuring seamless interoperability and a consistent flow of information. The architecture also includes a knowledge base and a global optimization module, allowing agents to share data in real-time, explore various scenarios, and choose the best strategies together. This configuration, which is both decentralized and coordinated, enhances the flexibility of the system, allowing it to dynamically adapt to changing market and operational conditions, while ensuring a seamless connection between product development goals and downstream supply chain performance as shown in Fig. 1.

The specific agents that make up this architecture, as well as their roles and objectives, are detailed in the next subsection.

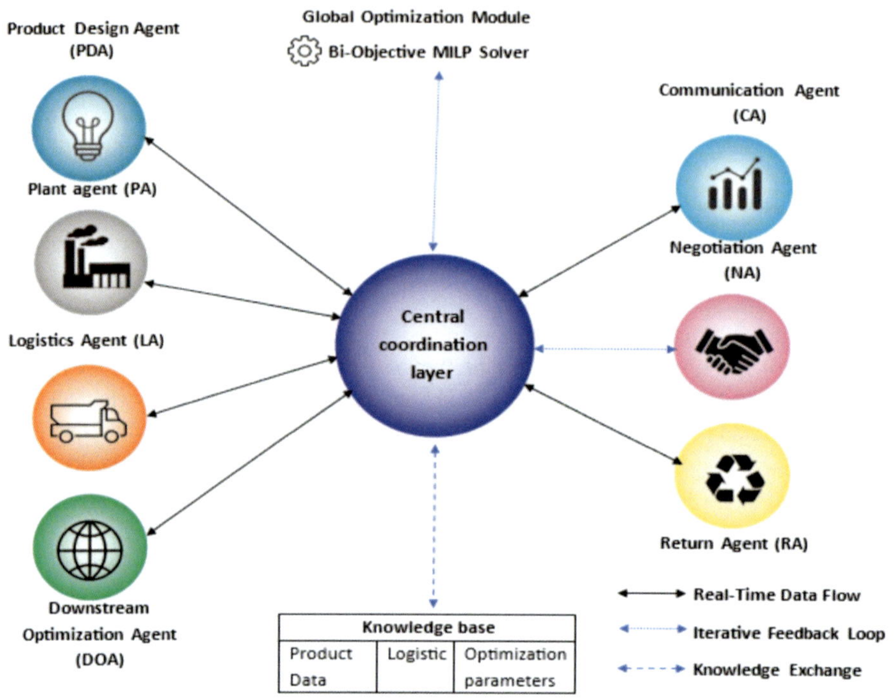

Fig. 1. Multi-Agent Decision Support System (MADSS) architecture for integrated product and downstream supply chain optimization

3.2 Key Agents

The architecture we propose is both hierarchical and collaborative composed of autonomous agents who represent the main entities of the product-downstream supply chain integration. The Product Design Agent (PDA) is responsible for generating the various product configurations, such as bills of materials, modular designs, and sustainability parameters. The plant agent (PA) evaluates product design alternatives according to manufacturing constraints. The Logistics Agent (LA) designs and manages warehouses and transport networks and organizes distribution flows with the aim of reducing costs and environmental impact. A new contribution in this system is the Downstream Optimization Agent (DOA), which uses a dual-objective mixed integer linear programming (MILP) model, as explained in Sect. 4, to determine the optimal distribution network configuration based on product characteristics. Reverse logistics and end-of-life management are handled by the return agent, who plans recycling and reconditioning strategies. Finally, two mediation agents ensure the coherence of the system: the communication agent, who facilitates the exchange of information between the agents, and the negotiation agent, who resolves conflicts when local constraints are violated. Through continuous collaboration, agents arrive at overall optimal solutions balancing profitability, service level and sustainability [14].

3.3 Decision Making Process

The decision-making process within the Multi-Agent Decision Support System (MADSS) is structured as an iterative and collaborative sequence [15]. It starts with the Product Design Agent, who proposes one or more design alternatives. These options are then passed on by the Communication Agent (CA) to the Plant Agent (PA), the Logistics Agent (LA) and Return Agent (RA), who evaluate the proposal according to their own constraints and objectives. The Downstream Optimizer (DOA) uses its built-in Mixed Integer Linear Programming (MILP) model to analyze the feasibility and overall efficiency of the proposed configuration. Then, the Communication agent aggregates the local evaluations and sends an overall performance evaluation back to the Product Design Agent. If the solution meets the predefined thresholds, whether related to cost, operational, or sustainability, it is validated as an acceptable configuration. Otherwise, a new iteration is launched: either the Product Design Agent proposes a revised alternative, or the Negotiation Agent (NA) intervenes to resolve conflicts and facilitate arbitration, for example by relaxing delivery constraints to reduce transport costs. This process, which is based on negotiation and optimization, makes the system particularly suitable for dynamic planning in contexts where demand fluctuates, where there are supply uncertainties or cost variations.

3.4 Describing the MILP Model as a "Decision Support Tool"

In this work, we introduce the MILP formulation not only as a mathematical way to represent the optimization problem, but also as a valuable basic decision-making tool [16]. In MAS, the MILP model is not a static entity but is dynamically invoked by the Downstream Optimizer Agent (DOA) during the negotiation process. By bringing together technical, logistical, and operational constraints in a single model focusing on downstream operations, the MILP framework allows agents to weigh trade-offs between competing objectives such as cost, time, and resource utilization [17]. This method transforms the model from a simple computational task to a practical support system for strategic and tactical decision-making [18]. Essentially, the MILP model provides a structured and quantitative basis for guiding choices in downstream supply chain configuration, ensuring that solutions are not only feasible but also consistent with the overall performance objectives of the product-supply chain system [19].

4 Integrated Mathematical Model

This section details the bi-objective Mixed-Integer Linear Programming (MILP) model that serves as the computational engine for the Downstream Optimization Agent (DOA) within the proposed MADSS. The model is adapted from the framework proposed by Labbi and Ahmadi (2024) [13] to be seamlessly integrated into the agent-based decision-making process described in Sect. 3.

4.1 Model Overview

The mathematical model used in our MADSS system is based on a two-objective mixed-integer linear programming (MILP) framework. It was designed to consider only the downstream levels of the supply chain: distribution centers, transportation modes, customers, and reverse logistics flows from customers to collection centers. Within this framework, several sets and parameters are involved, representing costs, capacities, demand, distances, and emissions. The model's two main objectives can be summarized in two functions: first, minimize the total cost of the downstream supply chain (DCS), which includes distribution, reverse logistics, and infrastructure installation costs; and second, reduce the environmental impact by limiting facility operations and transportation CO_2 emissions. As part of our multi-agent system, this model is embedded in the DOA. This later can interact with the model via a function like solve_MILP(), passing real-time data (such as demand forecasts, product attributes from the PDA, and current logistics costs from the LA) as parameters. The model then returns optimized decisions regarding facility selection, product allocation, and recovery strategies, which the DOA uses to evaluate product design proposals.

4.2 Model Formulation

- **Model assumptions and definitions**

The model is based on key assumptions that aim to simplify its complexity:

- The demand is known.
- The logistics network is made up of several distribution centers (DCs), customers, return collection centers, and possible modes of transport, which allows for flexibility in the configuration of the network.
- Transportation costs are modeled as linear functions that depend on the characteristics of the product, while activation costs for facilities (such as distribution centers) are fixed.
- The planning horizon is multi-period, divided into discrete time intervals $t \in T$.

The essential elements of the downstream supply chain are represented by the following sets:

- Candidate distribution centers (D).
- Customers (R).
- Modes of transport (M).
- Collection centers for returned products (L).
- Discrete planning periods (T).

Model parameters which quantify system characteristics include:

- Demand levels for each customer and each period.
- Storage capacities of the distribution centers and transport modes.
- $C_{\text{transp}}^{i,m,t}$: Transport costs from Plant to distribution center i using transportation mode m during period t

- $C_{\text{transp}}^{i,j,m,t}$: Variable transport costs from distribution center i to customer j using transportation mode m during period t.
- $C_{\text{returns}}^{j,l,t}$: Costs of managing returned products from customer j to collection center l during period t.
- F_i: Fixed costs related to the opening of the distribution centers i.
- F'_l: Fixed cost related to the opening of the collection center l.
- $CO_{2\text{transp}}^{i,m,t}$: CO_2 emission factors for transportation from plant to distribution center i using transportation mode m in period t.
- $CO_{2\text{transp}}^{i,j,m,t}$: CO_2 emission factors for transportation from distribution center i to customer j using transportation mode m in period t.
- $CO_{2\text{returns}}^{j,l,t}$: CO_2 emission factors for reverse logistics flows from customer j to collection center l in period t.
- $CO_{2\text{op}}^{i}$: CO_2 emissions associated with distribution center i operations.
- $CO_{2\text{op}}^{l}$: CO_2 emissions associated with collection center l.

The model is based on several decision variables that determine the most effective actions to take. Here are the main ones:

- X_{ijm}^{t}: flow of product from distribution center i to customer j using transport mode m during period t.
- W_{jl}^{t}: quantity of returned products from customer j to a collection center l during period t.
- Y_i: Binary variable equal to 1 if distribution center i is opened, 0 otherwise.
- L_l: Binary variable equal to 1 if collection center l is opened, 0 otherwise.
- P_{im}^{t}: Quantity of products transported from plant to distribution center i using transport mode m during period t.
- **Objective Function and Constraints**

Objective 1: Minimization of total downstream supply chain (DSC) costs.

$$\min Z_1 = \sum_{t=1}^{T}\sum_{i=1}^{I}\sum_{j=1}^{J}\sum_{m=1}^{M}\left(C_{\text{transp}}^{i,j,m,t}\cdot X_{ijm}^{t}\right) + \sum_{t=1}^{T}\sum_{j=1}^{J}\sum_{l=1}^{L}\left(C_{\text{retours}}^{j,l,t}\cdot W_{jl}^{t}\right) + \sum_{i=1}^{I}(F_i\cdot Y_i) + \sum_{l=1}^{L}\left(F'_l\cdot L_l\right) + \sum_{t=1}^{T}\sum_{i=1}^{I}\sum_{m=1}^{M}\left(C_{\text{transp}}^{i,m,t}\cdot P_{im}^{t}\right) \quad (1)$$

This objective function includes costs of transport from Plant to distribution centers, costs of transport from distribution centers to customers, costs of managing product returns, and the fixed costs of opening new distribution centers.

Objectif 2: Minimization of total CO_2 emissions from downstream operations.

$$\min Z_2 = \sum_{t=1}^{T}\sum_{i=1}^{I}\sum_{j=1}^{J}\sum_{m=1}^{M}\left(CO_{2\text{transp}}^{i,j,m,t}\cdot X_{ijm}^{t}\right) + \sum_{t=1}^{T}\sum_{j=1}^{J}\sum_{l=1}^{L}\left(CO_{2\text{retours}}^{j,l,t}\cdot W_{jl}^{t}\right) + \sum_{i=1}^{I}\left(CO_{2\text{op}}^{i}\cdot Y_i\right) + \sum_{l=1}^{L}\left(CO_{2\text{op}}^{l}\cdot L_l\right) + \sum_{t=1}^{T}\sum_{i=1}^{I}\sum_{m=1}^{M}\left(CO_{2\text{transp}}^{i,m,t}\cdot P_{im}^{t}\right) \quad (2)$$

This function aims to minimize CO_2 emissions generated by transportation, reverse logistics flows, and the operations of distribution centers.

The model must meet several constraints to ensure the feasibility of the solution:

- Demand Satisfaction Constraints: For each customer and at each time, the demand for products must be fully met.
- Flow and Capacity Constraints: Product flows must respect network capacities and variable domains.
- Reverse Flow Constraints: Returned products must not exceed the recovery potential.
- Environmental Constraints: Total CO_2 emissions must respect operational and environmental limits.

5 Negotiation Agent Within MADSS

The negotiation process of the proposed Multi-Agent Decision Support System (MADSS) is illustrated by the UML sequence diagram in Fig. 2.

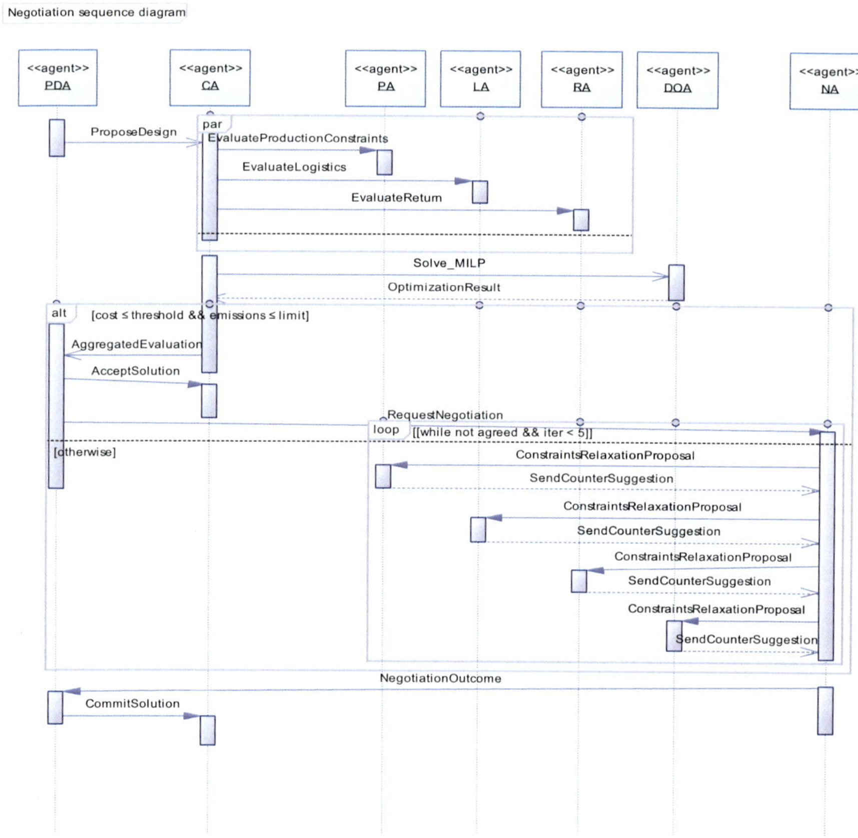

Fig. 2. UML sequence diagram of the negotiation process within the MADSS framework

It starts when the Product Design Agent (PDA) submits a new configuration. This proposal is then forwarded by the Communication Agent (CA) to the Plant Agent (PA), the Logistics Agent (LA) and the Return Agent (RA) for evaluation. Simultaneously, the Downstream Optimization Agent (DOA) runs a dual-objective Mixed Integer Linear Programming (MILP) model, leveraging real-time data from other agents to determine the optimal downstream configuration, balancing costs with environmental goals. Once the assessments are compiled, the results are sent back to the PDA. If the proposed design meets all defined thresholds (costs, CO_2 emissions), it is validated. Otherwise, the Negotiation Agent (NA) intervenes to manage a negotiation process between the agents. During this phase, the negotiator proposes potential constraints 'relaxations, collects counterproposals from the product design, plant, logistics, returns and optimization agents, and collaborates with the communication agent to make any necessary adjustments. This negotiation process continues until a common agreement is reached between the agents or until the maximum number of negotiation cycles is reached. This structured negotiation protocol ensures overall consistency, optimal Pareto coordination and smooth collaboration between product design choices and downstream logistics optimization within the MADSS system.

6 Conclusion

This paper presents a multi-agent decision support system (MADSS) designed to optimize product design and downstream supply chain configuration. This framework intelligently combines decentralized negotiation between agents and centralized bi-objective optimization, ensuring consistent and efficient decision-making at both the design and logistics levels. By integrating a mixed integer linear programming (MILP) model with the Downstream Optimization Agent (DOA), the system can simultaneously afford economic and environmental objectives, resulting in Pareto-optimal trade-offs between cost and sustainability.

Through structured communication and negotiation protocols, agents collaborate to evaluate design options, streamline logistics flows, and manage reverse operations, enhancing the adaptability and resilience of downstream supply chains. This hybrid architecture effectively connects product design choices to supply chain operational performance, paving the way for more robust and sustainable industrial systems.

Our future research will focus on the implementation and validation of the proposed MADSS using multi-agent simulation tools such as NetLogo. In addition, it will be applied to real-world industrial case studies to assess its scalability and practical effects on collaborative decision-making within complex supply chain networks.

References

1. Ivanov, D., Dolgui, A.: Viability of intertwined supply networks: extending the supply chain resilience angles towards survivability. A position paper motivated by COVID-19 outbreak. Int. J. Prod. Res. **58**(10), 2904–2915 (2020)
2. Saeed, K.A., Malhotra, M.K., Abdinnour, S.: How supply chain architecture and product architecture impact firm performance: an empirical examination. J. Purch. Supply Manag. **25**(1), 40–52 (2019)

3. Khan, O., Stolte, T., Creazza, A., Hansen, Z.N.L.: Integrating product design into the supply chain. Cogent Eng. **3**(1), 1210478 (2016)
4. Ramezanian, R., Khalesi, S.: Integration of multi-product supply chain network design and assembly line balancing. Oper. Res. Int. J. **21**(1), 453–483 (2021)
5. Aguila, J.O., ElMaraghy, W.: Simultaneous global supply chain and product architecture design considering natural hazard exposure and geographical facility location. Procedia CIRP **72**, 533–538 (2018)
6. Baud-Lavigne, B., Agard, B., Penz, B.: Simultaneous product family and supply chain design: an optimization approach. Int. J. Prod. Econ. **174**, 111–118 (2016)
7. Terrada, L., Khaïli, M.E., Ouajji, H.: Multi-agents system implementation for supply chain management making-decision. Procedia Comput. Sci. **177**, 624–630 (2020)
8. Sun, L., et al.: Multi-agent coordination across diverse applications: a survey. arXiv preprint arXiv:2502.14743 (2025)
9. Giannakis, M., Louis, M.: A multi-agent based system with big data processing for enhanced supply chain agility. J. Enterp. Inf. Manag. **29**(5), 706–727 (2016)
10. Djennas, M., Benbouziane, M., Djennas, M.: Agent-based modeling in supply chain management: a genetic algorithm and fuzzy logic approach. Int. J. Artif. Intell. Appl. **3**(5), 13–30 (2012)
11. Plinere, D., Aleksejeva, L., Merkuryev, Y.: Multi-agent system development and application for supply chain management tasks. Inf. Technol. Manag. Sci. **24**, 1–7 (2021)
12. Malviya, R.K., Mandala, V., Reddy, M.S.: Reinforcement learning for collaborative decision-making in multi-agent systems: applications in supply chain optimization. (Unpublished manuscript)
13. Labbi, O., Ahmadi, A.: A dynamic bi-objective optimization model for a closed loop supply chain design under environmental policies. Stat., Optim. Inf. Comput. **12**(6), 1725–1744 (2024)
14. Attajer, A., Mecheri, B., Hadbi, I., Amoo, S.N., Bouchnita, A.: Sustainable supply chain strategies for modular-integrated construction using a hybrid multi-agent–deep learning approach. Sustainability **17**(12), 5434 (2025)
15. Sharko, P.A., Burlutskaya, Z.V., Zubkova, D.A., Gintciak, A.M., Pospelov, K.N.: AI-supported decision making in multi-agent production systems using the example of the oil and gas industry. Appl. Sci. **15**(10), 5366 (2025)
16. Wang, S., Li, Y., Wang, J.: Production decision optimization based on a multi-agent mixed integer programming model. Mathematics **13**(11), 1827 (2025)
17. Alshababi, N.M., Matar, A.-H., Abdelati, M.H.: Multi-objective mixed-integer linear programming for dynamic fleet scheduling, multi-modal transport optimization, and risk-aware logistics. Sustainability **17**(10), 4707 (2025)
18. Lopez-Landeros, C.E., Valenzuela-Gonzalez, R., Olivares-Benitez, E.: Dynamic optimization of a supply chain operation model with multiple products. Mathematics **12**(15), 2420 (2024)
19. Labbi, O., Ouzizi, L., Douimi, M., Ahmadi, A.: Genetic Algorithm combined with Taguchi method for optimization of supply chain configuration considering new product design. Int. J. Logist. Manag. Syst. **31**(4), 531–561 (2018)
20. Ballouki, I., Labbi, O., Douimi, M., Ouzizi, L.: A multi-agent approach for simultaneous design of a product and its supply chain: suppliers' selection example. In: Proceedings of the Global Operations and Logistics (GOL) Conference, pp. 1–6 (2016)
21. Li, N., Li, X., Shen, Y., Bi, Z., Sun, M.: Risk assessment model based on multi-agent systems for complex product design. Inf. Syst. Front. **16**(2), 311–327 (2013)
22. Chiu, M.-C., Tsai, C.-H.: Design a personalised product service system utilising a multi-agent system. Adv. Eng. Inform. **43**, 101036 (2020)

Evaluating the Tribal Intelligent Evolution Optimization (TIEO) for Travel Route Planning: A Case Study of the Traveling Salesman Problem

Ton Nguyen Trong Hien(✉)

Faculty of Business Administration, Van Lang University, Ho Chi Minh City, Vietnam
hien.tnt@vlu.edu.vn

Abstract. Tribal Intelligent Evolution Optimization (TIEO) is a novel metaheuristic algorithm introduced in early 2025. As a relatively new method, TIEO has yet to undergo extensive empirical validation, particularly for solving classic combinatorial optimization problems such as the Traveling Salesman Problem (TSP). This study evaluates the performance of TIEO on a constrained TSP case study, benchmarking it against Modified Particle Swarm Optimization (MPSO), a well-established and robust algorithm in this domain. The experimental results reveal a significant trade-off between speed and accuracy. While TIEO exhibits a substantially lower runtime (approximately 6.7 times faster than MPSO) and high temporal stability, it suffers from a large optimality gap and premature convergence. In contrast, MPSO consistently achieves superior solution quality and demonstrates effective local search behavior. Our findings suggest that TIEO's current architecture prioritizes exploration but lacks sufficient exploitation capabilities required for complex combinatorial landscapes. Consequently, while TIEO offers a low computational footprint, it requires stronger local refinement operators to become a competitive alternative to established methods, such as MPSO, for high-precision optimization tasks.

Keywords: Metaheuristics · Routing · TIEO

1 Introduction

Travel route planning is essential for effective logistics, field service, delivery, and transportation operations. At the heart of this challenge lies the TSP, a computational problem that poses difficulties for optimization algorithms when applied to real-world issues constrained [1, 2]. Incorporating constraints into a larger problem space leads to a combinatorial explosion, rendering exact algorithms computationally intractable for scalable instances. Therefore, exact algorithms fail to address the combinatorial explosion caused by scalability issues [3]. To overcome limitations, metaheuristic-related algorithms have established themselves as a viable approach to finding solutions within realistic time bounds. Among these methods, nature-inspired algorithms have proven successful to some extent, as they simulate biological or social processes to explore a

Y. Benadada et al. (Eds.): GOL 2026, LNNS 2034, pp. 179–188, 2027.
https://doi.org/10.1007/978-3-032-29677-1_17

multidimensional solution space [4, 5]. A notable recent development is the TIEO algorithm, introduced in January 2025. As a member of the social-inspired metaheuristics family, TIEO leverages a metaphorical representation of human societal mobility and interactions to efficiently explore the solution space [6].

The TIEO stands out for its unique social structure, in which the search population is divided into rival tribes, and knowledge is shared through migration at longer intervals. Intra-tribe refinement and inter-tribe diversification comprise this two-part mechanism, which helps address a critical issue in all population-based approaches, namely, the delicate balance between exploration and exploitation. TIEO does not follow a standard population-based optimization approach, as it applies social dynamics based on tribal systems. Within these structures, subgroups develop a degree of semi-independence and then engage in structured migration to exchange knowledge. Algorithms are frequently developed based on natural or societal metaphors, often lacking rigorous scientific validation, leading to suboptimal designs [7]. TIEO has already been shown to be far superior to many other algorithms for calculating the optimal control of cooling plants [6]. Despite promising findings, the capabilities of TIEO have not yet been fully explored in transportation, an area where the tribal structure could be effectively applied to optimize spatially clustered systems. It would represent a notable gap if newly developed metaheuristic approaches were not assessed using benchmark transportation problems.

To address this gap, this study will employ a validation approach. To begin, the TIEO algorithm will be evaluated for its effectiveness in optimizing travel routes connecting the continental United States. "TSP is one of the most important issues in combinatorial optimization problems used in many engineering sciences" [8]. The TSP case study on the 48 continental United States capitals is a well-established benchmark that has been extensively utilized in prior research, as exemplified by studies [8, 9]. The results will be benchmarked against the MPSO algorithm, which has been demonstrated to outperform several other algorithms in experimental studies on the TSP [8], as documented in prior research. To date, no empirical study has directly compared the performance of TIEO and MPSO. Consequently, this research aims to fill this gap by applying both algorithms to a TSP case study, representing a problem domain that has not yet been explored with these methods.

2 Methodology

Using a case study on a particular real-life situation, TSP, each method is applied independently, and the corresponding outcomes are recorded. A rigorous and standardized evaluation procedure is then conducted to validate these results. TIEO and MPSO are evaluated within the selected case study to assess their capabilities in terms of solution quality, exploration-exploitation balance, and overall performance. Each algorithm was executed 10 times independently, each with 100 iterations, to ensure statistical reliability.

2.1 Data

This study uses classical TSP, a well-known combinatorial optimization problem. The situation involves a network of locations, with geographic coordinates for each location that establish distances between them (Illustrated in Listing 1). The goal is to determine a tour that visits each location and returns to the starting point, thereby creating a closed route. The transmission between the two locations is calculated based on their coordinates. The algorithms discussed here, TIEO and MPSO, are accused of not only finding the shortest route but also one that meets this restriction, allowing us to understand what each of these two algorithms is capable of when faced with a challenging route optimization problem. It is important to note that the shape of the geographical area does not have a direct meaning for the algorithms. What matters is the structure of the distance matrix, which is generated from the city data. A complex distance matrix will create a difficult problem, regardless of the geographical shape [10].

```
data = {
    "Montgomery": (32.3668, -86.3000), # Alabama
    "Phoenix": (33.4484, -112.0740), # Arizona
    "Little Rock": (34.7465, -92.2896), # Arkansas
    "Sacramento": (38.5816, -121.4944), # California
    #...
}
```

Listing 1. Plotting city coordinates.

TSP is a problem in which a salesman must visit all cities, each city is visited only once, and the tour starts and ends at the origin city. The goal is to determine the route with the minimum total distance. Such constraints serve as standardized, rigorous benchmarks for testing the solution quality, constraint-handling ability, and optimization effectiveness of newly developed metaheuristics [11, 12].

2.2 TIEO

Introduced by Yao et al. [6], TIEO simulates the intelligent behavior of tribal communities through a novel combination of classification mechanisms and reinforcement learning strategies.

Step 1: Initialization

Define bounds for each variable $x_{ij} \in [x_{\min}, x_{\max}]$, where i denotes the i-th individual in the population ($i = 1,\ldots, N$), and j denotes the j-th dimension or city index ($j = 1,\ldots, D$). The population is randomly initialized as follows:

$$X_i = x_{\min} + \text{rand}(0, 1)^* (x_{\max} - x_{\min}) \quad (1)$$

where rand(0,1) is a uniform number in the range [0,1].

Step 2: Tribal clustering

The population is partitioned into k tribes using an adaptive clustering mechanism.

Splitting: If a tribe's diversity (e.g., variance in fitness) exceeds a threshold $\sigma_{\max}$, split it into two tribes by re-clustering its individuals.

$$T_m \rightarrow \{T_{m1}, T_{m2}\} \text{ if } \text{var}(f(X_i) \in T_m) \geq \sigma_{\max} \quad (2)$$

Unification: If two tribes are too similar (e.g., average fitness difference below), merge them into one tribe:

$$T_m \cup T_n \rightarrow T_{new} \text{ if } f_{avg}(T_m) - f_{avg}(T_n) < \delta_{\min} \quad (3)$$

For each tribe T_m, compute the best solution $\left(X_{best,m}\right)$ and average fitness.

Step 3: Autonomy phrase

For each $X_i \in T_m$, update its position toward the tribe's best solution using a permutation-based movement operator:

$$X_{\text{new},i} = \text{move\_toward}\left(X_i, X_{\text{best},\ m}, \alpha\right) \quad (4)$$

where α represents the learning rate affecting the segment lengths. The move_toward function is implemented as a segment-based crossover: A segment of cities is randomly selected from the tribal best solution $X_{\text{best},\ m}$. The length of this segment is proportional to α. This segment is copied directly into $X_{\text{new},i}$ at the same positions. The remaining positions in $X_{\text{new},i}$ are filled with the missing cities in the order they appear in the original individual X_i, ensuring a valid TSP permutation without duplicates. Applying mutation with probability p_m:

$$X_{\text{new},i} = \text{mutate}\left(X_{\text{new},i}\right) \text{ if rand}(0, 1) < p_m \quad (5)$$

$$\text{If } f\left(X_{\text{new},i}\right) < f(X_i),\ \text{update } X_{\text{i}} = X_{new,\ i}$$

Step 4: Diplomacy phase

Simulate cultural exchange by allowing individuals to adopt features from other tribes. For each $\text{X}_i \in T_m$, select a random individual $\text{X}_j \in T_n$ $(n \neq m)$ and perform a cooperative update:

$$X_{new,i} = \text{crossover}\ (X_i,\ X_i,\ \beta) \quad (6)$$

where β controls the extent of crossover, crossover is an order-based crossover that ensures valid TSP paths. Update X_i if $f\left(X_{new,\ i}\right) < f(X_i)$

Step 5: War phase

Randomly select t individuals from the tribe and replace the weakest with a new individual created via crossover with the global best $\left(X_{global}\right)$ or a top-ranked tribe's best $\left(X_{best,m}\right)$.

$$X_{new,i} = \text{crossover}\left(X_i,\ X_{global}\right),\ F \quad (7)$$

where F is the influencer factor that determines the proportion of genes (cities) inherited from X_{global}.

Step 6: Reinforce learning

Define a reward function to guide parameter adaptation:

$$R = \frac{f_{\text{prev,global}} - f_{\text{global}}}{\max\left(f_{\text{prev, global}}, \varepsilon\right)} \quad (8)$$

where f_{global} is the current global best fitness, $f_{\text{prev, global}}$ is the previous global best, and ε is a small constant to prevent division by zero. Update parameter α, β, p_m based on R:

- If $R < R_{\text{min}}$, increase exploration:

$$\alpha \leftarrow \min(1, \alpha \times 1.1), \beta \leftarrow \min(1, \beta \times 1.1), p_m \leftarrow \min(0.5,\ p_m \times 1.2) \quad (9)$$

- If $R \geq R_{\text{min}}$, focus on exploitation:

$$\alpha \leftarrow \max\,(0.1, \alpha \times 0.9), \beta \leftarrow \max(0.1, \beta \times 0.9), p_m \leftarrow \max(0.05,\ p_m \times 0.8) \quad (10)$$

Adjust the tribe count k if diversity or convergence metrics indicate stagnation.

Step 7: Update global best
Update X_{global} if a better solution is found across all tribes.

Step 8: Termination check
If the maximum iteration is reached or the fitness level falls below a threshold, return X_{global} and $f\left(X_{global}\right)$. Otherwise, return to step 2.

2.3 MPSO

Inspired by social swarms, the Particle Swarm Optimization (PSO) algorithm is a powerful metaheuristic for solving optimization problems. In the standard PSO, particles traverse a continuous search space, updating their position based on their best experience (*pbest*) and the global best experience (*gbest*). However, to tackle discrete combinatorial problems such as the TSP, a specialized hybrid approach is often required.

The new position of a particle, $X_i(t+1)$, is derived from a movement operation that takes the particle's current position, $X_i(t)$, and guides it towards a target position (e.g., $pbest_i$ or $gbest$). This movement is often implemented as a series of permutation operations (swaps, insertions, etc.).

When a new $pbest_i(t+1)$ or $gbest(t+1)$ is identified, it is immediately refined by a local search algorithm. This can be expressed as:

$$pbest_i(t+1) = \text{LocalSearch}(X_i(t+1)) \text{ if } \mathrm{X_i}(t+1) \text{ is better than } pbest_i(t) \quad (11)$$

$$gbest(t+1) = \text{LocalSearch}(X_i(t+1)) \text{ if } \mathrm{X_i}(t+1) \text{ is better than } gbest_i(t) \quad (12)$$

2.4 Validation

A validation process was conducted by comparing its performance with the well-established MPSO algorithm. This comparison includes Solution quality, where the effectiveness of both algorithms in finding optimal solutions was evaluated.

Exploration and exploitation capabilities assessment: Using Mean Gap, a metric that quantifies the average percentage deviation from the known global optimum of the benchmark problem. This method provides a normalized, robust measure for comparing the final solution quality of the two algorithms. The convergence behavior was analyzed using convergence curves, which plot the best-found solution over generations. This visual representation enables a direct comparison of the algorithms' search efficiency and their ability to steadily improve solutions throughout the optimization process.

Performance analysis: Each algorithm was executed under the same conditions, including hardware specifications, stopping criteria, and problem size. TIEO and MPSO were independently run to account for randomness and ensure statistical validity. The average execution time and standard deviation for each algorithm were recorded and visualized in a bar chart. Error bars represent the standard deviation across runs. This analysis aims to highlight TIEO's computational efficiency relative to a well-established metaheuristic baseline.

3 Results and Discussion

3.1 Solution Quality

The figure provides a comparative visualization of the routes generated by the TIEO algorithm (Fig. 1a) and the MPSO algorithm (Fig. 1b) for solving the TSP after 10 independent runs, each consisting of 100 iterations, to identify the optimal solution.

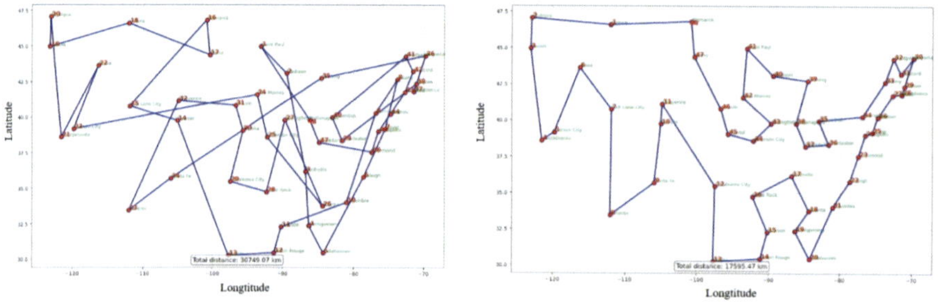

(a) TIEO (b) MPSO

Fig. 1. Visualization of the pathfinding obtained.

The TIEO solution yields a total route distance of about 30,749.07 km. In contrast, MPSO achieves a substantially shorter distance of nearly 17,595.47 km. This discrepancy of over 13,150 km indicates MPSO's superior optimization capability in this case study.

Visually, the TIEO-generated route exhibits multiple unnecessary crossings and geographically inconsistent detours. Such patterns suggest that TIEO, under the tested configuration, may have struggled to effectively exploit promising regions of the search space, leading to premature convergence or entrapment in local optima. On the other hand, the MPSO solution demonstrates a more coherent and systematically organized trajectory, with cities visited in a generally progressive geographic sequence. This reflects

MPSO's stronger balance between exploration and exploitation, enabling it to refine high-quality candidate solutions while maintaining population diversity. From a methodological standpoint, the significant performance disparity highlights a critical limitation in the current TIEO architecture: the tribal interaction mechanisms prioritize diversity but lack the sufficient local search intensity required to refine solutions in complex, combinatorial landscapes, such as the TSP, unlike the more balanced search behavior observed in MPSO.

3.2 Exploration and Exploitation Capabilities Assessment

Figure 2 The bar chart illustrates the mean optimality gap (expressed in percentage) between the solutions obtained by TIEO and MPSO. The TIEO algorithm shows a significantly larger mean gap of 108.11%, with a standard deviation of 9.23%, indicating greater deviation from the optimal solution and greater variability in performance across runs. In contrast, the MPSO algorithm exhibits a significantly smaller mean gap of 1.64% and a standard deviation of 1.19%, reflecting its ability to approximate near-optimal solutions with minimal fluctuation consistently. The error bars further highlight the statistical robustness of MPSO, as its narrow spread suggests stable convergence, whereas TIEO's wider spread indicates inconsistent performance and potentially premature convergence.

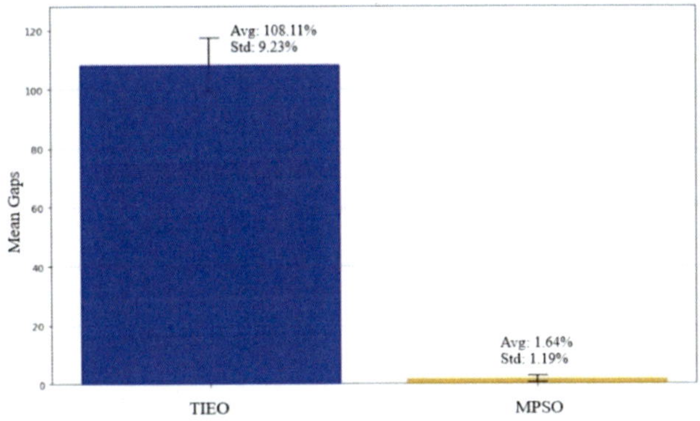

Fig. 2. Mean gap distance among groups.

The red curve (Fig. 3), representing TIEO, shows a gradual and slow decline in the objective value. The algorithm starts with an extremely high initial distance (over 65,000 km) and progressively decreases; however, even after 100 iterations, it remains above 35,000 km. This slow convergence indicates limited exploitation capability and suggests that TIEO may struggle to refine high-quality solutions within a reasonable number of iterations.

In contrast, the blue curve (Fig. 3) corresponding to MPSO demonstrates rapid convergence within the first few iterations, dropping from over 30,000 km to below 18,000 km almost immediately. After this steep decline, the curve stabilizes, indicating that MPSO reaches near-optimal solutions early and maintains consistent performance without significant fluctuation. This reflects strong convergence speed and effective local search behavior, suggesting a good balance between exploration and exploitation in the algorithm's design.

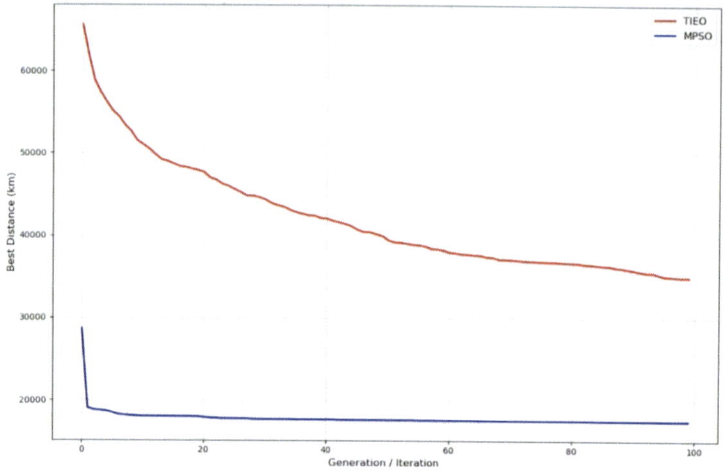

Fig. 3. Convergence curve.

The stark contrast between the two curves emphasizes that MPSO achieves better final solution quality with far greater efficiency, requiring fewer iterations to converge. On the other hand, TIEO's slower, less stable convergence trajectory suggests a need for further tuning or algorithmic enhancements to improve its search dynamics and convergence reliability.

3.3 Performance Analysis

MPSO recorded an average runtime of 1877.60 s with a standard deviation of 60.60. TIEO achieved the same number of iterations with an average runtime of only 280.98 s and a much lower standard deviation of 1.98 (Fig. 4).

While TIEO exhibits a significantly lower runtime (approximately 6.7 times faster than MPSO), this metric must be interpreted in conjunction with the solution quality outcomes. The reduced execution time reflects the lower computational complexity of TIEO's operators per iteration compared to MPSO. However, given the substantial optimality gap observed, this speed advantage does not translate to algorithmic efficiency. Instead, it indicates that the current TIEO configuration converges prematurely, failing to explore the solution space adequately. In contrast, MPSO's longer runtime is justified by its ability to perform the intensive local search operations necessary to achieve near-optimal solutions.

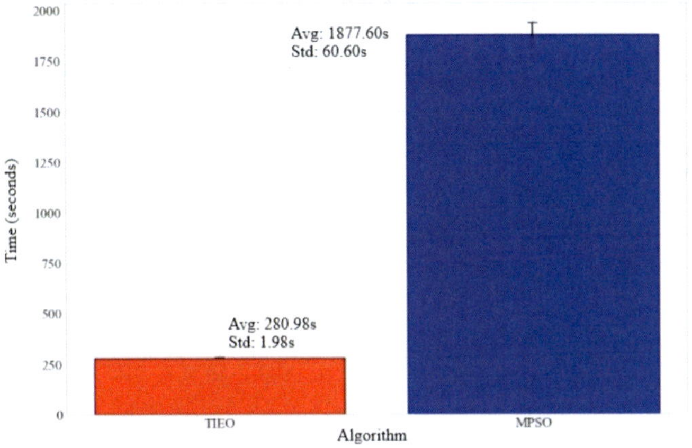

Fig. 4. Comparison of computational time between TIEO and MPSO algorithms.

4 Conclusion

The case study on the constrained TSP shows that MPSO achieves better separation between optimal and sub-optimal solutions, suggesting superior robustness and precision in its final outputs. Additionally, MPSO consistently exhibits steeper convergence toward lower objective values, particularly during the early iterations, highlighting its strong capacity to exploit the search space and avoid premature convergence effectively. However, TIEO, despite its lower convergence accuracy, proves significantly more efficient when computational cost is considered, with a remarkably low standard deviation in runtime, reflecting its high temporal stability. Although introduced in 2025 with claims of computational efficiency, our findings suggest that TIEO offers performance efficiency rather than computational superiority compared to well-established algorithms, such as MPSO. Our study also reinforces Yousefikhoshbakht's [8] findings, affirming that MPSO holds considerable potential for solving TSP problems with high accuracy. Nonetheless, we contribute an additional insight: this performance comes at the cost of higher computational time. Future research should focus on hybridizing the strengths of both algorithms, for instance, integrating the rapid convergence and accuracy of MPSO with the lightweight computational footprint of TIEO to design more balanced and adaptive metaheuristics for solving constrained combinatorial problems. Furthermore, testing on larger-scale TSP instances and other problem domains would provide more generalizable insights into algorithmic behavior.

Acknowledgement. The author would like to thank Van Lang University, Ho Chi Minh City, Vietnam, for their support. Special thanks are also extended to the anonymous reviewers for their insightful comments and valuable suggestions, which have greatly improved the quality of this manuscript.

References

1. Halim, A.H., Ismail, I.: Combinatorial optimization: comparison of heuristic algorithms in the travelling salesman problem. Arch. Comput. Methods Eng. **2**(5), 367–380 (2017)
2. Sui, J., Ding, S., Huang, X., Yu, Y., Liu, R., Xia, B., Ding, Z., Xu, L., Zhang, H., Yu, C., Bu, D.: A survey on deep learning-based algorithms for the traveling salesman problem. Front. Comput. Sci. **19**(6) (2024)
3. Prosser, P.: Exact algorithms for maximum clique: a computational study. Algorithms **5**(4), 545–587 (2012)
4. Shandilya, S.K., Datta, A., Nagar, A.K.: Nature-inspired algorithms. In: Studies in Computational Intelligence, pp. 3–36 (2023)
5. Chandra, S.V., Anand, H.S.: Nature-inspired meta-heuristic algorithms for optimization problems. Computing **104**(2), 251–269 (2021)
6. Yao, Y., Hong, X., Xiong, L.: Study on a new metaheuristic algorithm—Tribal intelligent evolution optimization and its application in optimal control of cooling plants. Appl. Energy **383**, 125339 (2025)
7. Wang, C., Hu, K., Wu, X., Ou, Y.: Rethinking Metaheuristics: unveiling the myth of "Novelty" in metaheuristic algorithms. Mathematics **13**(13), 2158 (2025)
8. Yousefikhoshbakht, M.: Solving the traveling salesman problem: a modified metaheuristic algorithm. Complexity **2021**(1) (2021)
9. Lin, M., Schaposnik, L.P.: A carbon aware ant colony system for the sustainable generalized traveling salesman problem. Sci. Rep. **15**(1) (2025)
10. Cook, W.J.: In pursuit of the traveling salesman: mathematics at the limits of computation. Choice Rev. Online **49**(10), 49–57 (2012)
11. Lin, S.: Computer solutions of the traveling salesman problem. Bell Syst. Tech. J. **44**(10), 2245–2269 (1965)
12. Bajaj, A.S., Dhodiya, J.M.: Sustainable multiple travelling salesman problem solved by reference point aspiration level based multi objective quasi oppositional Jaya algorithm under uncertain environment. Evol. Intel. **17**(4), 2577–2616 (2024)

Multi-agent System for Supply Chain Management

Manel Neffati(✉), H. Murat Afsar, and Leila Merghem Boulahia

LIST3N Laboratory, University of Technology of Troyes, Troyes, France
{manel.neffati,murat.afsar,leila.merghem_boulahia}@utt.fr

Abstract. This paper presents a multi-agent system for supply chain management in the carpentry industry. The proposed model simulates interactions between client, producer, and multiple suppliers. In situations where customer orders cannot be fulfilled due to raw material shortages, the producer initiates a call for tenders. Supplier proposals are then evaluated using a hybrid decision-making system. To enhance negotiation outcomes, suppliers adapt their pricing and delivery strategies through Q-learning, learning from experience over time. Experimental results demonstrate improvements in both profit margins and customer satisfaction. The trained agents develop strategic bidding behaviors that prioritize system-wide efficiency over individual gains, leading to consistent service quality. Overall, this work contributes to the integration of AI into supply chain operations and provides a scalable framework for intelligent procurement in make-to-order manufacturing contexts.

Keywords: Multi-agent systems · Supply chain management · Negotiation · Q-learning · MCDM

1 Introduction

Supply chains in the carpentry industry face significant challenges due to variable material quality, product customization, and unpredictable demand [7]. Traditional static decision-making approaches are poorly suited to such dynamic conditions, particularly when balancing cost, delivery reliability, quality, and customer satisfaction. Recent advances in artificial intelligence and multi-agent systems address these limitations by enabling autonomous agents to adapt their strategies based on experience.

The trade-off between centralized and decentralized decision-making remains critical. Centralized approaches offer global optimization but limited responsiveness, while decentralized systems provide flexibility with reduced coordination [3]. Multi-agent systems overcome these limitations by supporting autonomous negotiation and learning, making them well suited for adaptive supply chains [13].

Y. Benadada et al. (Eds.): GOL 2026, LNNS 2034, pp. 189–199, 2027.
https://doi.org/10.1007/978-3-032-29677-1_18

To address these challenges, this work integrates Q-learning, AHP, and TOPSIS. Q-learning enables suppliers to adapt negotiation strategies [9], while AHP-TOPSIS ensures transparent and consistent supplier evaluation through systematic weighting and ranking [4]. This combination balances adaptability with decision auditability.

This paper proposes a multi-agent system for carpentry supply chain management, integrating intelligent supplier selection, adaptive negotiation, and coordinated production planning in make-to-order environments.

2 Related Work

Multi-agent systems (MAS) have gained increasing attention for supply chain optimization. Xu et al. [13] proposed autonomous, self-orchestrating supply chains combining MAS with foundation models, later validated through an implementation case [14]. Terrada et al. [12] developed MAS-based decision-making frameworks, highlighting the importance of communication and coordination among agents.

Multi-criteria decision-making (MCDM) methods have been widely applied to supplier selection, with AHPTOPSIS being one of the most adopted combinations. Sequeira et al. [11] demonstrated its effectiveness through a fuzzy AHPTOPSIS model for manufacturing relocation decisions. Jain et al. [4] validated the robustness of integrated AHPTOPSIS for supplier selection in the automotive industry, while Paredes Rodríguez et al. [8] and Kumar et al. [6] extended its application to inventory classification and sustainable supplier selection, respectively. However, most existing studies treat supplier selection as a static optimization problem, overlooking the dynamic nature of supplier behavior and market conditions.

Reinforcement learning (RL) has shown strong potential for adaptive supply chain optimization. Rolf et al. [9] identified Q-learning as the most widely used RL algorithm in supply chain applications, particularly for inventory management. Chaharsooghi et al. [2] pioneered Q-learning-based ordering mechanisms, and more recent work by Zhou et al. [15] focused on efficiency optimization through RL. Nevertheless, the integration of RL with MCDM in multi-agent negotiation contexts remains limited, especially in manufacturing environments requiring transparency and traceability.

Agent-based negotiation protocols have evolved from simple auctions to multi-round frameworks capable of handling complex preferences. The Automated Negotiating Agent Competition (ANAC) has served as a benchmark for negotiation strategy evaluation since 2010 [1]. Recent studies have introduced more advanced mechanisms, such as trust-based and token-based negotiation protocols [5].

Supply chain optimization research in the carpentry and wood manufacturing sector remains scarce despite its high product customization, diverse material requirements, and lead time variability. Molinaro et al. [7] explored Industry 4.0 adoption in the wood sector, but supply chain optimization and adaptive coordination were not addressed. Most existing studies rely on traditional optimization methods and do not consider adaptive agent behavior or learning-based coordination.

Despite advances in MAS, MCDM, and RL, existing approaches typically emphasize either adaptability or transparency, but rarely both. **To the best of our knowledge, this is the first work to integrate Q-learning with AHPTOPSIS within a multi-agent negotiation framework**, enabling adaptive supplier strategies while preserving decision auditability in a make-to-order carpentry supply chain.

3 Problem Definition

The proposed system represents a small carpentry manufacturer with one producer, one client, and two competing suppliers. The environment is stochastic, with Poisson-distributed demand (λ) and partial observability due to incomplete information on supplier strategies.

3.1 Overall System Design

The system adopts a hierarchical multi-agent architecture in which autonomous agents represent key supply chain stakeholders, as illustrated in Fig. 1. Three main agent types are defined:

- **Client Agent**: Generates customer orders specifying wood type, product characteristics, and delivery deadlines according to a Poisson distribution with parameter λ

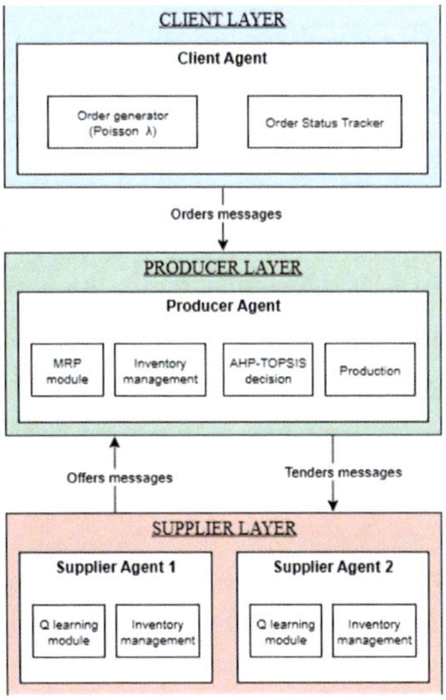

Fig. 1. Multi-agent system architecture for supply chain management

- **Producer Agent**: Acts as the central coordinator, managing inventory, production planning (MRP), supplier negotiations, and order fulfillment
- **Supplier Agent**: Implements Q-learning to adapt negotiation strategies and improve offer acceptance rates

3.2 Agent Interaction

The system follows a multi-phase interaction protocol to ensure consistent communication and synchronized decision-making, as illustrated in Fig. 2.

The sequence is as follows:

1. **Order Creation**: The client agent generates orders with specific requirements and deadlines, triggering the supply chain process.
2. **Call for Tender**: If inventory is insufficient, the producer broadcasts tender specifications to suppliers.
3. **Iterative Negotiation**: Suppliers submit offers evaluated via AHPTOPSIS. The process iterates up to 7 rounds, allowing suppliers to adapt strategies using Q-learning.
4. **Delivery and Production**: The selected supplier delivers materials, enabling the producer to complete manufacturing and fulfill orders.

3.3 Q-Learning Implementation for Suppliers

Each supplier agent uses Q-learning to adapt its bidding based on past negotiation outcomes. The state space is defined as $s = (X, Y, Z)$, where X is order urgency (days until deadline), Y is the stock ratio (current stock/requested quantity), and Z indicates the negotiation round.

The action space a consists of four strategies: (1) competitive pricing, (2) standard pricing, (3) premium pricing, and (4) fast delivery for urgent orders.

The Q-value $Q(s, a)$ represents the expected cumulative reward for taking action a in state s, balancing immediate acceptance probability with long-term profit. Rewards are positive only when the offer is accepted by the producer after AHPTOPSIS evaluation. This adaptive mechanism allows suppliers to refine strategies according to the standard Q-learning procedure detailed in Algorithm 1.

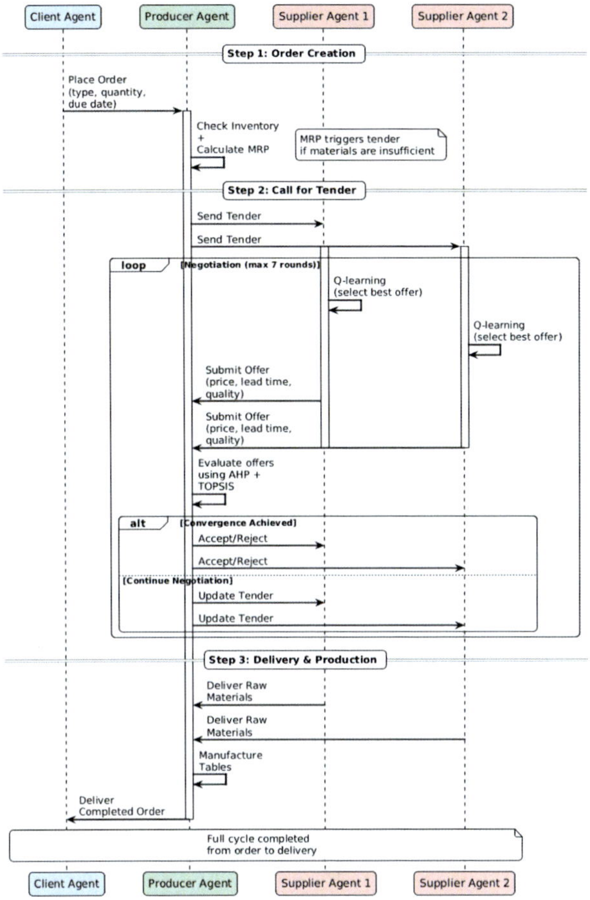

Fig. 2. Agent interaction workflow

Algorithm 1 Q-Learning learning process

Require: α: learning rate, γ: discount factor, ϵ: exploration rate
Require: episodes: number of training episodes
Ensure: Trained Q-table for optimal supplier strategies
Initialize Q-table $Q(s, a) \leftarrow 0$ for all state-action pairs
for episode = 1 to episodes **do**
    $s \leftarrow$ initial state (urgency, stock_level, round)
    while episode not terminated **do**
        if random() $< \epsilon$ **then**
            $a \leftarrow$ random action from action space
        else
            $a \leftarrow \arg\max_a Q(s, a)$
        end if
        Execute action a, observe reward r and next state s'
        $Q(s, a) \leftarrow Q(s, a) + \alpha[r + \gamma \max_{a'} Q(s', a') - Q(s, a)]$
        $s \leftarrow s'$
    end while
    Decay ϵ linearly towards minimum value
end for
return trained Q-table

The reward function provides binary feedback:

$$R = \begin{cases} +100 & \text{if offer accepted} \\ -10 & \text{if offer rejected} \end{cases}$$

This binary reward (+100/−10) emphasizes acceptance probability over order size. In make-to-order contexts, consistent reliability is preferred, while order value is implicitly captured through TOPSIS evaluation and state representation (stock levels, urgency).

3.4 Negotiation and Decision Making

The negotiation process uses a multi-criteria decision-making approach for consistent supplier evaluation. AHP [10] determines criteria weights through pairwise comparisons, covering cost, delivery time, and service quality, while ensuring consistency and handling qualitative and quantitative data. Suppliers are ranked using TOPSIS by measuring closeness to ideal and anti-ideal solutions, ensuring balanced performance. The integrated AHPTOPSIS framework aligns strategic priorities with robust mathematical rankings. This is implemented in the negotiation subsystem (Algorithm 2), which manages multi-round bidding and combines Q-learning-based supplier adaptation with the producer's evaluation. The supplier with the highest overall score across all criteria j is selected as optimal.

Algorithm 2 Iterative Negotiation with AHP-TOPSIS Evaluation

Require: tender, max_rounds = 7, convergence_threshold = 0.05
Ensure: Selected supplier or negotiation failure

round ← 1, previous_best_score ← 0
while round ≤ max_rounds **do**
    offers ← Collect supplier offers using Q-learning strategies
    Construct decision matrix D_{ij} from offers
    Normalize: $r_{ij} = \frac{x_{ij}}{\sqrt{\sum_{k=1}^{m} x_{kj}^2}}$
    Apply weights: $v_{ij} = w_j \times r_{ij}$
    Determine ideal solutions:
    $A^+ = \{\max_i v_{ij}\}$ for service quality, $\{\min_i v_{ij}\}$ for cost/time
    $A^- = \{\min_i v_{ij}\}$ for service quality, $\{\max_i v_{ij}\}$ for cost/time
    for each supplier i **do**
        $S_i^+ = \sqrt{\sum_j (v_{ij} - v_j^+)^2}$
        $S_i^- = \sqrt{\sum_j (v_{ij} - v_j^-)^2}$
        $C_i^* = \frac{S_i^-}{S_i^+ + S_i^-}$
    end for
    best_score ← $\max(C_i^*)$
    if round > 1 **then**
        relative_gap ← $\frac{|best\_score - previous\_best\_score|}{previous\_best\_score}$
        if relative_gap < convergence_threshold **then**
            Accept offer with highest C_i^*
            return selected supplier
        end if
    end if
    Update Q-learning rewards
    previous_best_score ← best_score
    round ← round + 1
end while
return negotiation failure

where i denotes the supplier index and j denotes the criterion index:

- x_{ij}: raw criterion value for supplier i on criterion j
- r_{ij}: normalized value using vector normalization
- w_j: AHP-derived weight for criterion j
- v_{ij}: weighted normalized value
- m: total number of suppliers submitting offers
- S_i^+, S_i^-: Euclidean distances to ideal and anti-ideal solutions
- C_i^*: closeness coefficient determining final supplier ranking

4 Results and Discussion

4.1 Sensitivity Analysis

Before comparing system-wide performance, we conducted systematic sensitivity analysis to identify optimal Q-learning configurations. Three critical parameters were examined across realistic ranges: learning rate $\alpha \in \{0.05, 0.1, 0.2\}$, discount factor $\gamma \in \{0.85, 0.9, 0.95\}$, and training episodes $\in \{100, 300, 500\}$.

Table 1 summarizes optimization results for three distinct objectives, demonstrating that optimal parameter selection depends fundamentally on stakeholder priorities. The profit-maximizing configuration ($\alpha = 0.20$, $\gamma = 0.95$, episodes = 500) achieves 13,123e compared to 6774e for satisfaction-maximizing parameters, revealing fundamental trade-offs between economic performance and service quality.

Table 1. Optimization results summary

Objective	α^*	γ^*	Ep.*	Profit (e)	Satisf. (%)	Negot. (%)
Profit max	0.20	0.95	500	13,123	95.7	38.2
Satisf. max	0.20	0.95	100	6774	99.3	37.4
Negot. max	0.05	0.95	300	8185	97.4	41.4

This analysis validates that our original parameter selection ($\alpha = 0.1$, $\gamma = 0.95$, episodes = 500) achieves near-optimal performance. Training convergence analysis reveals that agents exhibit stable strategic behavior after approximately 200–250 episodes, corresponding to 6,000–7,500 learning experiences, with the remaining episodes ensuring policy robustness without over-fitting.

4.2 System-Wide Performance Comparison

Our experiments show that adding Q-learning to the multi-agent system makes a real difference. Figure 3 presents the comparative analysis between the intelligent Q-learning system and the non-learning approach across four key performance dimensions.

System Profit Analysis (Fig. 3a) reveals the economic benefits accruing to the producer through supplier learning. Using the profit formula:
Profit = $\sum$(Order_Revenue) – $\sum$(Storage_Costs + Production_Costs + Procurement_Costs), results show the Q-learning system achieves peak producer profits of approximately 14,000e at high demand ($\lambda = 1.0$), while the non-learning system approaches near-zero profitability. This substantial improvement occurs because trained suppliers learn to balance competitive pricing with acceptance probability, ultimately reducing procurement costs for the producer.

Customer Satisfaction (Fig. 3b), measured as the percentage of on-time deliveries, demonstrates the system's service quality. The learning-enabled system achieves near-perfect satisfaction (99%) under low demand compared to 92% for non-learning systems, though both approaches experience degradation under high-demand conditions.

Negotiation Effectiveness (Fig. 3c) is quantified through success rates. The Q-learning system maintains acceptance rates between 95–100% across low to moderate demand levels, demonstrating that supplier learning enables better alignment with producer evaluation criteria. Under high demand, the learning system maintains a competitive advantage (75%) over non-learning approaches (73%).

Delivery Performance (Fig. 3d), measured as average delays beyond promised deadlines, shows the learning system achieving superior performance at moderate demand levels with delays below 1.5 days compared to 2 days for non-learning systems, indicating that adaptive supplier strategies contribute to more reliable delivery commitments.

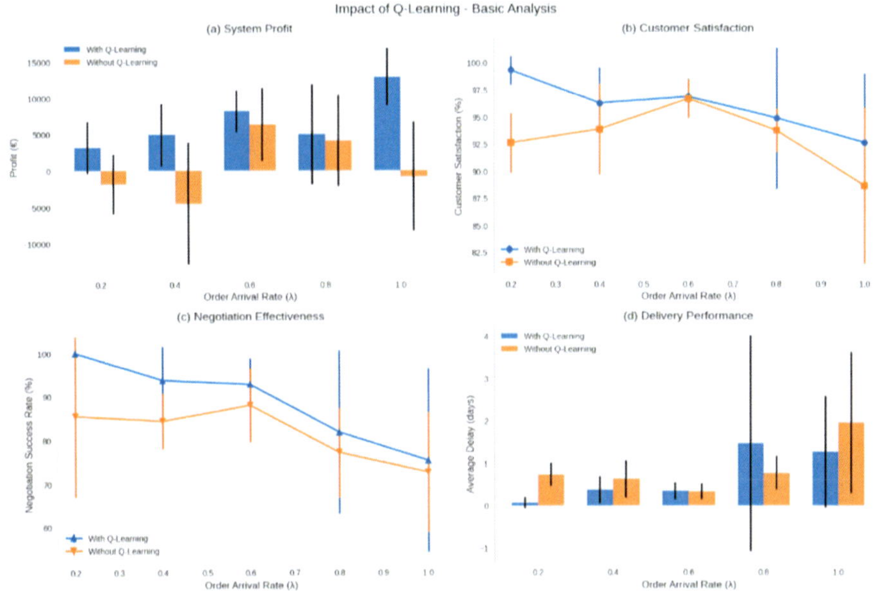

Fig. 3. Impact of Q-learning on system performance

4.3 Supplier Behavior Analysis

The supplier revenue analysis, shown in Fig. 4, reveals an unexpected pattern despite the system's overall profit improvements. The suppliers trained with

Q-learning consistently generate lower revenues compared to their non-trained counterparts, with this gap becoming more pronounced at higher demand levels. At $\lambda = 1.0$, non-trained suppliers achieve revenues around 35,000e while trained agents reach approximately 30,000e.

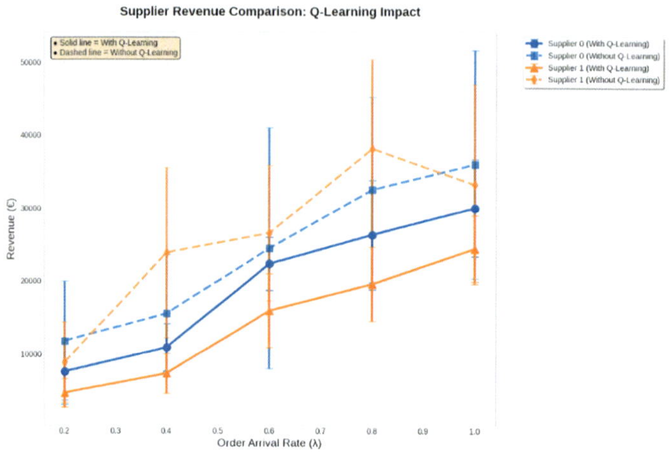

Fig. 4. Supplier revenue comparison across demand levels

This result reflects that Q-learning agents adopt more selective bidding strategies, focusing on achievable, profitable contracts aligned with producer evaluation criteria. While individual revenue decreases, system-wide profitability and negotiation efficiency improve, particularly under high demand, demonstrating a strategic trade-off that prioritizes overall supply chain performance.

5 Conclusion

This paper presents a multi-agent system integrating Q-learning with AHPTOPSIS for carpentry supply chain management. Experiments show up to 14,000e profit gains and 95% negotiation success rates. Trained suppliers prioritize offer acceptance over revenue maximization, improving system efficiency at the cost of individual profits.

Limitations include trade-offs between supplier revenue and customer satisfaction under high demand, simplified simulation assumptions (limited quality criteria, static market conditions), and potential need to refine the reward structure to better balance individual and system-wide goals.

Future work includes: (1) scaling to larger multi-agent networks, (2) integrating additional supply chain functions such as transportation, warehousing, and demand forecasting, (3) validating with real industrial data, and (4) developing multi-objective reward structures that balance individual and collective objectives.

References

1. AAMAS: The 23rd International Conference on Autonomous Agents and Multi-agent Systems—Competitions (2024)
2. Chaharsooghi, K., Heydari, J., Zegordi, S.H.: A reinforcement learning model for supply chain ordering management: an application to the beer game. Eur. J. Oper. Res. **269**(3), 955–983 (2008)
3. Giannoccaro, I.: Centralized vs. decentralized supply chains: the importance of decision maker's cognitive ability and resistance to change. Ind. Mark. Manage. **73**, 53–64 (2018)
4. Jain, V., Sangaiah, A.K., Sakhuja, S., Thoduka, N., Aggarwal, R.: Supplier selection using fuzzy AHP and TOPSIS: a case study in the Indian automotive industry. Neural Comput. Appl. **29**(7), 555–564 (2018)
5. Keskin, M., Cantürk, F., Eran, C., Aydoğan, R.: Decentralized multi-agent path finding framework and strategies based on automated negotiation. Auton. Agent. Multi-Agent Syst. **38**, 4985 (2024)
6. Kumar, A., Zavadskas, E.K., Mangla, S.K., Agrawal, V., Sharma, K., Gupta, D.: Using AHP-TOPSIS methodologies in the selection of sustainable suppliers in an electronics supply chain. Clean. Log. Supply Chain **5**, 100076 (2022)
7. Molinaro, M., Orzes, G., Poklemba, R., Continenza, R.: From forest to finished products: the contribution of industry 4.0 technologies to the wood sector. J. Manuf. Technol. Manag. **33**(4), 689–710 (2022)
8. Paredes-Rodríguez, A.M., Eras, J.J.C., Santos, V.S.: Fuzzy AHP TOPSIS methodology for multicriteria ABC inventory classification. J. Eng. **2023**, 7661628 (2023)
9. Rolf, B., Jackson, I., Müller, M., Lang, S., Reggelin, T., Ivanov, D.: A review on reinforcement learning algorithms and applications in supply chain management. Int. J. Prod. Res. **61**(20), 7151–7179 (2023)
10. Saaty, T.L.: The Analytic Hierarchy Process. McGraw-Hill, New York (1980)
11. Sequeira, M., Adlemo, A., Hilletofth, P.: A hybrid fuzzy-AHP-TOPSIS model for evaluation of manufacturing relocation decisions. Oper. Manag. Res. **16**, 164–191 (2023)
12. Terrada, L., El Khaïli, M., Ouajji, H.: Multi-agents system implementation for supply chain management making-decision. Proced. Comput. Sci. **177**, 624–630 (2020)
13. Xu, L., Almahri, S., Mak, S., Brintrup, A.: Multi-agent systems and foundation models enable autonomous supply chains: opportunities and challenges. SSRN Electron. J. **12**, 5986 (2023)
14. Xu, L., Chen, H., Zhang, Y., Wang, K.: On implementing autonomous supply chains: a multi-agent system approach. Comput. Ind. Eng. **189**, 109435 (2024)
15. Zhou, T., Xie, L., Zou, C., Yang, T.: Research on supply chain efficiency optimization algorithm based on reinforcement learning. Adv. Contin. Disc. Models **2024**, 51 (2024)

New Clustering Methods and Hybridization with ACO Algorithm for Energy Minimization Vehicle Routing Optimization

Nezha Jaaifi(✉) and Majda Fikri

System Computer Engineering, Mathematics and Applications Laboratory, Polydisciplinary Faculty of Taroudant, Ibnou Zohr University, 83000 Taroudant, Morocco
nezha.jaaifi@gmail.com

Abstract. In this paper, we focus on the Vehicle Energy Minimization Problem (EMVRP) since it has a significant economic and ecological impact in the logistics world. We propose a new approach that aims to overcome the limitations of existing resolution methods for this problem by integrating advanced techniques. Our goal is to develop an efficient and adaptable resolution method to handle different scenarios where geographical distributions and customer demands are varied. To do so, we propose two new clustering techniques that will be hybridized with the ant colony algorithm to better minimize vehicle energy consumption. Experimental results show the effectiveness of our approach, in terms of solution quality and computation time, on varied data with different customer distributions.

Keywords: Vehicle routing problem · Energy consumption · Heuristics · Metaheuristics · Clustering

1 Introduction

The Capacitated Vehicle Routing Problem (CVRP), introduced by Dantzig and Ramser [1], is a fundamental challenge in logistics, which aims to plan the most efficient routes for a fleet of vehicles, each with a fixed capacity, to deliver or collect goods from a set of customers.

Solving CVRP efficiently, essentially means reducing operational costs, improving service level and optimizing resource utilization. This type of problem is classified as NP-hard. Various methods have been proposed to address CVRP focusing mainly on minimizing distance travelled, reducing costs and the number of vehicles used. These approaches, encompassing exact methods and approximate methods namely metaheuristics such as Genetic Algorithms (GA), Ant Colony Optimization (ACO). Hybrid approaches have also been explored, combining the strengths of different metaheuristics to improve optimization performance [3, 4].

Despite the great contribution of CVRP in logistics optimization, it could not ensure ecological transport. Indeed, according to [5] road transport represents a major source of CO_2 emissions, the main cause of global warming. This has prompted researchers to

Y. Benadada et al. (Eds.): GOL 2026, LNNS 2034, pp. 200–224, 2027.
https://doi.org/10.1007/978-3-032-29677-1_19

consider the environment as a new dimension in transport problems, particularly CVRP. Kara et al. [6] have enriched CVRP with a new cost function and energy constraints with the aim of optimizing routes by taking into account not only the capacity but also the energy impact, which makes it possible to better respond to ecological issues. Thus, instead of minimizing only the distance travelled, the objective becomes reduced the total energy consumption related to the trips made. This new problem is called the Energy Minimization Vehicle Routing Problem (EMVRP). In the following we present a state of the art on the methods for solving the latter. The rest of this paper is organized as follows: Sect. 2 reviews related work. Section 3 presents problem description and the mathematical formulation. Section 4 introduces the proposed contributions: an improved initialization for k-means and k-means++, and a clustering adjustment method and full approach to solve the EMVRP, including the clustering, adjustment, and optimization phases using ACO. Section 5 reports the experimental results, Sect. 6 concludes the paper.

2 Literature Review

In recent years, the EMVRP problem has attracted considerable literature. Initially published by Kara et al. [6] who propose a novel cost function based on distance and load to minimize energy consumption, which has led to further research in this area. As an example, Wang and Liu [7] who consider EMVRP with a heterogeneous vehicle fleet and not all vehicles are used for better energy consumption, two mixed programming models are formulated for the problem: a cumulative energy-based formulation and a parallel machine scheduling-based model. Wang and Wu [8], also considered the case of heterogeneous vehicles and proposed a modified GA for solving EMVRP. This was also very evident in the work of Cooray and Rupasinghe [9] where a machine learning enhanced GA is used to show the interest of machine learning for adjusting metaheuristics, by suggesting approaches to develop more efficient solutions. In fact, EMVRP has practical applications in emergency situations. For example, from the study of Yi et al. [10] in their study on the problem of routing emergency vehicles with relief equipment during sudden disasters, a modified Monarch Butterfly Optimization (MBO) algorithm was proposed to provide an efficient solution to these problems. Funabashi et al. [11], formally defined this problem for energy minimization in delivery drone routing, proposing a dynamic programming algorithm as a solution. From our literature review, we noticed that most of the optimization approaches applied to EMVRP are: GA, Tabu Search (TS), machine learning, and hybrid algorithms, which have contributed to developing efficient solutions for energy minimization in vehicle routing which is illustrated by the results obtained namely [9, 12, 13]. Hybrid approaches, such as that of Frias et al. [13] that associate clustering with ant colony optimization, also show the current trend towards the integration of advanced methods to be able to address energy challenges in vehicle routing. All these advances show that the field of EMVRP is continuously evolving and is moving, on the one hand, towards multi-objective solutions adapted to modern logistics requirements, and on the other hand towards the integration of machine learning. Recent research on VRP focused on learning-based and adaptive optimization methods, including deep reinforcement learning, supervised learning, and hybrid approaches combined with clustering or large neighborhood search strategies. Notably, reinforcement learning and attention-based neural architectures have been successfully applied to complex VRP variants such as pickup and delivery problems, dynamic routing, and column generation frameworks [20, 21, 25]. Bibliometric analyzes further confirm that these learning-driven methods represent a major emerging trend in VRP research [22]. Comprehensive surveys also highlight the growing integration of machine learning techniques, particularly reinforcement learning, into routing problems to enhance adaptability and scalability [23, 24]. Indeed, to improve the methods for solving some NP-hard problems, several researchers [13–17] have bet on the hybridization of metaheuristics with machine learning methods, namely k-means as a clustering method. Indeed, this method consists of grouping the data into k clusters by starting by randomly selecting k centroids. Then, assign the points to the cluster whose centroid is closest. Afterwards, the centroids are updated by calculating the average of the points in each cluster. The assignment and update process is repeated until the stopping criterion is reached [18]. From our literature review, whether for EMVRP or others, we have noticed a wide use of k-means as a clustering method given its simplicity, speed, efficiency and easy interpretation of its results. Thus, we propose in Sect. 3.1 a new version of k-means to improve its performance, in Sect. 3.2 we propose a cluster balance method that we exploit in Sect. 3.3 to present a new approach to solving EMVRP.

3 Problem Description and Mathematical Formulation

In Kara et al. [6] defined EMVRP as follows: Each vehicle starts its journey from the depot, visits a number of customers to collect a given quantity of products, and then returns to the depot, with the condition that the maximum capacity of the vehicle is not exceeded. The cost to be minimized is not limited to the distance traveled, but also on the weight of the loaded goods; the weight of the vehicles changes during the delivery (or collection) which impacts the energy consumption.

Mathematically this consists of considering the graph G = (V, A) with V = {0, 1, …, n} is the set of nodes (customers) where 0 represents the depot and A = {(i, j): i, j ∈ V, i ≠ j} is the set of arcs. The variables are given as follows:

- d_{ij}: *The distance between* I *and* J, i, j ∈ V
- q_i: *The demand (the quantity to be collected) of* i, i ∈ V
- m: *The number of vehicles*
- Q_0: *The weight of the vehicle when empty*
- Q: *The maximum capacity of the vehicle*
- Y_{ij}: *The weight of the vehicle traveling from* i *to* j

The decision variable is: $x_{ij} = 1$ If the arc (i,j) belongs to the road and 0 otherwise.

Objective function

$$\boxed{\min \sum_{i=0}^{n} \sum_{j=0}^{n} d_{ij} Y_{ij} x_{ij}} \tag{1}$$

Constraints

$$\sum_{j=1}^{n} x_{0j} = m \tag{2}$$

$$\sum_{i=1}^{n} x_{i0} = m \tag{3}$$

$$\sum_{i=0}^{n} x_{ij} = 1, \quad \forall j = 1, 2, \ldots, n \tag{4}$$

$$\sum_{j=0}^{n} x_{ij} = 1, \quad \forall i = 1, 2, \ldots, n \tag{5}$$

$$\sum_{\substack{j=0 \\ j \neq i}}^{n} Y_{ij} - \sum_{\substack{j=0 \\ j \neq i}}^{n} Y_{ji} = q_i, \quad \forall i = 1, 2, \ldots, n \tag{6}$$

$$Y_{0j} = Q_0 x_{0j}, \quad \forall j = 1, 2, \ldots, n \tag{7}$$

$$Y_{ij} \leq \left(Q + Q_0 - q_j\right) x_{ij}, \quad \forall (i, j) \in A \tag{8}$$

$$Y_{ij} \geq (Q_0 + q_i) x_{ij}, \quad \forall (i, j) \in A \tag{9}$$

$$x_{ij} \in \{0, 1\} \quad \forall i = 1, 2, \ldots, n; \quad \forall j = 1, 2, \ldots, n \tag{10}$$

The objective function given in (1) represents the cost of the trip taken into account the distance and the vehicle load, since transporting heavier loads over longer distances results in higher energy consumption. Constraints (2) and (3) guarantee the use of all vehicles. Constraints (4) ensures that each customer is visited exactly once (is served by a single vehicle). Constraint (5) ensures that each vehicle departs from the depot, delivers customers, and then returns to the depot. Constraint (6) ensures the correct management of vehicle load during the routing process, it guarantees that the transported quantity is consistently updated as the vehicle visits each customer, maintaining the feasibility of the solution with respect to capacity limits. Constraints (7) initializes the flow in the first arc of each route by Q0. Constraint (8) verifies that the vehicle's load before arriving at a customer does not exceed its remaining capacity after serving that customer. Constraint (9), on the other hand, ensures that the vehicle carries at least the quantity required to serve the current customer before proceeding. The integrity of the constraints is given in (10).

4 Proposed Approach

K-Means and K-Means++ are popular clustering algorithms. K-Means assigns points to initial centroids and updates these centroids by calculating the mean of points in each group. K-Means++ improves this by selecting centroids that are more spread out, leading to faster convergence and better results. However, the choice of initial centroids is crucial as it directly impacts the convergence quality and final clusters.

To improve both methods, we propose to select centroids based on their distance from a global center of gravity, while also considering weight. This ensures better distribution and alignment with the dataset's structure.

Our method differentiates itself by selecting centroids that combine spatial distribution with specific criteria, leading to improved convergence and higher-quality clustering.

In this work, we present three contributions, the first is a new version of k-means to improve the performance of this method, the second is a proposal of a cluster balance method and the last is an improved approach to solving the EMVRP.

4.1 Contribution 1: Improvement of k-Means and k-Means++ Method by New Initialization Method

Consider V = {1, 2, …, n} a set of nodes where each point i(ai, bi) is characterized by a weight α that reflects its importance (mass for optimizing transport loads, energy consumption for balancing networks, population density for urban planning, access frequencies for depot management, etc.). To consider both geographic proximity and this weight, we developed a modified version of the k-means method where the initial centroids are not chosen randomly but are chosen according to their distances from the center of gravity weighted by the weights of the nodes. Then each node will be assigned to the cluster whose priority pi is maximum defined by the following formula:

$$p_i = \max_k \left(\frac{d_{ik}}{\alpha_i} \right) \tag{11}$$

where d_{ik} is the distance between node i and the k-th centroid, k = 1, 2, …, K and α is the weight of node i.

Afterwards, as in the k-means, we recalculate the position of each centroid which is the average of the coordinates of all the points of the cluster. The update step and the point assignment step are repeated until convergence. The following algorithm presents the detailed steps of our approach.

Algorithm 1 Center-of-Gravity-Based k-Means Clustering
Input:
Set of nodes $V = \{1,2,\ldots,n\}$;
Weight (demand) α_i of node i located at (a_i,b_i);
Number of clusters K.
Output:
A partition of V into K clusters.

1. **Initialization**
   - Compute the center of gravity $G(G_a,G_b)$ of all nodes:
     $$G_a = \frac{\sum_{i=1}^{n} a_i\alpha_i}{\sum_{i=1}^{n} \alpha_i}, G_b = \frac{\sum_{i=1}^{n} b_i\alpha_i}{\sum_{i=1}^{n} \alpha_i}.$$
   - Select the K nodes farthest from G as the initial centroids.
2. **Assignment Step**
   - For each node $i \in V$ and each centroid $k = 1,\ldots,K$, compute the priority:
     $$p_i^k = \frac{d_{ik}}{\alpha_i},$$

where d_{ik} denotes the distance between node i and centroid k.
   - Assign node i to the cluster with the maximum priority value.
3. **Update Step**
   - Recompute each centroid as the mean position of the nodes assigned to the corresponding cluster.
4. **Stopping Criterion**
   - Repeat Steps 2 and 3 until convergence, i.e., when no further changes in cluster assignments occur.

4.2 Contribution 2: Clustering Adjustment Method

In our previous work, we considered the weights of the points in the clustering, aiming to obtain better results in less time than k-means. Some clustering applications require not only a maximum similarity of the weights of the same cluster, but also require a total cluster weight to be respected, such as the case of client-server or groups of similar companies having a total turnover not to exceeded per group, or the group of similar households having a maximum total income per group. Therefore, applying the classic k-means or k-means++ method or even our first contribution will not necessarily ensure such a clustering so we propose in this section an algorithm to adjust a classic clustering in order to have clusters with a limit total weight.

After applying a clustering method, whether k-means, k-means++, or other, we sort the points in each obtained cluster in descending order according to their weights, we keep in each cluster those that can be included without exceeding its maximum capacity and the other points must be removed and reassigned to suitable clusters. The reassignment is based on priority (11).

Our following algorithm summarizes the steps to follow in Algorithm 2.

Algorithm 2 Cluster Adjustment Procedure
Input:
List of clusters obtained by a clustering method (e.g., k-means);
Maximum cluster capacity Q.
Output:
Adjusted clusters satisfying the capacity constraint.

Initialization
Create an empty list **RemovedNodes.**
Capacity Adjustment for Each Cluster
For each cluster C_k:

1. Create an empty list **LIST**.
2. Compute the total weight of the cluster: $S_k = \sum_{i \in C_k} \alpha_i$.
3. If $S_k > Q$:
   - Sort the nodes of $C_k$in descending order of their weights.
   - Set $S = 0$.
   - For each node $i$in the sorted list:
     - If $S + \alpha_i \leq Q$, then $S \leftarrow S + \alpha_i$and add node $i$to **LIST**.
     - Otherwise, add node $i$to **RemovedNodes**.
   - Update the cluster: $C_k \leftarrow$LIST.

Reassignment of Removed Nodes
For each node $i$in **RemovedNodes**:

- Identify all clusters whose remaining capacity allows the insertion of node i.
- Compute the priority of node $i$with respect to the centroids of the feasible clusters.
- Assign node $i$to the cluster with the highest priority value.

4.3 Contribution 3: New Approach for Solving EMVRP

Our literature review reveals that there aren't many studies focused on solving the EMVRP. The approach put forth by [13] is one of the few approaches specifically designed to address this issue among the existing work. In this framework we exploit our contributions 1 and 2 hybridized with the ACO in order to better optimize the energy.

Our resolution approach builds on the mathematical formulation described in Sect. 3, and we propose an algorithm that combines the previous contributions with an ant colony optimization (ACO). This algorithm works in three successive phases:

Clustering Phase

We begin our resolution approach by building homogeneous groups of customers formed by taking into account their demands and the capacity of the vehicles. This phase applies Contribution 1 by considering V the set of customers who are grouped into clusters by taking into account their geographical positions and their demands (the weights of the customers) so that the customers of each cluster can be served by a single vehicle. The number of clusters, noted k, corresponds to the number of vehicles is calculated by the following formula:

$$k = \sum_{i=0}^{n} \frac{\alpha_i}{Q} \tag{12}$$

Adjustment Phase

Once the clusters are formed, we adjust the clusters to meet the vehicle capacity constraints. This phase implements our Contribution 22, where the limit capacity of the clusters is equal to vehicle capacity. This ensures a balanced distribution while minimizing additional trips.

Optimization Phase with ACO

In this phase, we apply the Ant Colony Optimization (ACO) algorithm to optimize the routes within each cluster. ACO is a metaheuristic inspired by the foraging behavior. of ants, where artificial ants build solutions based on pheromone trails and heuristic information. Several works have successfully applied ACO to the Vehicle Routing Problem (VRP), demonstrating its efficiency in handling combinatorial structures and constraints.

In our method, ACO is applied independently within each cluster to solve the corresponding Traveling Salesman Problem (TSP).

5 Numerical Results

To demonstrate the effectiveness of the proposed methods, we evaluate the quality of the solutions using three key metrics: similarity, dissimilarity, and dispersion. Each metric plays a specific role: similarity assesses how compact the clusters are, dissimilarity evaluates how well-separated the clusters are from each other, and dispersion measures the overall spatial distribution of customers across the service area. A good clustering solution should achieve low similarity (high internal cohesion), high dissimilarity (good separation between clusters), and an appropriate level of dispersion depending on the problem instance.

Similarity is defined as the average distance between each customer and the centroid of its cluster. Dissimilarity is calculated as the average distance between the centroids of different clusters. Dispersion is computed as the standard deviation of the distances between customers and the global center of gravity. A low dispersion indicates that customers are closely grouped, while a high dispersion suggests a more widespread distribution.

In the results shown below, the model assumes a fixed vehicle capacity.

5.1 Contribution 1

We tested this improved k-means, improved k-means++ algorithm and k-means with center of gravity in the case of customers for random data: geographical positions and demands and they have been tested on different sets of randomly generated customers with different dispersion indices. Similarity measures the similarity between customers of the same clusters, while dissimilarity measures it between clusters. These two metrics are very important in clustering according to Jain and Dubes [19].

5.2 Contribution 2

To show the importance of the adjustment phase and its impact on the performance of the results, we compare the following three approaches: K-means with center of gravity initialization, K-means random and K-means++. These approaches have been tested on different sets of randomly generated customers with different dispersion indices. The following Table 1 illustrates the results of the comparison of these three approaches before and after the adjustment considering the similarity (intra-cluster), dissimilarity(inter-cluster), time and total requests metrics in each cluster.

Generally, regardless of the method used, the adjustment improves the similarity and corrects demand imbalances in the clusters, especially for the random k-means method which shows a remarkable difference before and after the adjustment. On the other hand, for the k-means method based on the center of gravity, the adjustment brings modest but consistent gains which shows that this method manages the clusters well from the start. On the other hand, the k-means++ method stands out more for high dispersion indices. Moreover, the adjustment time remains low for all methods, with a slight increase for high dispersion indices.

These results show the importance of the adjustment because it allows to avoid exceeding the overloads of the vehicles and optimizes the use of the vehicles. This shows the interest of combining a good initialization with an efficient adjustment for the minimization. These solid bases thus allow to move on to the resolution of the routing problem in each cluster, by optimizing the routes to minimize the distances traveled while respecting the defined constraints.

5.3 Contribution 3

In an applied context, we propose the generation of customers with geographic coordinates and random requests illustrated by the dispersion index. The Table 2 presents a detailed comparison of three clustering methods applied to different datasets, by criteria such as clustering time, similarity, dissimilarity, total distance and execution time.

Analyzing the results in the table above, we notice that our approach stands out for its ability to generate well-homogeneous clusters while maintaining relatively low and stable clustering times, even with large data (number of customers, dispersion). Although this method may result in slightly longer clustering times, this is justified by the quality of the results obtained. Indeed, using the center of gravity as a reference for the initialization of the centroids guarantees more homogeneous clusters that are better adapted to the geographical characteristics and customer demand. This results, in a reduction of the total distances traveled, which is essential for applications such as the EMVRP problem where the optimization of distance and energy is crucial.

Additionally, our approach has better control over the distribution of customers in each cluster. This method is particularly beneficial for datasets where customers are distributed non-uniformly, as it takes into account the density of customers in the area around each centroid, thus ensuring a more efficient distribution.

Although our approach shows competitive total distance and running time results, it can slightly outperform or equal k-means++, depending on the data used, especially when there is moderate distribution of customers. These features make it a robust and efficient method for applications requiring fast initialization and good clustering quality, especially for the EMVRP problem.

K-means++ method, on the other hand, often offers a good compromise between the first two, with reasonable similarity and acceptable execution time. It combines the advantages of random K-means and center of gravity, while seeking to avoid bad initializations, making it a reliable method for medium to large client sets.

In summary, the choice of method will depend on specific priorities, such as speed of execution, quality of clusters or scale of data. These results highlight the importance of selecting the clustering method based on the needs and characteristics of the dataset.

5.4 Experimental Study

In order to assess the effectiveness of our approach, we conducted a series of experiments on a set of generated EMVRP instances automatically thanks to an internal generator designed in Python. The instances are in the format EMVRP m n Q, where m is the number of vehicles, n is the number of customers, and Q is the maximum capacity of one vehicle (set at 100 units).

5.4.1 Validation of Our Proposed Approach

Initially, we evaluated our method by testing it on various instances in order to compare it with the exact technique of IBM ILOG CPLEX Optimization Studio (v22.1.2).

The results obtained are illustrated in Table 3.

The “OFV” and “t(s)” columns under “CPLEX” and “Our approach” correspond to the optimal solution and the corresponding computational time, respectively.

The results indicate that for instances with up to ten clients, our method provides similar energy values to CPLEX in a shorter time. However, CPLEX encountered memory limitations from instance EMVRP_2_13_2 because of the exponential increase in complexity for the exact method. On the other hand, our approach retains computational stability and satisfactory scalability. This experiment confirms the validity of our approach and demonstrates that it represents a robust method compared to exact resolution for small instances.

5.4.2 Comparative Analysis of Our Approach

Since CPLEX can only solve small instances, we performed a second series of experiments to compare global ACO with our method. Table 4 presents the results obtained.

Table 1. Comparison of total demand, total distance, and execution time before and after adjustment for different numbers of clients and dispersion indices.

n	Indices	Methods	Befor adjustment				After adjustment			
			Total demands in cluster	Similarity	Dissimilarity	Clustering time	Total demands in cluster	Similarity	Dissimilarity	Adjustment time
20	22.39	Cof	[48, **58**, 41]	90.33	27.32	**0.0012**	[48, 49, 50]	103.35	25.51	**0.000120**
		k-means rand	[50, **63**, 34]	86.71	32.10	0.0017	[50, 50, 47]	101.26	**26.00**	0.000156
		k-means++	[10, **64**, 37]	**134.77**	**35.11**	**0.0012**	[50, 47, 50]	**106.53**	25.56	0.000314
	134.36	Cof	[31, 43, 41]	541.95	163.92	**0.0012**	[31, 43, 41]	541.95	163.92	**0.000015**
		k-means random	[48, 46, 21]	**566.22**	158.73	0.0015	[48, 46, 21]	566.22	158.73	0.000019
		k-means++	[**66**, 18, 31]	563.01	**170.01**	0.0016	[50, 18, 47]	**671.18**	**170.22**	0.000873
	223.93	Cof	[37, **54**, 37]	903.25	273.20	**0.0012**	[42, 49, 37]	**908.50**	**273.13**	0.000471
		k-means random	[**55**, 33, 40]	**913.23**	273.03	**0.0012**	[49, 37, 42]	**908.50**	**273.13**	0.000143
		k-means++	[34, **57**, 37]	824.48	**298.83**	0.0015	[34, 50, 44]	903.93	272.15	**0.00014**

(continued)

Table 1. *(continued)*

n	Indices	Methods	Befor adjustment				After adjustment			
			Total demands in cluster	Similarity	Dissimilarity	Clustering time	Total demands in cluster	Similarity	Dissimilarity	Adjustment time
50	51.55	Cof	[**64**, **76**, **61**, 46, 14, 36, **51**]	**120.19**	**60.19**	0.0054	[50, 50, 48, 50, 50, 49, 47]	187.42	55.74	0.000986
		k-means random	[47, 17, **61**, **60**, 45, **58**, **60**]	115.32	59.10	**0.0031**	[50, 50, 50, 50, 49, 49, 50]	**197.05**	**55.61**	**0.000511**
		k-means++	[**52**, **59**, 47, **62**, 46, 21, **61**]	116.59	56.46	0.0054	[50, 49, 47, 50, 50, 49, 47]	174.51	57.09	0.000580
	154.66	Cof	[**57**, **73**, 49, 42, 22, 19, 41]	360.57	**180.57**	0.0156	[50, 50, 49, 42, 45, 26, 41]	404.93	171.55	0.001965
		k-means random	[**76**, 25, **70**, 8, 32, 50, 42]	**364.32**	179.77	**0.0064**	[49, 35, 50, 19, 50, 50, 50]	**641.59**	170.13	0.000824

(continued)

Table 1. *(continued)*

n	Indices	Methods	Befor adjustment				After adjustment			
			Total demands in cluster	Similarity	Dissimilarity	Clustering time	Total demands in cluster	Similarity	Dissimilarity	Adjustment time
		k-means++	[50, 30, 32, 45, 50, 49, 47]	339.87	175.42	0.0122	[50, 30, 32, 45, 50, 49, 47]	339.87	**175.42**	**0.000039**
	257.77	Cof	[**72**, **64**, **59**, **60**, 35, 7, 26, 35]	493.88	302.02	0.0145	[48, 50, 50, 49, 35, 28, 50, 48]	643.35	277.33	**0.000904**
		k-means random	[48, 34, 36, 41, **66**, 13, 36, **84**]	484.78	**302.66**	**0.0035**	[49, 50, 36, 48, 50, 25, 50, 50]	670.36	**280.92**	0.001544
		k-means++	[**65**, **90**, **71**, 19, 28, 46, 35, 4]	**615.14**	300.93	0.0062	[50, 50, 50, 43, 46, 50, 49, 20]	**765.38**	275.51	0.001137

(continued)

Table 1. *(continued)*

n	Indices	Methods	Befor adjustment				After adjustment			
			Total demands in cluster	Similarity	Dissimilarity	Clustering time	Total demands in cluster	Similarity	Dissimilarity	Adjustment time
100	51.08	Cof	[45, 44, 43, **90**, **62**, 33, **64**, **101**, 13, 35, **64**, 19, 47, 29]	90.50	**57.20**	**0.0237**	[50, 50, 50, 50, 50, 48, 50, 50, 44, 50, 50, 48, 49, 50]	**189.50**	53.21	0.003142
		k-means random	[27, 43, **116**, 36, 12, 22, **68**, **59**, **61**, 47, **71**, 18, 45, **64**]	85.47	54.74	0.1393	[50, 43, 50, 50, 50, 47, 50, 50, 50, 49, 50, 50, 50, 50]	166.96	53.68	0.002855
		k-means++	[**52**, **60**, **64**, 39, 33, **53**, 37, 4, 41, **64**, **90**, 44, 34, **74**]	**92.90**	55.94	0.0335	[50, 50, 50, 47, 50, 50, 48, 44, 50, 49, 50, 44, 50, 50]	139.94	**54.62**	**0.002148**

(continued)

Table 1. *(continued)*

n	Indices	Methods	Befor adjustment				After adjustment			
			Total demands in cluster	Similarity	Dissimilarity	Clustering time	Total demands in cluster	Similarity	Dissimilarity	Adjustment time
	153.24	Cof	[47, **83**, **78**, 12, **97**, **107**, 20, 10, 37, 25, 44, 22]	**411.23**	**176.82**	0.0167	[50, 50, 50, 50, 50, 50, 32, 50, 50, 50, 50, 50]	**661.44**	159.04	0.004230
		k-means random	[**55**, **56**, 36, 27, **52**, 15, 50, **64**, 23, **98**, **53**, **53**]	331.09	160.20	**0.0088**	[50, 50, 49, 50, 50, 33, 50, 50, 50, 50, 50, 50]	499.38	165.44	0.001942
		k-means++	[**56**, 45, 41, 24, 28, **64**, **54**, **76**, 50, **51**, 37, **56**]	319.63	168.30	0.0526	[50, 49, 48, 48, 38, 50, 50, 50, 50, 50, 50, 49]	416.81	**165.93**	**0.001359**

(continued)

Table 1. *(continued)*

n	Indices	Methods	Befor adjustment				After adjustment			
			Total demands in cluster	Similarity	Dissimilarity	Clustering time	Total demands in cluster	Similarity	Dissimilarity	Adjustment time
	255.40	Cof	[34, **58**, 44, **58**, **100**, **67**, 7, **53**, **114**, 14, 26, **71**, 20, 25, 36]	421.83	**285.62**	0.1265	[50, 49, 47, 50, 50, 50, 50, 48, 50, 35, 50, 50, 49, 49, 50]	**838.86**	267.84	0.010802
		k-means random	[38, **67**, 44, 49, **69**, 39, 40, **63**, 38, 47, 32, 28, **58**, **84**, 31]	353.26	274.55	**0.0134**	[50, 50, 50, 49, 50, 49, 50, 50, 42, 50, 50, 42, 49, 48, 48]	533.95	**268.60**	**0.002281**
		k-means++	[**58**, 31, 49, **126**, 35, **51**, **69**, 18, 38, 33, 16, 5, 33, **75**, **90**]	**422.49**	280.17	0.2332	[50, 50, 49, 50, 50, 50, 50, 50, 50, 49, 30, 49, 50, 50, 50]	725.99	262.43	0.003187

Table 2. Comparisons of execution times and distances traveled for the three methods for various numbers of customers and dispersion indices

n	Index	Methode	Time of clustering	Similarity	Dissimilarity	Total distance	Execution time
20	25.499	K-means cof	**0.0677**	176.9738	**489.2544**	**2508.6734**	1.4936
		K-means random	0.0678	**201.1687**	408.8469	2758.1980	1.6212
		K-means++	0.0699	160.8465	484.6576	**2508.6734**	**1.4892**
	63.748	K-means cof	0.0685	44.2435	**122.3136**	**627.1684**	**1.4591**
		K-means random	**0.0669**	**50.2922**	102.2117	689.5495	1.5694
		K-means++	0.0743	40.2116	121.1644	**627.1684**	1.5086
	254.993	K-means cof	**0.0677**	176.9738	**489.2544**	**2508.6734**	1.4936
		K-means random	0.0678	**201.1687**	408.8469	2758.1980	1.6212
		K-means++	0.0699	160.8465	484.6576	**2508.6734**	**1.4892**
50	18.9	K-means cof	0.3750	5.2363	**29.4306**	427.8656	4.1544
		K-means random	**0.2147**	**5.5761**	26.3720	**420.7724**	**3.8525**
		K-means++	0.2211	4.9184	27.2369	436.3677	4.0924
	75.61	K-means cof	**0.2072**	20.9454	**117.7224**	1711.4623	**4.1305**
		K-means random	0.2496	**22.3044**	105.4880	**1683.0897**	4.3856
		K-means++	0.2192	19.6736	108.9477	1745.4709	4.9407
	264.64	K-means cof	**0.2068**	73.3089	**412.0285**	5990.1182	**3.8367**
		K-means random	0.2070	**78.0654**	369.2081	**5890.8141**	4.1926
		K-means++	0.2273	68.8576	381.3171	6109.1481	4.0845
100	18.708	K-means cof	**0.4494**	**4.1730**	**30.2789**	901.3164	11.5371
		K-means random	0.5650	3.6969	27.0351	895.0837	**11.1294**

(continued)

Table 2. (*continued*)

n	Index	Methode	Time of clustering	Similarity	Dissimilarity	Total distance	Execution time
		K-means++	0.5642	3.7213	27.3021	**888.8793**	8.9189
	187.083	K-means cof	0.5174	**41.7297**	**302.7891**	9013.1641	10.2184
		K-means random	**0.4940**	36.9687	270.3510	8950.8369	10.5707
		K-means++	0.5543	37.2131	273.0213	**8888.7928**	**10.1653**
	299.332	K-means cof	0.4492	**66.7676**	**484.4625**	14421.0625	8.9292
		K-means random	**0.4207**	59.1500	432.5617	14321.3391	**8.8730**
		K-means++	0.6213	59.5410	436.8341	**14222.0684**	10.5900
150	19.23	K-means cof	0.6513	3.1382	**30.7065**	1375.7414	13.7752
		K-means random	**0.6263**	**3.2154**	26.4043	**1358.6591**	12.7048
		K-means++	0.6540	3.0517	26.5092	1365.7448	**12.6078**
	115.35	K-means cof	0.6121	18.8291	**184.2391**	8254.4484	12.4685
		K-means random	**0.6007**	**19.2924**	158.4259	**8151.9548**	12.5796
		K-means++	0.6535	18.3100	159.0550	8194.4690	**12.3029**
	307.62	K-means cof	0.7197	50.2110	**491.3042**	22011.8624	12.7688
		K-means random	**0.5969**	**51.4463**	422.4690	**21738.5461**	12.4340
		K-means++	0.6434	48.8268	424.1468	21851.9174	**12.3477**
200	19.12	K-means cof	0.9031	2.9425	**30.8869**	1940.9276	26.9444
		K-means random	**0.8495**	**3.2849**	25.9472	1928.4568	**19.4078**
		K-means++	1.1966	2.9515	27.6479	**1886.2664**	19.6844
	95.60	K-means cof	0.8262	14.7126	**154.4343**	9704.6381	17.8856

(*continued*)

Table 2. *(continued)*

n	Index	Methode	Time of clustering	Similarity	Dissimilarity	Total distance	Execution time
		K-means random	**0.7891**	**16.4243**	129.7358	9642.2841	**17.4614**
		K-means++	0.8534	14.7577	138.2395	**9431.3320**	19.2437
	305.94	K-means cof	0.8123	47.0805	**494.1898**	30977.1237	16.9989
		K-means random	**0.7741**	**52.5578**	415.1547	30855.3093	17.4666
		K-means++	0.8929	47.2245	442.3665	**30180.2625**	**16.9105**
250	19.31	K-means cof	3.7924	2.4493	**32.0373**	2379.4045	41.3585
		K-means random	**1.1068**	**2.4527**	26.3328	2418.9389	27.9082
		K-means++	1.4433	2.3152	26.0639	**2313.7075**	**25.7733**
	115.88	K-means cof	1.4800	14.6955	**192.2241**	14276.4270	23.8917
		K-means random	1.2711	**14.7159**	157.9967	14513.6336	23.8809
		K-means++	**1.2311**	13.8914	156.3835	**13876.4444**	**22.4602**
	347.65	K-means cof	1.2369	44.0866	**576.6723**	42829.2811	22.9224
		K-means random	**1.0936**	**44.1478**	473.9901	43540.9008	**21.9875**
		K-means++	1.4053	41.6742	469.1506	**41629.3332**	22.3978
300	19.45	K-means cof	1.5857	2.2376	**32.4591**	2963.0168	28.8451
		K-means random	1.7037	**2.2836**	26.9144	**2816.9632**	29.4803
		K-means++	**1.6626**	2.1637	27.1826	2840.0394	**27.9729**
	97.27	K-means cof	1.3463	11.1881	**162.2954**	14815.0841	25.9691
		K-means random	**1.2939**	**11.4179**	134.5719	**14071.0899**	26.0998
		K-means++	1.5054	10.8184	135.9129	14200.1971	**25.6840**

(continued)

Table 2. *(continued)*

n	Index	Methode	Time of clustering	Similarity	Dissimilarity	Total distance	Execution time
	311.26	K-means cof	1.4350	35.8020	**519.3452**	47408.2690	25.9133
		K-means random	**1.3564**	**36.5373**	430.6300	**44993.6937**	25.9744
		K-means++	1.6000	34.6189	434.9214	45440.6308	**25.5300**
500	19.23	K-means cof	7.9568	**1.7428**	**32.2162**	4734.6344	63.3864
		K-means random	5.3073	1.7255	27.1912	**4723.1449**	50.8933
		K-means++	**3.8055**	1.6409	27.2466	4809.8454	**46.1097**
	115.38	K-means cof	**2.6834**	**10.4566**	**193.2973**	28391.3347	**47.6589**
		K-means random	2.7974	10.3532	163.1469	**28303.9244**	50.1284
		K-means++	4.1349	9.8454	163.4798	28859.0727	56.8062
	307.67	K-means cof	**2.7544**	**27.8843**	**515.4595**	75738.6824	**45.8186**
		K-means random	3.9153	27.6085	435.0585	**75477.1318**	56.2112
		K-means++	3.9334	26.2543	435.9460	76957.5271	48.8864
1000	39.19	K-means cof	10.2850	**2.4991**	**66.7256**	19157.8577	99.7742
		K-means random	**6.9611**	2.4802	54.2274	18940.7205	**87.3185**
		K-means++	11.9695	2.3087	53.1420	**18878.3655**	94.2814
	117.58	K-means cof	9.7675	**7.4973**	**200.1767**	57457.1016	107.5421
		K-means random	**7.2130**	7.4406	162.6823	56809.4887	**88.5482**
		K-means++	11.7099	6.9261	159.4260	**56593.2794**	93.8826
	313 54	K-means cof	11.8057	**19.9927**	**533.8046**	153151.3494	102.9859
		K-means random	**6.8886**	19.8415	433.8194	151491.9698	**86.9359**

(continued)

Table 2. *(continued)*

n	Index	Methode	Time of clustering	Similarity	Dissimilarity	Total distance	Execution time
		K-means++	11.6479	18.4697	425.1360	**150833.5328**	93.9359
1500	37.07	K-means cof	16.0380	**1.6796**	**53.5080**	22996.9956	146.8441
		K-means random	**13.9421**	1.6622	42.4414	**22779.0464**	**136.3139**
		K-means++	27.7026	1.5532	42.6171	22875.3635	169.1989
	97.11	K-means cof	21.8258	**5.2487**	**167.2124**	71836.0452	180.3258
		K-means random	**14.6959**	5.1944	132.6293	**71184.5201**	**156.3959**
		K-means++	33.4744	4.8539	133.1784	71456.8384	191.1636
	310.76	K-means cof	42.6512	**16.7958**	**535.0796**	229893.6711	194.2432
		K-means random	**16.3516**	16.6222	424.4138	**227741.2026**	194.0677
		K-means++	34.7412	15.5325	426.1709	228628.0887	**191.9278**
3000	38.45	K-means cof	50.1699	**1.4058**	**66.7713**	56955.5612	304.9625
		K-means random	**40.6235**	1.3865	52.3015	55978.3144	**285.1402**
		K-means++	4203.0442	1.2990	52.2407	**55965.0717**	4450.6411
	115.33	K-means cof	58.3546	**4.2174**	**200.3139**	170872.4840	310.8122
		K-means random	**49.5332**	4.1596	156.9044	167919.7313	**290.5075**
		K-means++	148.4654	3.8970	156.7220	**167853.3979**	387.8253
	422.9	K-means cof	50.2353	**15.4637**	**734.4844**	626443.4383	298.5389
		K-means random	**41.0758**	15.2518	575.3162	615495.8987	**283.8261**
		K-means++	150.9286	14.2890	574.6473	**615462.4590**	390.2917

Table 3. Comparative results between CPLEX and our approach for different EMVRP instances

EMVRP_m_n_Q	Cplex		Our approach	
	OFV	t(s)	OFV	t(s)
EMVRP_1_3_2	1960	0.39	1960	0.003
EMVRP_2_3_2	3870	0.12	3870	0.003
EMVRP_1_5_2	4360	0.25	4360	0.009
EMVRP_2_5_2	4460	0.14	4775	0.004
EMVRP_3_6_2	8111	0.49	8111	0.005
EMVRP_4_6_2	9099	0.12	9099	0.005
EMVRP_1_8_2	3453	0.78	3463	0.047
EMVRP_2_8_2	5050	0.27	5065	0.008
EMVRP_4_8_2	11,710	0.13	11,710	0.005
EMVRP_1_10_2	4694	368.68	4897	0.063
EMVRP_2_10_2	8900	0.49	8936	0.007
EMVRP_3_10_2	10,226	0.89	10,419	0.009
EMVRP_4_10_2	9012	0.17	10,090	0.006
EMVRP_2_11_2	5797	3.97	5894	0.026
EMVRP_3_11_2	9687	1.56	9763	0.008
EMVRP_1_12_2	7297	364.42	8934	0.083
EMVRP_3_13_2	12,058	3.48	12,950	0.046

The columns in Table 4 show: EMVRP_m_n_Q instances, Customer Dispersion Index, Energy, and the iteration at which this value has been obtained and execution time(s). This information is presented for both methods in order to assess the quality of the solutions. We established a stopping criterion of 200 iterations for both configurations.

The table shows that our approach leads to a significant reduction in energy consumption, ranging from 30 to 60%, while also reducing execution time. Indeed, the optimal solution is obtained well before the maximum number of iterations. For example, in the case of the EMVRP_3_20_2 instance, we managed to reduce energy consumption from 14,260 (with global ACO) to 10,151, representing a decrease of 28.8%. Similarly, for EMVRP 4 50 2, there is a reduction of over 40% in energy consumption, and the computation time is halved.

These improvements stem from the use of weighted center of gravity-based clustering, which we developed to harmonize the distribution of clients before the optimization process on each cluster, thus promoting better energy performance and more efficient convergence.

We have demonstrated that it is possible to reduce combinatorial complexity while maintaining high solution quality. Compared to global ACO, our approach offers more

Table 4. Comparison of results between global ACO and our proposed approach

Instance	Index	ACO			Our approach		
		OFV	Iteration	Time(s)	OFV	Iteration	Time(s)
EMVRP 3_20_2	14.6315	14,260	20	0.452	10,151	42	0.319
	30.444	28,454	100	0.443	21,223	35	0.314
	183.9059	120,053	35	0.449	45,406	31	0.318
EMVRP_4_50_2	16.3862	20,773	45	1.912	12,007	34	0.947
	33.4428	41,585	126	1.851	23,344	57	0.908
	183.7104	201,465	106	1.862	144,541	36	0.946
EMVRP_4_100_2	16.4966	25,716	144	6.251	17,988	81	2.403
	36.8845	53,955	195	6.857	36,997	71	2.373
	179.6069	241,742	90	6.412	187,297	67	2.385
EMVRP_5_200_2	16.2254	36,139	96	46.511	23,671	75	22.189
	36.9674	79,686	106	37.54	50,830	88	15.48
	185.1295	394,163	101	32.828	275,547	72	16.642
EMVRP_6_500_2	15.9842	68,020	63	159.547	48,912	175	25.362
	35.7129	143,732	187	141.225	113,372	172	26.126
	193.2096	726,807	181	141.884	616,363	116	25.716

energy-efficient and faster solutions, particularly for cases characterized by large spatial dispersion.

6 Conclusion

In this work we have targeted two objectives.

First, we have proposed an improvement of the K-means method in terms of better similarity and in terms of clustering time, as well as a method of adjusting clusters in the case of limited weight or capacity of clusters.

Secondly, we have proposed an approach to solving the EMVRP based on a hybridization of our clustering approach with the ACO algorithm.

The experimental results showed the effectiveness of our method in reducing energy consumption under capacity constraints, which also improved the quality of the obtained clusters. By mutually comparing the clustering methods (random initialization, center of gravity, k-means++), we were able to highlight the strengths of our demand weighted initialization method.

In future work, we aim to introduce new constraints in the EMVRP to integrate other constraints influencing energy consumption depending on the nature of the roads used.

References

1. Dantzig, G.B., Ramser, J.H.: The truck dispatching problem. Manage. Sci. **6**(1), 80–91 (1959)
2. Elshaer, R., Awad, H.: A taxonomic review of metaheuristic algorithms for solving the vehicle routing problem and its variants. Comput. Ind. Eng. **140**, 106242 (2020)
3. Amrouche, H.: Hybridization of Metaheuristics. PhD thesis, Mouloud Mammeri University (2012)
4. Hachimi, H.: Hybridizations of Metaheuristic Algorithms in Global Optimization and Their Applications. PhD thesis, INSA Rouen; Mohammadia School of Engineers (Rabat, Morocco) (2013)
5. Jacob, B., Raynal, M.: ERS for the Decarbonization of Road Transport (2022)
6. Kara, I., Kara, B.Y., Yetis, M.K.: Energy minimizing vehicle routing problem. In: Combinatorial Optimization and Applications (COCOA 2007), LNCS, vol. 4616, pp. 62–71. Springer, Xi'an (2007)
7. Wang, S., Liu, X.: Energy minimization vehicle routing problem with heterogeneous vehicles. In: Proceedings of the 13th International Conference on Service Systems and Service Management (ICSSSM 2016), pp. 1–5. IEEE, Kunming (2016)
8. Wang, S., Wu, Y.: A genetic algorithm for energy minimization vehicle routing problem. In: International Conference on Service Systems and Service Management (ICSSSM 2017), pp. 1–5. IEEE, Dalian (2017)
9. Cooray, P.L.N.U., Rupasinghe, T.D.: Machine learning-based parameter tuned genetic algorithm for energy minimizing vehicle routing problem. J. Ind. Eng. **2017**(1), 3019523 (2017)
10. Yi, J.-H., Wang, J., Wang, G.-G.: Using monarch butterfly optimization to solve the emergency vehicle routing problem with relief materials in sudden disasters. Open Geosci. **11**(1), 391–413 (2019)
11. Funabashi, Y., Shibata, A., Negoro, S., et al.: A dynamic programming algorithm for energy-aware routing of delivery drones. In: Advances in Artificial Intelligence and Data Engineering: Select Proceedings of AIDE 2019, pp. 1217–1226. Springer, Singapore (2021)
12. Rathnakumara, G.W.H.H.P., Rupasinghe, T.D.: Clustering-based augmented tabu search for energy minimizing vehicle routing problem (EMVRP). In: SLAAI International Conference on Artificial Intelligence, p. 38. SLAAI, Colombo (2017)
13. Frías, N., Johnson, F., Valle, C.: Hybrid algorithms for energy minimizing vehicle routing problem: integrating clustering and ant colony optimization. IEEE Access **11**, 1–13 (2023)
14. Yang, F., Sun, T., Zhang, C.: An efficient hybrid data clustering method based on k-harmonic means and particle swarm optimization. Exp. Syst. Appl. **36**(6), 9847–9852 (2009)
15. Pandi, M., Sivakumar, T., Madasamy, N.S., et al.: Hybrid genetic algorithm and simulated annealing for clustering microarray gene expression data. In: Journal of Physics: Conference Series, p. 012034. IOP Publishing, Tamil Nadu (2021)
16. Kumar, A., Kumar, D., Jarial, S.: A novel hybrid k-means and artificial bee colony algorithm approach for data clustering. Dec. Sci. Lett. **7**(1), 65–76 (2018)
17. Yaghini, M., Ghazanfari, N.: Tabu-KM: a hybrid clustering algorithm based on tabu search approach. Appl. Soft Comput. **10**(3), 1023–1030 (2010)
18. Jain, A.K.: Data clustering: 50 years beyond k-means. Pattern Recogn. Lett. **31**(8), 651–666 (2010)
19. Jain, A.K., Dubes, R.C.: Algorithms for Clustering Data. Prentice-Hall, Englewood Cliffs (1988)
20. Li, J., Xin, L., Cao, Z., Lim, A., Song, W., Zhang, J.: Heterogeneous Attentions for Solving Pickup and Delivery Problem via Deep Reinforcement Learning, arXiv:cs.LG (2021)

21. Catapang, J.K., Solano, G.A.: A Floyd–Warshall-based reoptimization of Q matrix on the single DVRPPD with on-demand cancellations. In: Proceedings of the International Conference on Information (2021)
22. Ni, Q., Tang, Y.: A Bibliometric Visualized Analysis and Classification of Vehicle Routing Problem Research, Sustainability (2023)
23. Shahbazian, R., Pugliese, L., Guerriero, F., Macrina, G.: Integrating machine learning into vehicle routing problem: methods and applications. IEEE Access **4**, 785 (2024)
24. Ammon, S., Phillipson, F., Almeida, R.: A supervised machine learning approach for the vehicle routing problem. In: Proceedings of the International Conference on Operations Research (2024)
25. Abouelrous, A., Bliek, L., Gabor, A. F., Wu, Y., Zhang, Y.: Reinforcement Learning for Solving the Pricing Problem in Column Generation: Applications to Vehicle Routing, arXiv:cs.LG (2025)

Blockchain and the Future of Smart Ports

A Double Machine Learning Approach to Estimate the Operational Efficiency of Container Ports

Jose Humberto Ablanedo-Rosas(✉)

The University of Texas at El Paso, El Paso, TX 79968, USA
jablanedorosas2@utep.edu

Abstract. This research introduces a framework based on a double machine learning and bootstrap data envelopment approach to estimate the future operational efficiency of container ports when improvements to either the ports' infrastructure or the scale of operations are implemented. The double machine learning approach includes an unsupervised machine learning algorithm to classify container ports based on their infrastructure characteristics, and an artificial neural network trained to estimate bias-corrected efficiencies computed via a bootstrap data envelopment analysis. The proposed robust framework is tested using a real ports data set that includes container ports from five different worldwide developing regions. The results proved that the proposed approach was very effective in estimating the operational efficiency of container ports. Hence, the proposed tool can be highly beneficial for practitioners, port authorities, and decision makers who can be able to estimate the effect of port changes on the operational efficiency and productivity of the ports. The ports' changes can be in two different domains; the first one corresponds to the ports' infrastructure, while the second one corresponds to the scale of operations. This tool will be highly beneficial for the strategic planning process of container ports.

Keywords: Bootstrap DEA · Machine learning · Double machine learning · Container port efficiency · Developing countries

1 Introduction

Global supply chains are experiencing multiple challenges imposed by the new regulations and tariffs recently implemented. A critical component of the global supply chain is the container ports. Today, huge cargo ships travel the oceans delivering thousands of containers that are vital for the countries' economies. Countries rely on productive ports to efficiently import and export goods. Hence, their permanent goal is to build and sustain a competitive ports' infrastructure. However, the expansion and improvement of container ports are mid and long-term decisions; their impact is difficult to assess. Therefore, an appropriate framework to estimate the impact of such decisions is needed. This research fills the gap and proposes a double machine learning approach to estimate the effect that improvements in ports' infrastructure can have on the ports' productivity.

Y. Benadada et al. (Eds.): GOL 2026, LNNS 2034, pp. 227–232, 2027.
https://doi.org/10.1007/978-3-032-29677-1_20

The network of container port terminals across the globe is complex and highly heterogeneous. There are large ports that emerge as role models for container operations, while other small ports are incipient factors of development, mainly in developing economies. The latter ports have been neglected in the research literature. Some research exceptions include [1], who studied ports in the Middle East and East Africa [2], who investigated African ports [3], who researched some Caribbean ports, and, more recently [4], who benchmarked ports in different geographic regions that included several developing countries in Africa and America. These studies confirm the need for developing countries to enhance their strategic planning and be able to predict the productive boosting effect when enhancing their physical infrastructure. On the other hand, the challenge for medium and large container port terminals is to continue their continuous improvement journey to keep growing and attaining their productivity, environmental, and financial goals. Therefore, the operational improvement of container ports in developing countries will play a critical role in boosting their national economies. This research proposes a double machine learning approach to estimate the future impact on ports' productivity given feasible enhancements in their infrastructure. The proposed approach combines unsupervised machine learning in an initial stage, with bootstrap Data Envelopment Analysis (DEA) in an intermediate stage, and Artificial Neural Networks (ANN) in the final stage. The double machine learning approach is tested using the panel data set from [4], which covers five developing regions: North Africa, West Africa, South America, North and Central America, and the Caribbean. The results confirm the usefulness of the approach.

2 Methodology and Empirical Analysis

Current global supply chains are integrated by highly heterogeneous port systems, which include ports with a large variety of infrastructure characteristics and scale of operations. The data set utilized in this research project is from [4], and it encompasses 40 container ports from five developing regions: 12 ports in North Africa, 10 ports in South America, 8 ports in West Africa, 6 ports in Central and North America, and 4 ports in the Caribbean. In an initial stage, an unsupervised machine learning algorithm is utilized to create groups of ports with similar characteristics. Figure 1 depicts the proposed approach for the case under analysis. Figure 1 shows five different geographic regions with several ports in each region. The K-Means algorithm is implemented, and three clusters are determined (elbow method). The second stage corresponds to a bootstrap Data Envelopment Analysis (DEA) approach [5, 6] to estimate the operational efficiency of each container port [1–4]. The Constant Returns to Scale (CRS) assumption [7] and the original data set are utilized for the bias-corrected efficiency estimation. The final stage includes the development of an Artificial Neural Network (ANN) model to predict the operational efficiency of the container ports in each cluster. The ANN approach is particularly useful for predicting improvements in the ports' operational efficiency based on future changes in either the port's infrastructure, scale of operations, or both. This constitutes the significant contribution of this research paper.

The CRS assumption in the bootstrap DEA method includes scale and technical efficiency. The inputs used in the DEA approach are related to ports' infrastructure: the

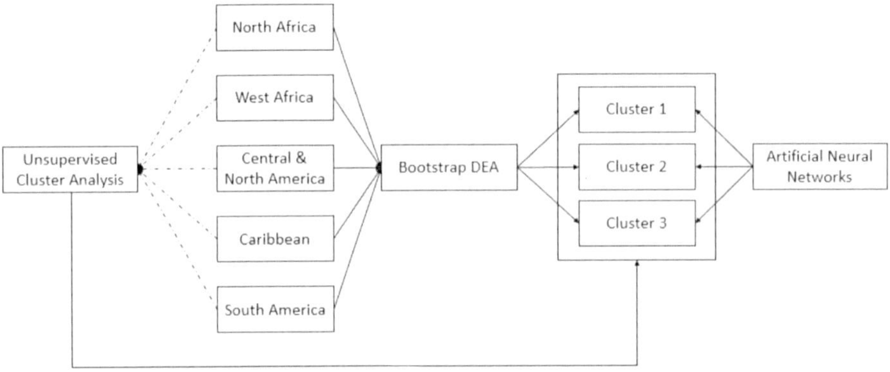

Fig. 1. Proposed double machine learning approach.

quay length, the land area, the draft length, and the number of berths. The number of transported containers and total cargo are the outputs [4]. After computing the bias-corrected CRS efficiency, three sets of container ports were defined, based on the results of the unsupervised machine learning algorithm. The ANN was then trained to estimate the bias-corrected efficiency scores based on the input and output values. A specific ANN was developed for each cluster of ports. As is common in ANN training, 80% of the original data was used for training purposes. The ANN configuration included twelve neurons in the first layer and seven neurons in the second layer. This configuration remained fixed across all the different clusters; furthermore, the same initial random seed was used across all runs. The ANN achieved a mean absolute error of $<10\%$ all cases. An immediate question that arises is whether the ANN provides an efficiency density distribution (a function that describes the distribution of the efficiency scores) that is statistically significant to the original efficiency distributions. Hence, a detailed kernel analysis was performed to compare the original efficiency distributions with the predicted density distribution.

Before comparing the original efficiency density distribution and the predicted efficiency density distribution, it is necessary to compare the efficiency distributions between clusters. Figure 2 depicts the kernel of the bootstrap CRS efficiencies per cluster. The simple inspection of the graphs suggests that Cluster 1 and Cluster 3 have similar means, whereas Cluster 2 differs. Hence, a test of the hypothesis is performed. The null hypothesis states that two given clusters have the same kernel density distribution, while the alternative hypothesis is that they differ. Table 1 shows the p-values corresponding to each compared pair of clusters. Cluster 1 and Cluster 3 have similar kernel densities, while Cluster 2 differs from Cluster 1 and Cluster 3. The initial interpretation of the results is that small and large container ports present similar kernel densities, while medium container ports present a different kernel density when compared with small and large container ports. More importantly, these results justify the training of a specific ANN for each cluster.

Figure 3 depicts the kernel densities of the bias-corrected CRS efficiencies for each cluster. It also depicts the kernel density estimated by the ANN and the corresponding confidence intervals. In all cases, the predicted kernel density falls within the confidence

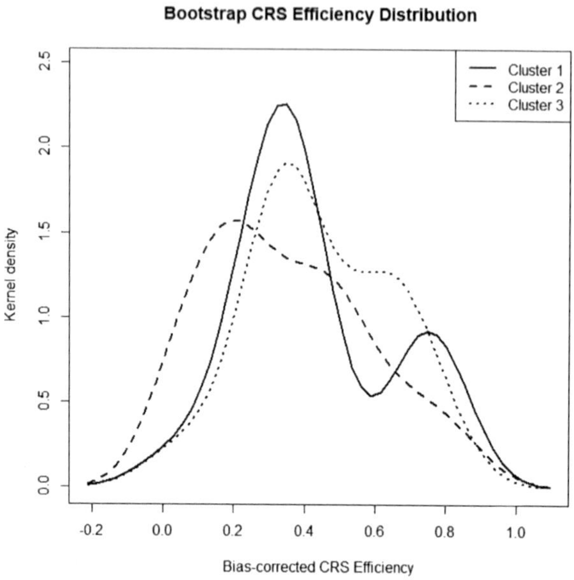

Fig. 2. Kernels of the bias-corrected CRS efficiency per cluster.

Table 1. *p*-values for the kernel test

p-values	Cluster 1	Cluster 2	Clusuter3
Cluster 1		0.05*	0.14
Cluster 2			0.00**

**p*-value ≤ 0.05, ***p*-value ≤ 0.01

interval, which shows a significant kernel density prediction performed by the ANN. Similarly, to Fig. 2, a Kernel test of equal densities is used to evaluate the statistical significance of the difference between the bias-corrected CRS efficiency density and the ANN predicted efficiency density. The null hypothesis posits that the efficiency densities are similar, while the alternative hypothesis suggests they differ. The very high *p*-values in Table 2 confirm that we failed to reject the null hypothesis. Hence, the kernel's original bias-corrected CRS efficiency densities and the ANN estimated efficiency densities are significantly similar. These major results support the usefulness of the proposed approach to estimate the efficiency of container ports. Decision makers can make inferences of the positive impact on efficiency and productivity when improving ports' infrastructure and scale of operations.

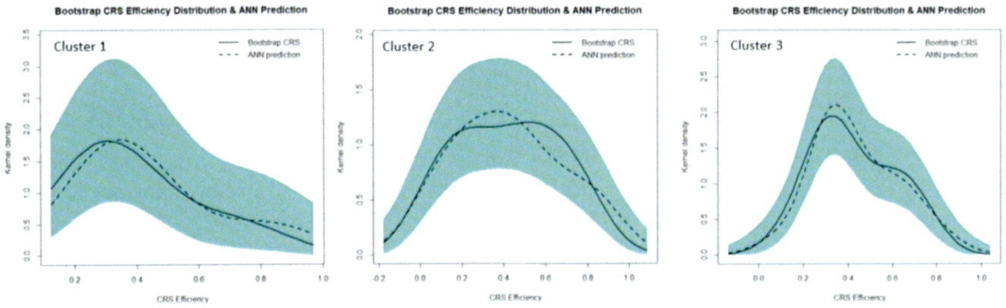

Fig. 3. Kernels of the bias-corrected and ANN estimated CRS efficiencies.

Table 2. p-values for the kernel test DEA-ANN

	Cluster 1	Cluster 2	Clusuter3
	Bootstrap DEA	Bootstrap DEA	Bootstrap DEA
ANN	0.77	0.92	0.96

3 Conclusions

Container port systems are highly heterogeneous by nature. Furthermore, they are in continuous transformation, as are the global supply chains that they support. Hence, predicting future operational efficiency based on ports' infrastructure improvements and size of operations changes is a complex challenge. This research presented a robust framework to tackle this challenge. The method based on a double machine learning approach combined with a bootstrap DEA proved to be effective when tested in a real data set containing several container ports from developing countries. Hence, the proposed method and its main findings have important implications for investors and port authorities, as the proposed approach can estimate the impact of infrastructure improvements and operational growth on the operational port efficiency. This methodology, adapted to each specific scenario, can serve as a strategic planning tool to enhance operational efficiency in container ports of developing countries. Immediate future research includes testing the approach in other container port systems and testing different configurations for the K-Means algorithm [8, 9], and the ANN.

References

1. Al-Eraqi, S.A., Khader, A.T., Mustafa, A.: DEA Malmquist index measurement in Middle East and East African container terminals. Int. J. Ship. Transp. Log. **1**(3), 249–259 (2009)
2. Suárez-Alemán, A., Trujillo, L., Cullinane, K.P.B.: Time at ports in short sea shipping: when timing is crucial. Mar. Econ. Log. **16**(4), 399–417 (2014)
3. Julien, S.A., Cowie, J., Monios, J.: Efficiency, productivity and returns to scale in ports: a comparison of data envelopment analysis and econometric estimation with application to Caribbean Small Island Developing States. Mar. Econ. Log. **22**(2), 1–26 (2018)

4. Hlali, A., Ablanedo-Rosas, J.H., Ruiz-Torres, A.J.: Operational efficiency of major cargo seaports in some developing regions: results from a bootstrap DEA. Int. J. Ship. Transp. Log. **16**(3–4), 290–319 (2023)
5. Simar, L., Wilson, P.W.: Sensitivity analysis of efficiency scores: how to bootstrap in nonparametric frontier models. Manage. Sci. **44**(1), 49–61 (1998)
6. Simar, L., Wilson, P.W.: A general methodology for bootstrapping in non-parametric frontier models. J. Appl. Stat. **27**(6), 779–802 (2000)
7. Charnes, A., Cooper, W.W., Rhodes, E.: Measuring the efficiency of decision-making units. Eur. J. Oper. Res. **2**(6), 429–444 (1978)
8. El Khattabi, M.Z., El Jai, M., Lahmadi, Y., Oughdir, L., Rahhali, M.: Understanding the interplay between metrics, normalization forms, and data distribution in K-means clustering: a comparative simulation study. Arab. J. Sci. Eng. **49**(3), 2987–3007 (2024)
9. El Khattabi, M.Z., El Jai, M., Lahmadi, Y., Oughdir, L.: Geometry-inference based clustering heuristic: new k-means metric for Gaussian data and experimental proof of concept. Oper. Res. Forum **5**(13), 1–26 (2024)

Automated Smart Contract Generation Using Multi-level Finite State Machines: A Service Agreement Case Study

Maxence Lambard(✉), Cyrille Bertelle, and Claude Duvallet

Université Le Havre Normandie, INSA Rouen Normandie, Université Rouen Normandie, Normandie Univ, LITIS UR 4108, F-76600 Le Havre, France
{maxence.lambard,cyrille.bertelle,claude.duvallet}@univ-lehavre.fr

Abstract. Traditional contracts face challenges in transparency, complexity, intermediaries, and costs. Blockchain and smart contracts offer decentralized, transparent, and tamper-proof mechanisms. However, their technical nature creates accessibility barriers for domain experts, like legal practitioners, hindering adoption. This paper introduces an approach to automatically generate smart contracts by modeling real-world agreements as multi level finite state machines. This model is generic, adaptable to any blockchain, and enables seamless on-chain tracking and execution. The methodology combines legal and technical perspectives ensuring regulatory compliance, operational efficiency, and security. A case study demonstrates the model's application on service agreements with financial elements. The results highlight the model's ability to translate clauses into formal, executable representations, advancing accessible and reliable smart contract deployment.

Keywords: Blockchain · Smart contract · Multi-level finite state machine · Code generation

1 Introduction

Traditional contract management faces challenges in transparency, administrative complexity, and high intermediary costs, often relying on trust, which can lead to disputes [4]. Blockchain and smart contracts address this by ensuring transparent, tamper-proof, and traceable automated execution without trusted third parties [3]. However, their potential is hindered by significant technical barriers, as the computational syntax of smart contracts remains difficult to understand for non-technical domain experts (e.g., legal professionals), increasing risks of errors and hindering adoption [12].

This paper proposes a new modeling of real-world contracts in the form of multi level Finite State Machine (FSM), enabling the automatic generation of smart contracts. The model is designed to be generic and adaptable to any blockchain. This modeling also allows for tracking and executing contracts directly on blockchain.

Y. Benadada et al. (Eds.): GOL 2026, LNNS 2034, pp. 233–242, 2027.
https://doi.org/10.1007/978-3-032-29677-1_21

This paper first reviews the foundations and challenges of smart contracts. It then introduces the multi-level FSM model, applies it to a concrete use case, and concludes with perspectives on future work.

2 Related work

Smart contracts, at the intersection of legal and technical challenges, are redefining contractual practices. From a technical standpoint, their automated generation relies on innovative approaches but must still overcome challenges related to security, complexity, and interoperability. First conceptualized by Szabo [1], smart contracts are computer protocols designed to automate, verify, or execute the terms of a contract. Blockchain, introduced by Satoshi Nakamoto in 2008 with Bitcoin [2], provided a decentralized and immutable infrastructure. However, Bitcoin's limited functionality led to the development of Ethereum by Buterin [11], a universal blockchain equipped with the Ethereum Virtual Machine (EVM), enabling the execution of complex smart contracts and paving the way for applications like DAOs and DeFi.

The automation of smart contract generation has gained popularity, leading to the emergence of various approaches. Mars et al. [8] propose a classification of different approaches: model-driven approaches (MD) [5,6], domain-specific languages (DSL) [7,9], and visual programming languages (VPL) [10]. The MD approach focuses on creating high-level abstract models (e.g., UML, BPMN) that are then transformed into blockchain source code [13]. For example, Caterpillar uses BPMN to generate Solidity code [14]. However, transforming models into executable code remains a challenge. DSLs offer a higher level of abstraction by using domain-specific syntax [8]. They can be classified based on their purpose or dependency [8]. Some, like DSL4SC [17], generate state diagrams, while others, like Ergo[1] focus on legal smart contracts. Finally, VPLs allow users to create programs using visual elements [8]. Blockly[2] for instance, has been used for smart contract development [15]. FSolidM [18], a framework based on FSM, provides a graphical editor for designing and generating Solidity code. Despite these advancements, challenges related to security and interoperability remain [16].

Existing technical approaches, such as Symboleo [19] and FSM synthesis [20], have made significant progress but present limitations. Symboleo models the lifecycle with statecharts but lacks real-time execution tracking and multi-level adaptation. FSM synthesis focuses on verification before deployment but doesn't offer a dynamic framework for legal obligations. Our contribution, a multi level FSM model, addresses these gaps. It enables hierarchical abstraction levels (contract, obligations, transactions) and incorporates dynamic tracking to adapt contract states in real-time based on execution events. This approach offers a more flexible and dynamic framework for managing and executing smart legal contracts.

---

[1] Accord Project, Ergo. Available at: https://accordproject.org/projects/ergo/.
[2] Blockly, Google Developers. Available at: https://developers.google.com/blockly.

3 Multi-level representation

This section introduces the Multi-Level Finite State Machine (MLFSM) model, describing its structure and central role in the generation.

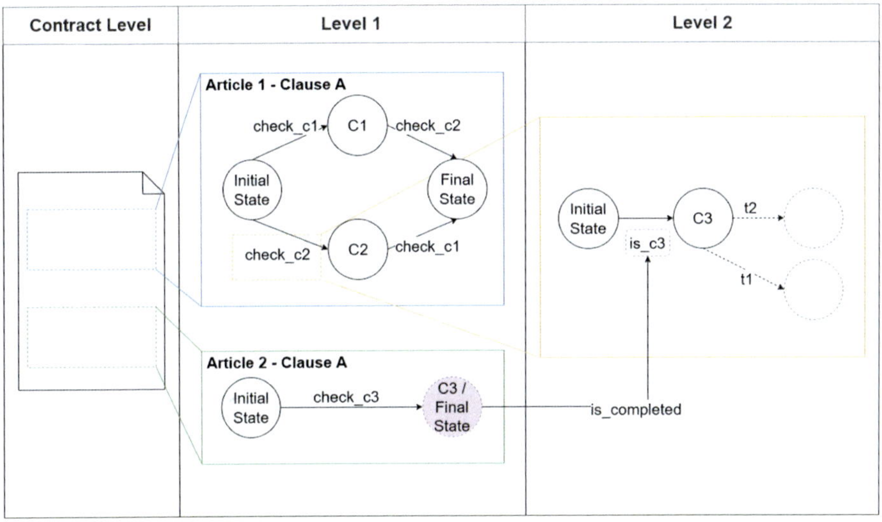

Fig. 1. Different levels of the model

As shown in Fig. 1, these FSM allow for modeling the contract clauses at different levels of granularity, providing a hierarchical view of the obligations and conditions. This approach presents several advantages:

- Global consistency: it ensures that transitions between different states respect the defined rules.
- Modularity: multi level representations allow for easy modification or adjustment of certain clauses without affecting the entire model.
- Adaptability: this structure facilitates the integration of specific requirements for each contract or blockchain platform.

Figure 1 illustrates our two-level FSM. The first level is a global FSM for validating general clause conditions. The second level focuses on execution, using a more granular FSM to model the specifics of each condition. These second-level FSMs operate interdependently. The result (e.g., state or output) of one FSM can serve as an input condition or trigger for another. This mechanism enables the creation of FSM chains to manage complex processes broken into interdependent steps.

By combining legal and technical expertise, this generic model produces reliable smart contracts that meet both legal requirements and technological imperatives for their intended use cases.

4 Use Case: The Service Agreement Contract

As part of the validation of our model, we conducted a study applied to a real-world case in order to test its feasibility and relevance in a realistic context. To do this, we collaborated within the framework of the SmartLogiLab[3] with the legal department of a service company to identify concrete case studies and evaluate our model's ability to be correctly implemented.

To this end, we based our analysis on a service agreement contract related to software, concluded between two fictional companies. This contract is structured into several articles, each subdivided into clauses and conditions. To simplify the implementation of the model, we chose to focus our study on a single article titled "Financial Elements."

4.1 Methodology

Before implementing the model, it was necessary to prepare the contract. This involved an in-depth analysis of each article to distinguish between the operational (transposable) contract clauses and the non-operational (non-transposable) clauses. Additionally, identifying the interactions between different clauses was essential to ensure the correct implementation of the model. In our study, two main clauses were identified: "Service Price" and "Billing Terms."

Table 1. Service prices and billing milestones

Services	Net amount to be received (currency unspecified)	Billing milestones
Software development and implementation	$1,000,000	– 30% on order – 25% on first deliverable – 25% on second deliverable – 20% on signature of the verification service rrovided (VSP)
Software maintenance	10% of the development and implementation cost. For this case, $100,000 per year	After signature of the VSP
Software hosting	$100,000 per year	After signature of the VSP

The first defines the service prices and associated payment milestones (see Table 1). The second outlines the applicable terms in case of payment delays, particularly the associated penalties:

[3] ANR, Smart Logistics Laboratory—SmartLogiLab, available at https://anr.fr/Projet-ANR-22-LCV2-0013.

- Must be paid within 30 calendar days from the date of invoice issuance;
- Any delay in payment will incur late penalties, set at an annual rate of 18%;
- Dispute resolution procedure in case of non-payment (not addressed in this article).

The simulation is based on the importance of validating the connection between the theoretical model and the practical requirements specified in the "Service Price" clause. This clause establishes billing milestones, which are key stages in the contract process. By associating each milestone with a distinct state of the FSM, it is possible to precisely represent the dynamics of the contract, ensure consistency with operational expectations, and verify that the FSM can correctly handle the planned transitions. This preliminary validation is crucial to ensure the applicability and relevance of the model in a real-world context. Following this validation, we proceeded to develop a proof of concept, based on the same principle, but this time applying the model rather than simulating it.

The next section focuses on the implementation of the model, which serves as a proof of concept in itself.

4.2 Implementation

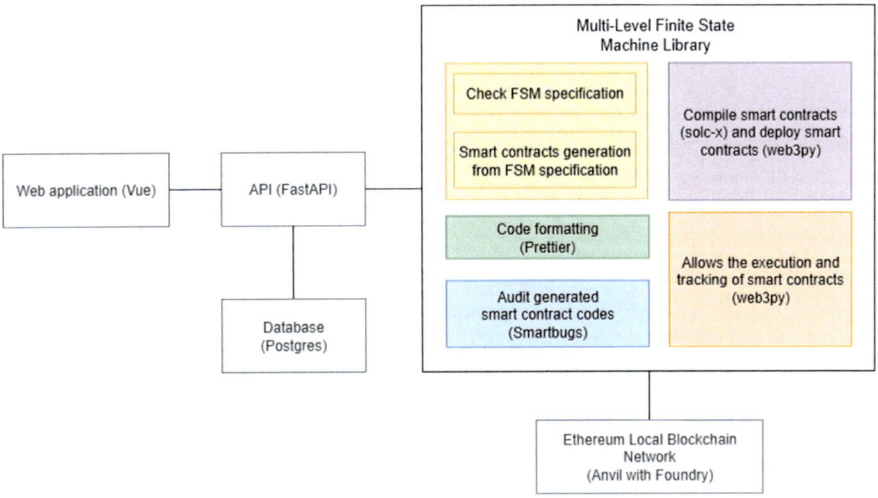

Fig. 2. Proof of concept architecture

The figure above illustrates the architecture of the proof of concept we have developed for the service contract use case. This architecture highlights the various technologies and components used for the generation and deployment of smart contracts based on multi-level finite state machines (Fig. 2).

To delve deeper into the specifics, the architecture comprises several key components:

- **Web Application**: The web application, developed with Vue.js, serves as the user interface for interacting with the system. It allows users to specify the conditions and clauses of contracts, visualize them, and also execute them.
- **API**: The API, built with the FastAPI framework, acts as an intermediary between the web application and the MLFSM library. It receives MLFSM specifications from the web application, processes them, and returns the appropriate results. FastAPI was chosen for its speed and ease of use, enabling smooth and efficient integration between the front-end and back-end.
- **Database**: The PostgreSQL database stores the FSM specifications, contract states, and other relevant information. It allows for efficient retrieval and updating of the data necessary for the generation and execution of smart contracts. PostgreSQL was selected for its robustness and ability to handle complex queries, ensuring reliable management of contractual data.
- **MLFSM Library**: The core component that handles the entire smart contract lifecycle. It validates FSM specifications; generates Solidity code; compiles and deploys the smart contracts; formats the code; audits for vulnerabilities using Smartbugs [21]; and manages the execution and real-time state tracking on the blockchain.
- **Ethereum Local Blockchain Network**: Anvil is used to simulate a local Ethereum network, enabling the deployment and testing of smart contracts in a controlled environment before their deployment on a main network. Anvil, part of the Foundry ecosystem, provides a fast and flexible local development environment, ideal for testing and debugging smart contracts.

A first JSON configuration file allows to implement the manual transposition of the contractual clauses. This file plays a central role in modeling the process, as it gathers the necessary information to define and execute the FSM corresponding to the various identified services. It includes the FSM associated with the three main services, as well as the one related to the Verification of Regular Service (VSP). These FSM translate the contractual specifications into a formal model, ready to be used in the implementation of the smart contract.

The configuration file is structured in a clear and organized manner, facilitating its use for subsequent stages. It relies on the use of several functions defined within the packages using a second JSON files, such as functions for state transitions, milestone validation, and managing dependencies between FSM. These functions formalize the interactions and specific constraints associated with each service.

Each service is represented in the configuration file as dedicated sections, detailing the associated milestones, the states defined in the FSM, and any potential relationships with other services. For example, for the service "Software development and implementation", the file specifies the key milestones (such as deliverables or billing stages) and their logical sequencing within the corresponding FSM. This level of detail ensures a rigorous representation of the contractual clauses and facilitates their use in subsequent stages of validation and deployment.

4.3 Human Machine Interface

Our platform/application empowers users to create, monitor, and execute contracts on the blockchain seamlessly. In the following sections, we will delve into the various functionalities of our Human-Machine Interface, showcasing its capabilities and user-friendly design.

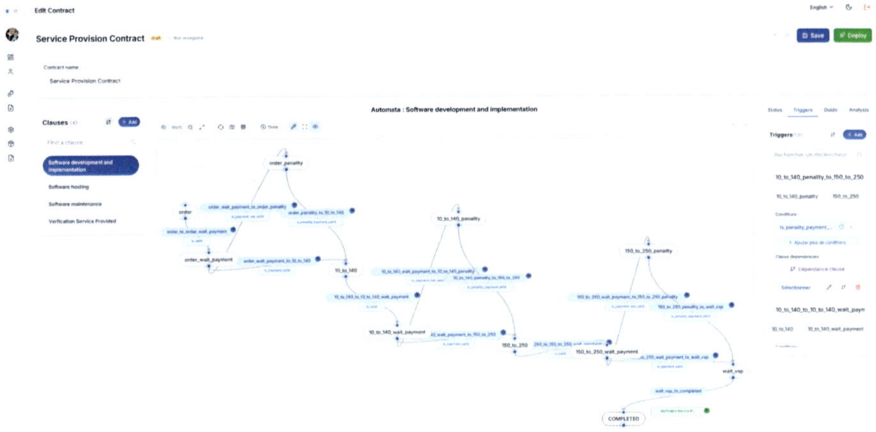

Fig. 3. Transposition of the contract to MLFSM

Figure 3 illustrates the contract creation interface for service provision, demonstrating the transposition of contract clauses into a Multi-Level Finite State Machine (MLFSM). This interface allows users to meticulously define the states, transitions, and conditions of various clauses within the contract. One of the standout features of this interface is its ability to notify users if a particular clause is not feasible. For instance, it can detect circular dependencies between clauses, ensuring that the contract logic remains sound and executable.

Figure 4 presents the contract execution interface, which becomes accessible once the contract clauses have been deployed on the blockchain. This interface is equipped with a variety of functions that facilitate the management and execution of the contract. Users can verify the completion status of clauses using functions such as is_completed. Additionally, the interface allows for the evolution of the contract through the use of triggers. For example, the trigger "order_to_order_wait_payment" can transition the contract from the "order" state to the "order_wait_payment" state. It is important to note that these triggers are governed by conditions defined during the contract creation phase. The interface also provides a comprehensive view of the overall contract progression, highlighting which clauses are pending.

In summary, our platform provides a robust and intuitive interface for creating, executing, and monitoring contracts on the blockchain. The detailed visu-

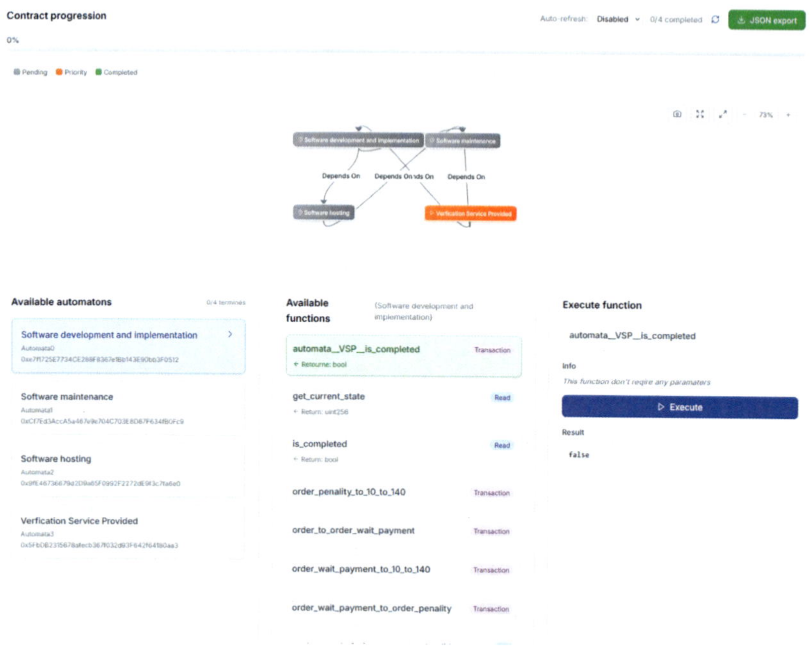

Fig. 4. Execution of the contract after deployment

alizations and user-friendly design ensure that users can manage their contracts with ease and precision.

5 Conclusion and Perspectives

This article presents an innovative approach for the automatic generation of smart contracts by modeling real-world agreements as multi level FSM. The proposed model, which combines legal and technical expertise, addresses a major challenge in the realm of smart contracts: transposing traditional contracts into an automated format while ensuring security, execution, and compliance. By offering a modular and flexible structure that is both technically and legally robust, this model overcomes the limitations of traditional smart contract generation approaches, which can sometimes face obstacles related to clause interpretation or integrating legal requirements into a decentralized environment. By merging legal rigor and technical formalism, this model meets the demands of contracts while simplifying their management and execution.

The case study presented in this article demonstrates the feasibility and effectiveness of this approach. It highlights the model's ability to precisely transpose complex contractual elements into a blockchain-applicable format. Through the validation of key clauses, such as those related to financial conditions and payment terms, the case study tested and refined the model, proving its capacity to simplify contract execution. The automated process not only accelerates the management of contractual transactions but also ensures that obligations and defined milestones are met, thus enhancing transparency and trust among stakeholders. Moreover, the integration of multi level FSM demonstrated that compliance with regulatory requirements can be guaranteed while offering flexibility to adapt contracts to different contexts.

These developments would not only enhance the flexibility and robustness of the model but also promote the widespread adoption of smart contracts in diverse legal and economic environments. Ultimately, the widespread use of this approach could lead to a transformation in contractual practices by facilitating contract management, reducing the risk of errors, and automating administrative tasks, all while ensuring legal compliance and contractual obligations. In the long run, this model is part of a future where smart contracts could become the norm, offering greater security and transparency in transactions while simplifying their management.

Acknowledgements. This work was funded by the Region of Normandy, Le Havre Seine Métropole, and the Agence National de la Recherche (ANR) as part of the SmartLogiLab project in the labcom program (ANR-22-LCV2-0013) which is a collaboration between SOGET and LITIS. Their contribution played a crucial role in the success of this study.

References

1. Szabo, N.: Smart Contracts: Building Blocks for Digital Markets (1996). http://www.truevaluemetrics.org/DBpdfs/BlockChain/Nick-Szabo-Smart-Contracts-Building-Blocks-for-Digital-Markets-1996-14591.pdf
2. Nakamoto, S.: Bitcoin: A Peer-to-Peer Electronic Cash System (2008). https://bitcoinwhitepaper.co/bitcoin.pdf
3. Prorok, J.: L'automatisation sur la blockchain des sanctions de l'inexécution du contrat. In: Structure et Usage de la Blockchain, pp. 105–116. Dalloz (2024)
4. Jaouen, M.: Le traitement de l'inexécution peut-il être totalement automatisé? In: Structure et Usage de la Blockchain, pp. 117–128. Dalloz (2024)
5. Pintado, O.L.: Caterpillar: A Blockchain-Based Business Process Management System (2017)
6. Frantz, R.Z., Corchuelo, R., Basto-Fernandes, V., Rosa-Sequeira, F., Roos-Frantz, F., Arjona, J.L.: A cloud-based integration platform for enterprise application integration: a model-driven engineering approach. Softw. Pract. Experi. **51**(4), 824–847 (2021). https://doi.org/10.1002/spe.2916
7. Dornelles, E., Parahyba, F., Frantz, R.Z., Roos-Frantz, F., Quintero, A.R., Molina-Jiménez, C., Bocanegra, J., Sawicki, S.: Advances in a DSL to Specify Smart Contracts for Application Integration Processes (2022). https://doi.org/10.5753/cibse.2022.20962
8. Mars, R., Cheikhrouhou, S., Kallel, S., Hadj Kacem, A.: A survey on automation approaches of smart contract generation. J. Supercomput. **79**(14), 16065–16097 (2023). https://doi.org/10.1007/s11227-023-05262-8

9. Mavridou, A., Laszka, A.: Designing Secure Ethereum Smart Contracts: A Finite State Machine Based Approach (2017). arXiv preprint arXiv:1711.09327. https://doi.org/10.48550/ARXIV.1711.09327
10. Mavridou, A., Laszka, A., Stachtiari, E., Dubey, A.: VeriSolid: Correct-by-Design Smart Contracts for Ethereum (2019). arXiv preprint arXiv:1901.01292. https://doi.org/10.48550/ARXIV.1901.01292
11. Buterin, V.: Ethereum: A Next-Generation Smart Contract and Decentralized Application Platform (2015). https://ethereum.org/en/whitepaper/
12. Frantz, C.K., Nowostawski, M.: From institutions to code: towards automated generation of smart contracts. In: Proceedings of the 2016 IEEE 1st International Workshops on Foundations and Applications of Self* Systems (FAS*W), pp. 210–215. IEEE (2016). https://doi.org/10.1109/FAS-W.2016.53
13. Amaral De Sousa, V., Burnay, C., Snoeck, M.: B-MERODE: a model-driven engineering and artifact-centric approach to generate blockchain-based information systems. In: Dustdar, S., Yu, E., Salinesi, C., Rieu, D., Pant, V. (eds.) Advanced Information Systems Engineering, pp. 117–133. Springer International Publishing (2020)
14. López-Pintado, O., García-Bañuelos, L., Dumas, M., Weber, I., Ponomarev, A.: CATERPILLAR: A Business Process Execution Engine on the Ethereum Blockchain (2018). arXiv preprint arXiv:1808.03517. https://doi.org/10.48550/ARXIV.1808.03517
15. Guida, L., Daniel, F.: Supporting reuse of smart contracts through service orientation and assisted development. In: Proceedings of the 2019 IEEE International Conference on Decentralized Applications and Infrastructures (DAPPCON), pp. 59–68. IEEE (2019)
16. Zupan, N., Kasinathan, P., Cuellar, J., Sauer, M.: Secure smart contract generation based on petri nets. In: Rosa Righi, R.D., Alberti, A.M., Singh, M. (eds.) Blockchain Technology for Industry 4.0, pp. 73–98. Springer (2020)
17. Tateishi, T., Yoshihama, S., Sato, N., Saito, S.: Automatic smart contract generation using controlled natural language and template. IBM J. Res. Develop. **63**(2), 6:1-6:12 (2019)
18. Mavridou, A., Laszka, A.: Tool Demonstration: FSolidM for Designing Secure Ethereum Smart Contracts (2018). arXiv preprint arXiv:1802.09949
19. Sharifi, S., Parvizimosaed, A., Amyot, D., Logrippo, L., Mylopoulos, J.: Symboleo: towards a specification language for legal contracts. In: Proceedings of the 2020 IEEE 28th International Requirements Engineering Conference (RE), pp. 364–369. IEEE (2020)
20. Suvorov, D., Ulyantsev, V.: Smart Contract Design Meets State Machine Synthesis: Case Studies (2019). arXiv preprint arXiv:1906.02906. https://doi.org/10.48550/ARXIV.1906.02906
21. di Angelo, M., Durieux, T., Ferreira, J.F., Salzer, G.: SmartBugs 2.0: an execution framework for weakness detection in ethereum smart contracts. In: Proceedings of the 38th IEEE/ACM International Conference on Automated Software Engineering (ASE), pp. 2102–2105. IEEE Computer Society (2023)

Enhancing Maritime Logistics with Blockchain: Securing and Tracing Dangerous Goods in Smart Ports

Rim Abdallah[1(✉)], Cyrille Bertelle[1], Claude Duvallet[1], Jérôme Besancenot[2], and Frédéric Gilletta[2]

[1] University of Le Havre, Le Havre 76600, France
rim.abdallah@etu.univ-lehavre.fr,
{cyrille.bertelle,claude.duvallet}@univ-lehavre.fr

[2] HAROPA PORT, 71 Quai Colbert, 76600 Le Havre, France
{jerome.besancenot,frederic.gilletta}@haropaport.com

Abstract. This paper investigates how permissioned blockchain can enhance traceability, integrity, and secure multi-party data exchange in maritime logistics, with a particular focus on the governance of dangerous goods. The research adopts a design and implementation methodology grounded in a real operational context, integrating a permissioned blockchain layer within an existing Port Community System (PCS). A proof of concept coupling TIMAD, a PCS component dedicated to dangerous-goods management, with Hyperledger Fabric is implemented to notarize safety-critical events, enforce role-based access control, and provide tamper-evident auditability without disrupting legacy workflows. The approach is validated through functional and performance testing in a multi-organization deployment, demonstrating regulatory traceability, interoperability, and acceptable operational latency. The results show that permissioned blockchain can effectively function as a complementary trust layer for existing port information systems, while laying the foundations for more automated and interoperable smart port ecosystems.

Keywords: Smart ports · Blockchain · Traceability · Dangerous goods · Hyperledger fabric · Port community systems · Maritime logistics

1 Introduction

Maritime logistics underpins roughly ninety percent of world trade [1]. The sector's increasing operational scale and complexity from vessel mega-sizing to just-in-time flows has amplified the need for high-fidelity traceability across heterogeneous information systems. Incidents such as the 2021 Suez Canal blockage [2] and large-scale industrial accidents involving hazardous cargo emphasize that fragmented visibility, latent data latency, and unclear accountability chains carry systemic risk.

Y. Benadada et al. (Eds.): GOL 2026, LNNS 2034, pp. 243–252, 2027.
https://doi.org/10.1007/978-3-032-29677-1_22

Ports have long invested in digitization, moving from paper manifests to EDI and, subsequently, to Port Community Systems/Cargo Community System that coordinate actors across customs, terminals, carriers, and authorities. Yet, centralization brings its own issues: cross-organization trust is mediated by a single operator; audit trails may be incomplete or mutable; and inter-port interoperability remains uneven. For sensitive processes like dangerous goods declaration, handling, and transfer authorization, stakeholders require both *assurance* (immutability, provenance, non-repudiation) and *continuity* (seamless fit with existing workflows and regulatory controls).

We examine permissioned blockchain as a complement to the Port Community System layer, focusing on integration patterns that (i) minimize disruption, (ii) preserve domain governance, and (iii) provide verifiable traces for safety-critical events. The work is organized around five research questions (opportunities, challenges, feasibility, adoption prospects, and integration methods) and is validated via a Hyperledger Fabric proof of concept (PoC) that augments TIMAD with notarized event logging and privacy-aware querying.

Contributions and research objectives This paper provides: (C1) a requirements-to-features mapping aligning dangerous goods traceability needs with permissioned-ledger capabilities; (C2) an integration architecture that augments a production-grade Port Community System (TIMAD) without replacing existing systems of record; (C3) a Hyperledger Fabric PoC with smart-contract logic for dangerous goods events and role-based access tied to port governance; and (C4) an outlook on scalability and federation, including sharding and inter-port channels. The objective of this research is to investigate whether and how a permissioned blockchain can be integrated into an existing Port Community System to enhance traceability, security, and inter-organizational trust in dangerous goods management. The study aims to design an integration architecture, implement a proof of concept in a real port environment, and evaluate its technical feasibility and operational relevance.

2 Background: From EDI to Smart Ports and the Role of Traceability

2.1 From EDI to Community Systems

The maritime information landscape evolved from bilateral EDI links toward community platforms that implement a "single window" for port stakeholders. Port Community Systems and Cargo Community System consolidate messaging, reference data, and process orchestration across public authorities (port, customs, safety) and private operators (terminals, shipping lines, forwarders). While implementations vary by governance model and national regulation, the architectural intent is consistent: reduce fragmentation, standardize exchanges, and increase operational predictability.

Despite these gains, three persistent limitations motivate exploring a cryptographically assured complement:

- **Trust Boundary Asymmetries**. A central operator mediates data custody and access. Cross-jurisdictional trust or inter-port federation can be hindered by differing operator mandates.
- **Auditability Constraints**. Traditional databases support change logs but are not append-only ledgers by design. Ensuring tamper-evident, end-to-end provenance across organizations is non-trivial.
- **Interoperability at Scale**. Community systems have matured within ports; inter-port, inter-vendor, and inter-nation scaling remains uneven, complicating cross-border traceability.

2.2 Traceability as a Safety and Compliance Primitive

Traceability couples *physical* and *digital* identities across events: declarations, transfers of custody, storage, and movement. Standards (e.g., ISO 6346 for container IDs [3]) and regulatory frameworks (e.g., IMDG, SOLAS [4]) provide the scaffolding for data semantics and safety rules. In the dangerous goods context, traceability serves four functions:

1. **Hazard Visibility**. Ensuring relevant authorities and operators can access accurate, timely data about cargo class, quantity, and location.
2. **Process Control**. Validating pre-arrival notifications, storage constraints, and transfer permissions against policy.
3. **Accountability**. Producing non-repudiable sequences of who did what, when, and under which authorization.
4. **Incident Response**. Speeding forensic reconstruction and targeted interventions when deviations or emergencies occur.

High-impact failures demonstrate that gaps in data availability, integrity, or governance exacerbate risk [5,6]. Conversely, ports with mature traceability and stakeholder coordination report improved compliance, reduced congestion, and better emergency preparedness [12].

2.3 Positioning Blockchain as a Complementary Layer

Rather than replacing existing port information systems, permissioned blockchain is positioned in this study as a complementary trust layer that enhances traceability, governance, and inter-organizational verifiability while preserving the operational role of legacy systems. This architectural choice reflects the reality of port ecosystems, where Port Community Systems (PCS), Terminal Operating Systems (TOS), and customs platforms remain the primary systems of record and cannot be disrupted without significant operational risk.

Figure 1 illustrates the proposed integration model, in which the blockchain layer operates as an augmented ledger backplane interfacing with existing operational systems through standardized APIs. In this architecture, core port applications continue to manage business processes and data storage, while the blockchain layer anchors safety-critical events, enforces shared governance rules, and provides tamper-evident auditability across stakeholders.

More specifically, the blockchain layer fulfills three complementary functions. First, it enables tamper-evident provenance by recording immutable traces of dangerous goods events such as manifest creation, custody transfers, and storage validations. Second, it supports a shared yet access-controlled representation of port data, allowing different stakeholders to access a common ledger while preserving confidentiality through channel-based segmentation and role-based permissions. Third, it introduces programmable governance through smart contracts, which automate policy enforcement and validation logic without altering existing workflows.

By decoupling operational processing from trust and verification mechanisms, this layered architecture allows blockchain to enhance the reliability and transparency of maritime information exchanges while maintaining the continuity and stability of established port systems.

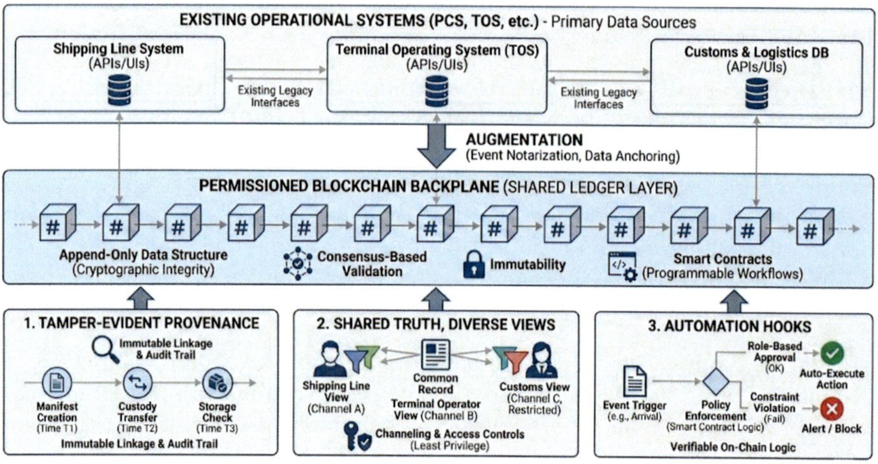

Fig. 1. Permissioned blockchain as an augmented "ledger backplane" for port operations, enhancing cross-party verifiability without replacing existing operational systems.

3 Research Framework and Scope

We formalize five research questions (RQ1–RQ5) covering (i) lessons learned from prior maritime blockchain attempts, (ii) use-case mapping and value streams, (iii) challenge taxonomy (technical, organizational, regulatory), (iv) adoption likelihood and barriers, and (v) a concrete integration method with a working PoC. The applied scope centers on HAROPA PORT (Le Havre) and TIMAD for dangerous goods processes, developed under a CIFRE collaboration bridging industry and academia.

3.1 Research Methodology

This study follows a design-science research methodology applied to a real industrial context. The research process consists of four steps: (1) analysis of maritime traceability requirements and existing PCS architecture, (2) design of a blockchain-based integration model, (3) implementation of a proof of concept using Hyperledger Fabric and TIMAD, and (4) functional and performance evaluation of the proposed solution. This approach ensures both technical validation and practical relevance for maritime stakeholders.

4 Blockchain for Maritime Logistics: Features Versus Requirements

The integration of blockchain into maritime logistics must align technological affordances with operational and governance realities. Building upon prior analyses of sectoral readiness and blockchain's typological diversity [13], this section maps the intrinsic properties of permissioned ledgers to the requirements of modern port ecosystems.

4.1 Sectoral Requirements

From a systems-engineering perspective, the maritime domain imposes four main requirements on any technological innovation:

1. **Interoperability and Data Standardization**. Community Systems rely on ISO/UN/CEFACT-based message formats and must ensure compatibility across customs, terminals, and regulatory APIs.
2. **Regulatory Traceability and Accountability**. Every operation involving hazardous materials must be reproducible for auditing and post-incident analysis under IMO and SOLAS mandates.
3. **Operational Continuity**. Ports operate on 24/7 cycles; any new component must integrate without introducing downtime or critical latency.
4. **Data Confidentiality and Governance**. Given the mix of public and private stakeholders, selective data disclosure and access control are essential.

These requirements guided the design decisions of the proposed blockchain layer, particularly the choice of a permissioned architecture.

4.2 Blockchain Capabilities and Alignment

Permissioned blockchain frameworks such as Hyperledger Fabric integrate modular architecture, distributed consensus, and identity-based governance through Membership Service Providers (MSPs), enabling immutable traceability, automated compliance via smart contracts, decentralized trust among stakeholders, and fine-grained access control.

4.3 Gap Analysis

The cross-analysis conducted in [13] highlighted partial overlaps between blockchain affordances and maritime needs. Table 1 summarizes the correspondence between key functional requirements—traceability, auditability, security, and interoperability—and the blockchain mechanisms that address them. The analysis showed that while immutability and notarization effectively ensure traceability and accountability, interoperability remains an open challenge that demands standardized data models and robust off-chain integration gateways.

Table 1. Mapping between functional requirements and blockchain mechanisms.

Requirement	Supporting blockchain feature
Traceability	Immutable ledger, event notarization
Auditability	Cryptographic provenance, consensus validation
Security	Role-based access control, TLS, MSP identity
Interoperability	API gateways, standard data schemas (UN/CEFACT, ISO)

4.4 Strategic Implications

Rather than introducing a disruptive replacement, the approach favors layered augmentation, whereby blockchain operates as a verifiable backbone beneath existing Port Community Systems, ensuring integrity of inter-organizational transactions, tamper-evident auditability for regulatory reporting, and programmable interfaces for secure automation. This positioning avoids redundancy, preserves stakeholder trust, and aligns with incremental digital-transformation roadmaps in the port sector [7].

5 Implementation and Proof of Concept: TIMAD + Hyperledger Fabric

Building upon the requirements–features alignment, a Proof of Concept (PoC) was implemented to integrate blockchain with the existing TIMAD platform at HAROPA PORT Le Havre. The objective was to validate the feasibility of embedding a permissioned ledger as a complementary traceability layer within an operational Port Community System.

5.1 Architecture Overview

The system was implemented on Hyperledger Fabric (v2.x), where maritime stakeholders operated organizational nodes and certificate authorities, an ordering service ensured consensus, and dangerous-goods events were recorded as immutable, X.509-signed transactions guaranteeing non-repudiation and traceability.

5.2 Smart-Contract Design

Two core smart contracts were implemented:

1. **Manifest SC**: handles submission, validation, and querying of hazardous-goods manifests.
2. **BAT SC**: manages Brokered Authoritative Transitions (BAT), governing access rights and trace logs.

Both contracts were written in Go, compiled, and deployed onto the Fabric chaincode environment. They exposed RESTful interfaces accessible through TIMAD's existing API gateway, maintaining continuity with legacy user interactions.

5.3 Integration with TIMAD

Integration followed an API-bridge model. TIMAD acted as the client application, while the blockchain operated as a backend micro-service. The key steps were:

1. Authentication via TIMAD's JWT tokens mapped to Fabric identities.
2. Submission of manifest data through POST requests, triggering smart-contract transactions.
3. Retrieval of notarized records through GET requests for audit or visualization.

This setup ensured transparent ledger operations without altering TIMAD's core database or front-end logic.

5.4 Network Deployment and Validation

The network was containerized using Docker Compose, with four peer organizations distributed across separate machines. Tests validated:

- successful peer-to-peer communication and block propagation;
- secure transaction endorsement and ordering;
- accurate query responses from the ledger;
- and resilience under concurrent write operations.

Results confirmed full functional interoperability between TIMAD and Hyperledger Fabric, demonstrating that blockchain can be embedded into live port-community infrastructure with minimal friction.

5.5 Security and Governance Considerations

Following recommendations from [10], TLS encryption and certificate rotation were enforced across all nodes. Role-based access policies limited endorsement rights to specific authority peers, while auditors operated read-only nodes for compliance verification.

6 Results, Outlook, and Conclusion

The proof of concept validated both the technical feasibility and strategic potential of embedding a permissioned blockchain layer within a live port environment. Results confirmed end-to-end interoperability between TIMAD and Hyperledger Fabric, with secure, auditable data exchange achieved without altering the operational logic of existing Port Community Systems. This outcome demonstrates that blockchain can be seamlessly anchored into port infrastructures as a verifiable transaction backbone, transparent to end users, yet fully traceable for authorities.

6.1 Results and Discussion

The experimental evaluation reproduced the operational cycle of dangerous-goods management within HAROPA PORT. Blockchain transactions successfully recorded the sequence of declarations, approvals, and custody transfers. Each operation produced a corresponding block event, ensuring a continuous, immutable audit trail across multiple organizations. Tests verified:

- transaction latency remaining within acceptable operational thresholds (<1 s for endorsement, <3 s for block commitment);
- resilience under concurrent manifest submissions by different peers;
- accurate retrieval of historical records through chain queries; and
- secure role segregation preventing unauthorized access to confidential cargo data.

The proposed architecture was validated through functional tests, performance measurements, and multi-organization deployment scenarios. The evaluation focused on transaction latency, interoperability with TIMAD, role-based access enforcement, and auditability of dangerous goods events. These criteria were selected to reflect operational constraints in port environments.

Beyond raw performance, the PoC highlighted blockchain's capacity to act as a programmable trust engine and added layer of trust. By encapsulating business logic within smart contracts, the system automated validation workflows traditionally performed through manual coordination or point-to-point data exchange. This finding supports prior conclusions in [11] that the value of blockchain in maritime logistics extends beyond notarization toward the orchestration of distributed decision-making.

6.2 From Complementary Layer to Automation Core

Although the initial objective was to integrate blockchain without intervening on existing Port Community Systems, the results reveal a broader architectural potential. Blockchain emerges as a core automation and security layer around which port information systems can be progressively reconfigured, with Port Community Systems and legacy databases interfacing with a shared, tamper-evident ledger governing permissions, data provenance, and event-driven interactions.

This perspective aligns with broader trends in smart port digitalization and distributed trust architectures, suggesting an evolution toward federated, blockchain-centric ecosystems that unify governance, traceability, and interoperability within a common technological framework. In this paradigm, blockchain transcends the role of a supplementary component and becomes a coordination kernel—or trust middleware—capable of automating regulatory controls, synchronizing multi-actor workflows, supporting AI-driven analytics through secure data provenance, and enabling interoperable digital corridors across ports.

6.3 Future Directions

Three research directions emerge from this architectural shift: enhancing scalability and federation through sharding and inter-chain communication, strengthening standardization and interoperability via cross-chain data schemas aligned with UN/CEFACT and IMO standards, and advancing governance automation through smart-contract templates that formalize port regulations and safety procedures. These directions echo ongoing discussions in [8] while extending them by positioning blockchain as a foundational automation kernel within port digital ecosystems.

6.4 Conclusion

This work demonstrates that permissioned blockchain technology, when carefully aligned with maritime operational and regulatory requirements, can enhance traceability, accountability, and process automation without disrupting existing systems. The TIMAD + Hyperledger Fabric prototype shows that blockchain can function as a transparent trust backbone capable of harmonizing data exchanges among diverse stakeholders.

Beyond technical feasibility, the broader implication is architectural: blockchain can serve as the core automation and security nucleus around which next-generation port information systems are re-architected. Such a nucleus supports secure interoperability, verifiable governance, and adaptive innovation—laying the groundwork for a resilient, connected, and intelligent maritime logistics ecosystem.

Acknowledgments. This work was conducted within the framework of a CIFRE collaboration between Université Le Havre Normandie and HAROPA PORT. The authors thank the LITIS Laboratory for its academic guidance and HAROPA PORT's digital division for operational insights and support during the implementation phase.

References

1. United Nations Conference on Trade and Development (UNCTAD): Review of Maritime Transport 2023: Towards a Green and Just Transition. United Nations Publication, Geneva (2023). https://doi.org/10.18356/9789213584569
2. Gast, J., Binsfeld, T., Marsili, F., Jahn, C.: Analysis of the Suez Canal blockage with queueing theory. In: European Commission Joint Research Centre Working Papers in Transport and Mobility (2021). https://www.econstor.eu/bitstream/10419/249644/1/hicl-2021-31-943.pdf
3. International Organization for Standardization (ISO): ISO 6346:2022—Freight Containers—Coding, Identification and Marking. ISO (2022). https://www.iso.org/standard/83558.html
4. International Maritime Organization (IMO): International Maritime Dangerous Goods (IMDG) Code. IMO (2024). https://www.imo.org/en/ourwork/safety/pages/dangerousgoods-default.aspx
5. Organisation for Economic Co-operation and Development (OECD): Management of Hazardous Substances in Port Areas. OECD Publishing (2024). https://doi.org/10.1787/5d09ac8a-en
6. Duan, L., Peng, W., Wang, R.: Causation analysis and governance strategy for hazardous cargo accidents at ports: case study of Tianjin Port's hazardous goods explosion accident. J. Loss Prevent. Process Ind. **60**, 58–69 (2019). https://doi.org/10.1016/j.jlp.2019.02.005
7. Fundación Valenciaport: Smart Ports Manual: Strategy and Roadmap. Lucenti, K. (Ed.) (2020). https://doi.org/10.18235/0002384
8. Guan, P.: Blockchain adoption in the port industry: a systematic literature review. Cogent Bus. Manag. **11**(1), 2431650 (2024). https://doi.org/10.1080/23311975.2024.2431650
9. Abdallah, R.: Enhancing Maritime Logistics with Blockchain Technology: Application to Secure and Trace Dangerous Goods in Smart Ports. Doctoral Dissertation, Université Le Havre Normandie, France (2024). https://theses.hal.science/tel-04652638
10. Lesavre, L., Varin, P., Mell, P., Davidson, M., Shook, J.: A taxonomic approach to understanding emerging blockchain identity management systems. In: NIST Computer Security White Paper (CSWP 0114-2020). National Institute of Standards and Technology (2020). https://doi.org/10.6028/NIST.CSWP.01142020
11. Alqarni, M.A., Mandurah, T., Zeitoun, A., Derhab, A.: Use of blockchain-based smart contracts in logistics and supply-chain management. Electronics **12**(6), 1340 (2023). https://doi.org/10.3390/electronics12061340
12. Tseng, P.-H., Yeh, C.-C.: Maintaining and researching port safety: a case study of port safety in Kaohsiung port. Taiwan. Eur. Transp. Res. Rev. **9**(31), 250 (2017). https://doi.org/10.1007/s12544-017-0250-z
13. Abdallah, R., Besancenot, J., Bertelle, C., Duvallet, C., Gilletta, F.: An extensive preliminary blockchain survey from a maritime perspective. Smart Cit. **6**(2), 846–877 (2023). https://doi.org/10.3390/smartcities6020041

Simulation of Miner Pool Behavior in Blockchains Using a Multi-agent System

Aicha Ferjani(✉), Niels Kerne, Claude Duvallet, and Cyrille Bertelle

LITIS (UR4108) Université Le Havre Normandie, 76600 Le Havre, France
{aicha.ferjani,niels.kerne,claude.duvallet, cyrille.bertelle}@univ-lehavre.fr

Abstract. Blockchain systems are inherently complex and decentralized, making them challenging to analyze using traditional analytical approaches. Among the key components influencing blockchain dynamics is miner behavior, particularly their strategic decisions regarding participation in mining pools. These choices directly affect the performance, security, and decentralization of proof of work blockchains. In this paper, we present a simulation framework based on a multi-agent modelling approach, where miners are represented as autonomous agents capable of making decisions in response to economic incentives. To capture the strategic nature of pool selection, the model integrates game-theoretic mechanisms that govern agent interactions and decision-making processes. The simulation is implemented using AnyLogic software, a platform that supports agent-based modelling and is widely used for simulating complex systems. Through this framework, we analyze how individual miner strategies aggregate to influence overall blockchain behavior, with particular attention to centralization risks and the distribution of computational power. The results demonstrate that miner incentives and pool selection algorithms play a critical role in shaping the resilience and decentralization of the network. This work contributes to a deeper understanding of the emergent properties of blockchain systems and highlights the value of agent-based simulation in studying strategic behaviors within decentralized environments.

Keywords: Blockchain · Simulation · Miners · Pools · Modelling · Multi-agents

1 Introduction

Born out of the need for a decentralized and transparent digital currency with the advent of Bitcoin in 2008 [1], blockchain quickly established itself as a major technological innovation. Its fundamental principle is based on the establishment of a peer-to-peer network enabling the sharing of a public, sequential and unalterable transaction ledger. Each new block of data is cryptographically linked to the previous one, thus forming a secure chain that can be audited by all participants. This intrinsic structure eliminates the need for a central authority to validate and record exchanges, establishing distributed trust within the network.

Y. Benadada et al. (Eds.): GOL 2026, LNNS 2034, pp. 253–261, 2027.
https://doi.org/10.1007/978-3-032-29677-1_23

Since its initial design, the blockchain ecosystem has expanded considerably, giving rise to a multitude of platforms and protocols. These developments are reflected in the diversification of consensus algorithms (proof of work, proof of stake, etc.), permission models (public, private, consortiums) and the functionalities offered. Blockchains such as Ethereum have introduced the concept of smart contracts [2], paving the way for decentralized applications (Dapps) and new forms of organization.

At the heart of a blockchain system lies the ability of its participants to collaborate in building a distributed ecosystem with economic, social and technological dimensions. Transactions are initiated by users and then propagated across the network, where they are validated by validator nodes (also known as miners) according to the rules of the protocol. This validation process, often incentivized by reward mechanisms (such as mining in some blockchains), guarantees the integrity and durability of the ledger. Notably, participants interact and carry out verified transactions without the need for complete interpersonal trust or a central trusted entity [3]. Their knowledge of the system is often fragmented, limited to the information necessary for their own interactions, and the topology of the network evolves dynamically as peer connections change. This dynamic is reminiscent of co-evolving knowledge networks, where individual understanding of the blockchain and the structure of the network influence each other over time.

A fundamental characteristic of blockchain systems is the coexistence of participants with distinct but interdependent roles and motivations. Some are simply users who utilize the blockchain as a transactional infrastructure, while others (validators or miners) are actively incentivized to contribute to its operation and to securing the network by providing validation services. Validators are incentivized through a mechanism that selects who will receive remuneration in cryptocurrency to compensate for their investment in computing resources. This remuneration is based on the transaction fees that users must pay for each deposit (called an anchor) on the blockchain. This mechanism creates a form of symbiosis between the actors, which is essential to the vitality of the system. However, the strong economic incentives induced by competition between validators can lead to opportunistic and divergent behavior, which may compromise the stability and integrity of the whole for the benefit of individual interests. In their quest to maximize their gains, validators may organize themselves into pools, increasing their chances of being selected, even if they then have to share the profits of the selection when they obtain it. Careful design of the objectives and incentive mechanisms for participants is therefore crucial to prevent such behavior.

The intrinsic complexity of blockchain systems exceeds human capabilities for prediction and manual problem solving. Challenges such as aligning incentives to avoid selfish behavior, managing behavioral deviations by malicious actors, and implementing effective regulation of the blockchain itself are particularly difficult. In this context, in-depth study of the behavior of miners within mining pools (i.e. groups of miners) is of paramount importance, particularly for proof of work based blockchains. By concentrating considerable computing power, these pools significantly influence block production and, consequently, the security and decentralization of the network. Their resource allocation strategies, participation choices, and responsiveness to economic incentives can impact power distribution and resistance to attacks. Excessive concentration could create vulnerabilities and encourage harmful strategic behavior.

Given the limitations of analytical approaches in capturing the emergent and strategic behaviors within mining pools, a simulation-based methodology becomes essential. To this end, we propose a multi-agent model that simulates the behavior of miners within a blockchain network, incorporating behavioral rules to analyze their collective impact on network performance and stability.

2 Fundamentals of Blockchain Systems

2.1 Overall Functioning of the Blockchain

Blockchain is a distributed ledger operating on a peer-to-peer (P2P) network, allowing agents to collaborate without a trusted intermediary. Each transaction added must comply with a consensus ensuring agreement among nodes on a shared ledger. The ledger is composed of blocks, each containing a list of transactions. The blocks are linked to each other via their fingerprints, forming a chain that ensures the integrity of transactions. Users create transactions, often accompanied by fees, and then submit them by broadcasting them across the blockchain network for confirmation. Miners or block validators receive transactions that they are responsible for validating and then anchoring on the blockchain through a consensus mechanism managed in a dedicated peer-to-peer network.

Once a certain number of transactions have been received, block validators select some of them to be confirmed and order them into a dedicated block, created using a consensus mechanism specific to the blockchain [4]. This consensus mechanism allows the validator(s) to be selected, who are then remunerated with the cryptocurrency associated with the blockchain and funded by transaction fees. This consensus mechanism is therefore essential to the proper functioning of a blockchain (particularly a public one); it requires the mobilization of validators, sometimes involving very costly resources, in order to be selected by these consensus mechanisms.

Proof of Work Consensus . In a blockchain, each node has a copy of the ledger. To ensure that all nodes have the same information and to avoid conflicts, a consensus protocol is used to validate and add blocks to the chain. Consensus is a mechanism that allows participants in the blockchain network to agree on a common state of data without the need for a central authority. The consensus algorithm is at the heart of the blockchain.

There are several consensus mechanisms. In this study, we focus specifically on proof of work consensus, which was used in the first blockchains, notably for Bitcoin. In this mechanism, block validators, known as miners, must solve a complex cryptographic problem before they can add a new block to the blockchain.

This process guarantees security but is energy-intensive [5]. The blockchain adjusts its 'difficulty' to maintain a constant rate of block addition (approximately 10 min for Bitcoin), thus avoiding disagreements in the P2P network despite fluctuations in the miners' computing power.

To solve this cryptographic problem, miners must use significant hardware resources. One strategy is to form pools. A miner pool is a group of miners working together under the direction of a leader, pooling their computing power to increase their chances of solving the problem. When a block is found, the reward is distributed among the pool members according to their contribution. This approach offers greater consistency in earnings and reduces uncertainty, although individual rewards are shared.

Mining Pools and Centralization Risks . Taking the case of the Bitcoin network in June 2024, the total computing power (hash rate) of the peer-to-peer network reached approximately 606 Exa Hash per second (EH/s). An individual miner equipped with an ASIC device delivering 195 Tera Hash per second (TH/s) at a cost of approximately e4350 would theoretically have to wait an average of 58 years to discover a block, assuming the mining difficulty remained constant. This estimate is based on the ratio between the miner's power and that of the network. However, by joining a mining pool, the same miner could earn an estimated daily income of e18.94, excluding pool fees. Added to this are the energy costs associated with mining, estimated at around e60,541 per bitcoin produced in France in 2023. This situation illustrates how, faced with scarce rewards and the high costs of individual mining, miners are encouraged to form pools, a phenomenon that could contribute to the gradual centralization of the blockchain despite its decentralized architecture [6]. This centralization poses security risks, as these pools can control the blockchain by reaching 51% of its computing power. Once this threshold is reached, the leader can then filter transactions added to the blockchain and make double expenditures [7].

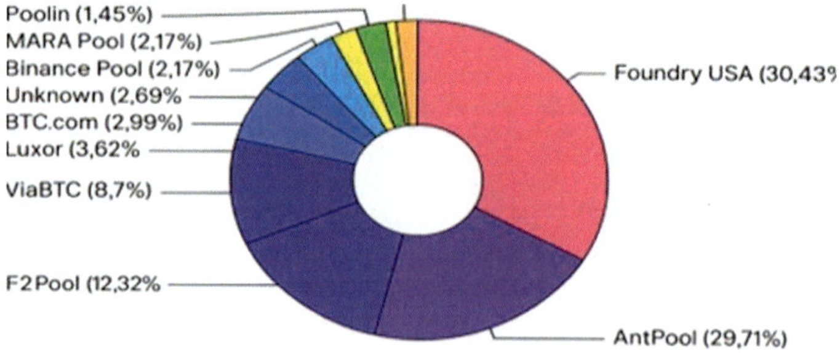

Fig. 1. Distribution of Bitcoin mining pools. *Source* http://mempool.space/

Figure 1 illustrates the centralization present in the Bitcoin blockchain through its proof of work consensus mechanism. According to a study conducted in March 2024 on the centralization of hash rate around pools, the Foundry USA and AntPool pools account for approximately 60% of the Bitcoin blockchain's computing power.

3 Implementation of a Multi-agent Simulator

Existing studies in the scientific literature have introduced a variety of paradigms for simulating blockchain systems (e.g. Object-Oriented [8], and Agent-Oriented approaches [9]). Each paradigm offers distinct perspectives and modelling capabilities. The Agent-Oriented paradigm is particularly well-suited for capturing the dynamic and heterogeneous interactions among autonomous entities within the network.

In light of these advantages, the present study adopts the agent-oriented paradigm to develop a simulation framework that models miner behavior and strategic decision-making within a blockchain environment. We particularly based on the AGR framework proposed by Roussille et al. [10] to model the behavior of block validators in the Bitcoin blockchain. Figure 2 presents the multi-agent architecture developed for the Bitcoin blockchain context. It highlights the different types of agents involved, such as the transaction proposer agent and the validator agent, as well as the behaviors associated with each in the transaction processing process.

To implement the AGR system, we used the AnyLogic simulation software, which supports multiple modelling paradigms, including discrete-event, system dynamics, and agent-based modelling. Its flexibility and extensibility make it particularly suitable for complex systems involving heterogeneous agents and dynamic interactions. Our simulator integrates the blockchain infrastructure:

- The representation of blocks composed of recorded transactions and the hash chaining mechanism with the cryptographic hashing mechanism used;
- Representation of the peer-to-peer network;
- The consensus mechanism used for block validation;
- Description of behavior.

The simulator also incorporates the behavior of miners, characterized in particular by their risk aversion and the computing power they can mobilize. It allows miner pools to be represented based on their profit redistribution algorithm. It is therefore necessary to characterize the behavior of miners, in particular their strategy for selecting a pool to join, which will be mainly motivated by their potential gain linked to the pool's redistribution algorithm, but also by:

- The ease or probability for the miner to find partial solutions within the pool;
- The reputation and reliability of the pool,
- The centralization of the pool, for example to avoid joining those that are becoming too powerful and may lead to risks of double spending or attacks.

Based on these characteristics, three pool selection algorithms were implemented using game theory [11].

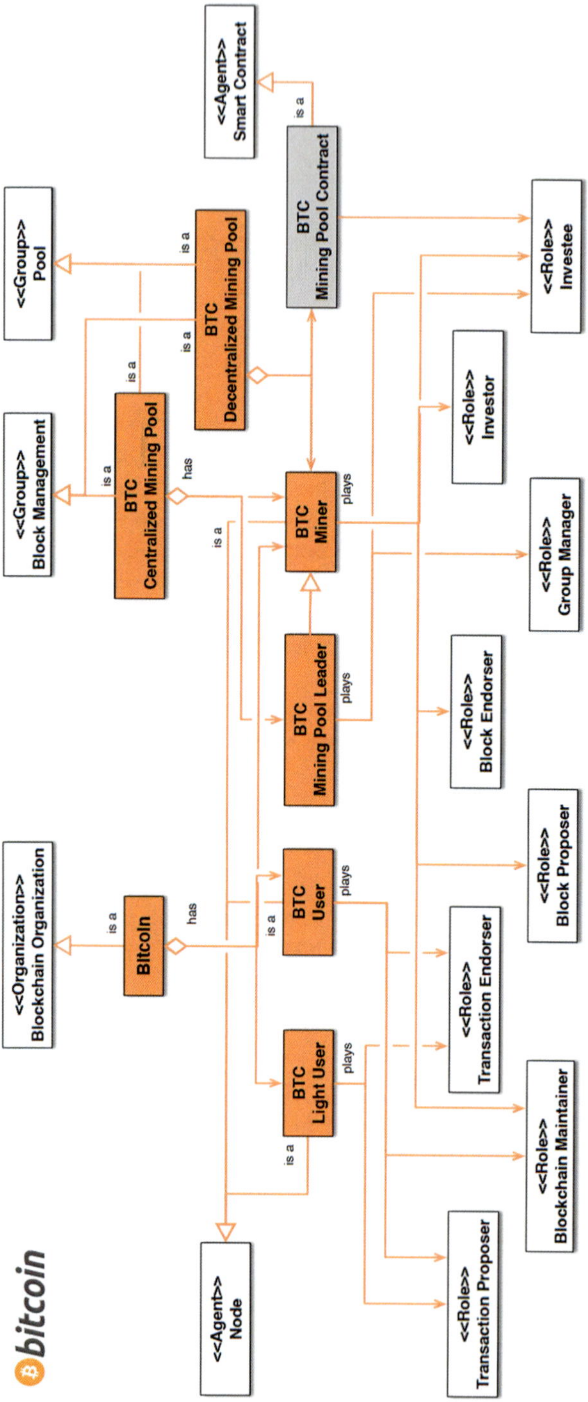

Fig. 2. AGR model for Bitcoin blockchain [10].

- **Algorithm 1**: based on 'simulated annealing' [12], which is used in our simulator to prevent massive and disorderly migration of miners between pools.
- **Algorithm 2**: designed to help miners choose a pool by taking into account not only its attractiveness, but also the risks associated with excessive centralization that could lead to double-spending attacks.
- **Algorithm 3**: This is based on Kantian ethics [7]. Miners assess the consequences of imitative behavior by all miners, which could lead them to choose the same pools. Miners then seek more balanced distributions between pools.

4 Results and Discussions

This study represents a preliminary step toward the systematic simulation of diverse decision-making algorithms and consensus mechanisms. The primary objective is to establish a comparative framework to evaluate blockchain network behavior and quantify the impact of local agent rules on global system properties, such as decentralization. We focus our analysis on two distinct approaches to pool selection: Algorithm 1, which utilizes a stochastic optimization method based on Simulated Annealing, and Algorithm 3, which is grounded in Kantian moral principles. By comparing these two paradigms, we aim to assess how different behavioral heuristics influence the structural evolution of the network.

To monitor the progression of the blockchain state, snapshots of the network were recorded at stochastically selected intervals, specifically at $t_0 = 0$, $t_1 = 15$, and $t_n = 30$ time units. The use of these random temporal points ensures that the observed state of the pools provides an unbiased representation of agent behavior and network distribution at different stages of the simulation.

The simulation results, visualized in Figs. 3 and 4, employ a graph-based representation where nodes represent individual agents within the blockchain ecosystem. In these graphs, green dots denote simple miners while purple dots identify pool leaders. The edges or arcs connecting these nodes represent the explicit affiliation of a miner to a specific pool leader at that given moment. Through this visualization, we can analyze the dynamic transition of agents and the resulting clustering effects, providing insights into the centralization risks associated with each algorithmic approach.

Figure 3 highlights an initially decentralized structure, characterized by dispersed connections and a relatively homogeneous distribution of agents within the network. During the simulation, the network reorganizes itself around a few dominant pools, while retaining a diversity of local substructures, which contrasts with the more uniformly centralized configuration induced by Algorithm 3.

In comparison, Fig. 4 reveals a highly centralized structure that remains relatively stable over time, with dominant pools retaining their influence over the entire graph, reflecting a more rigid and hierarchical network dynamic.

Fig. 3. Evolution of mining pools on blockchain network based on Algorithm 1.

Fig. 4. Evolution of mining pools on blockchain network based on Algorithm 3.

5 Conclusion and Outlook

In this study, we presented the development of a simulation tool designed to model the functioning of a blockchain system. The simulation is based on a multi-agent modelling framework. To analyze their pool selection strategies, we incorporated game-theoretic methods. Various behavioral scenarios have been implemented, and the simulation tool developed with Anylogic makes it possible to measure the impact of these behavioral choices. As future work, the proposed model could be extended to study a broader range of blockchain systems and alternative consensus mechanisms beyond proof of work. Additionally, it would be valuable to implement and compare various pool selection algorithms based on different strategic and economic criteria. In addition, future studies will focus on the introduction of quantitative performance indicators and the conduction of a more comprehensive statistical analysis of the simulation results, in order to provide a deeper and more rigorous evaluation of the proposed simulator.

References

1. Nakamoto, S.: Bitcoin: A Peer-to-Peer Electronic Cash System (2008). https://doi.org/10.2139/ssrn.3440802
2. Szabo, N.: Formalizing and securing relationships on public networks. First Monday **2**(9), 548 (1997). https://doi.org/10.5210/fm.v2i9.548
3. Bonneau, J., Miller, A., Clark, J., Narayanan, A., Kroll, J.A., Felten, E.W.: SoK: research perspectives and challenges for bitcoin and cryptocurrencies. In: Proceedings of the 2015 IEEE Symposium on Security and Privacy, pp. 104–121 (2015). https://doi.org/10.1109/SP.2015.14
4. Wang, W., Hoang, D.T., Hu, P., Xiong, Z., Niyato, D., Wang, P., Wen, Y., Kim, D.I.: A survey on consensus mechanisms and mining strategy management in blockchain networks. IEEE Access **7**, 22328–22370 (2019). https://doi.org/10.1109/ACCESS.2019.2896108
5. Chehade, I.: Essais sur l'économie des technologies blockchain et des cryptomonnaies. PhD Thesis (2023). http://www.theses.fr/2023NORMR057/document
6. Aponte-Novoa, F.A., Orozco, A.L.S., Villanueva-Polanco, R., Wightman, P.: The 51% attack on blockchains: a mining behavior study. IEEE Access **9**, 140549–140564 (2021). https://doi.org/10.1109/ACCESS.2021.3119291
7. Goldmann, L.: Immanuel Kant. Verso Books (2020)
8. Rocha, H., Ducasse, S.: Preliminary steps towards modeling blockchain oriented software. In: Proceedings of the 1st International Workshop on Emerging Trends in Software Engineering for Blockchain, pp. 52–57 (2018). https://doi.org/10.1145/3194113.3194123
9. Ciatto, G., Mariani, S., Omicini, A., Zambonelli, F.: From agents to blockchain: stairway to integration. Appl. Sci. **10**(21), 7460 (2020). https://doi.org/10.3390/app10217460
10. Roussille, H., Gürcan, Ö., Michel, F.: AGR4BS: a generic multi-agent organizational model for blockchain systems. Big Data Cognit. Comput. **6**(1), 1 (2021). https://doi.org/10.3390/bdcc6010001
11. Cao, T., Decouchant, J., Yu, J.: Leveraging the Verifier's Dilemma to Double Spend in Bitcoin (2023). arXiv:2210.14072. https://doi.org/10.48550/arXiv.2210.14072
12. Kirkpatrick, S., Gelatt, C.D., Jr., Vecchi, M.P.: Optimization by simulated annealing. Science **220**(4598), 671–680 (1983)

IoT Technologies for Industry 4.0

Aviation Workforce 4.0: Competences for a Sustainable and Digitalized Sector

Adina-Roxana Munteanu(✉), Adina Petruta Pavel, and Sorin Eugen Zaharia

National University of Science and Technology POLITEHNICA Bucharest, Bucharest, Romania
adina.munteanu@upb.ro

Abstract. The transition toward a sustainable and digital economy requires new workforce models that combine technical, digital, and transversal competences. This paper examines how aviation, as a strategic global industry, illustrates the broader challenges of Workforce 4.0. Drawing on data collected within the AVIONIC project, the study analyzes the effectiveness of training systems, the alignment of higher education with labor market demands, and the emerging need for green and digital skills. Findings reveal a strong demand for programming, managerial, and transversal competences, a continued preference for face-to-face and experiential training formats, and a limited emphasis on environmental sustainability skills. While the case of aviation highlights sector-specific dynamics, the implications extend to other industries undergoing similar twin transitions of digitalization and decarbonization. The paper argues that education systems and business organizations must adopt modular and work-integrated approaches to skill development in order to enhance competitiveness and resilience in a sustainable global economy.

Keywords: Workforce 4.0 · digital competences · green skills · sustainable economy · aviation industry · education and training · Industry 4.0

1 Introduction

The global economy is undergoing a profound transformation shaped by the dual imperatives of digitalization and sustainability. These structural changes, often described as the "twin transition," are redefining the nature of work, reshaping industries, and challenging education systems to deliver new forms of human capital. Workforce 4.0, a concept that captures the convergence of technological innovation, organizational change, and environmental responsibility, has become a critical priority for businesses, policymakers, and higher education institutions worldwide.

Within this context, the aviation sector represents a particularly relevant case study. As a highly regulated and technology-intensive industry, aviation has long relied on specialized competences in areas such as flight operations, maintenance, and safety management. However, the sector is now experiencing unprecedented pressures: the rapid digitalization of services, the introduction of automation and artificial intelligence, and the urgent requirement to reduce carbon emissions. These developments are reshaping

Y. Benadada et al. (Eds.): GOL 2026, LNNS 2034, pp. 265–273, 2027.
https://doi.org/10.1007/978-3-032-29677-1_24

occupational profiles, creating demand for interdisciplinary skills, and forcing companies to rethink both recruitment and training strategies.

Education and training systems play a decisive role in supporting this transition. Yet, evidence suggests a growing mismatch between traditional curricula and the dynamic requirements of modern industries. While aviation-specific studies highlight gaps in programming, digital fluency, and environmental competences, these findings mirror broader global challenges in aligning workforce capabilities with economic and societal transformation. The case of aviation therefore offers valuable insights into how education, business organizations, and policy frameworks must adapt to ensure competitiveness, resilience, and sustainable growth.

This paper contributes to this agenda by presenting results from the AVIONIC project (Creative Digital Teaching and Learning for Green Air Transport and Logistics), a European initiative focused on developing innovative educational tools and frameworks for the aviation sector. Using survey data collected from industry professionals, the study evaluates perceptions of training effectiveness, collaboration between companies and higher education institutions, and the emergence of new competence requirements. By situating these findings within the broader debate on Workforce 4.0, the paper highlights implications for workforce development strategies across sectors and provides recommendations for aligning education systems with the needs of a sustainable and digitalized global economy.

2 Literature Review

The transformation of the global economy under the twin pressures of digitalization and sustainability has intensified debates on the future of work and the competences required for long-term competitiveness. The concept of Workforce 4.0 has emerged as a framework to capture the intersection of technological innovation, organizational change, and human capital development in the context of Industry 4.0 [1, 2]. Workforce 4.0 emphasizes not only technical proficiency but also transversal capabilities such as adaptability, critical thinking, and systems integration, which are essential for navigating complex and rapidly evolving environments.

From an economic perspective, human capital development is widely recognized as a driver of growth and resilience in the digital age [3, 4]. However, traditional education-to-employment pipelines often fall short of delivering the agility required to address ongoing technological disruption. Scholars and policymakers increasingly highlight the need for lifelong learning, modular education, and microcredentials as mechanisms to ensure the continuous renewal of competences [5, 6]. These approaches allow individuals to adapt to new occupational profiles while maintaining alignment with labor market demand.

The green transition further complicates workforce development, as industries are called to integrate sustainability objectives into operational and strategic practices. While environmental skills are well established in sectors such as energy and construction, their integration into high-technology industries like aviation remains limited [7] The European Union's reports on "twin skills for the twin transition" emphasize that the lack of green competences may undermine progress toward climate neutrality if not embedded systematically into education and training [8].

Aviation provides a particularly illustrative case of these broader challenges. Traditionally skill-intensive and safety-driven, the industry is now compelled to combine its emphasis on technical expertise with new requirements in digitalization, data analytics, automation, and environmental responsibility [9, 10]. Studies show that while aviation education continues to focus on core operational skills, there is a growing mismatch with emerging demands such as programming, cybersecurity, and carbon management [11].

At the same time, collaboration between higher education institutions and industry has been identified as a critical success factor in bridging skill gaps [4]. Yet, evidence suggests that these partnerships often remain limited in scope, with insufficient co-design of curricula or integration of experiential learning. The adoption of microcredentials and modular programs offers an opportunity to strengthen this collaboration, providing flexible and labor-market-responsive pathways that can enhance both employability and sectoral competitiveness [12].

In summary, the literature highlights three key themes relevant for this study: (1) the emergence of Workforce 4.0 as a response to global economic transformation; (2) the critical role of education systems in fostering digital and green competences through innovative and flexible models; and (3) the persistent gaps in aligning higher education and industry needs, particularly in sectors undergoing rapid twin transitions. Against this backdrop, aviation serves as a case study for understanding how these dynamics unfold in practice and what lessons can be drawn for workforce development in other strategic industries.

3 Methodology

This study adopts a case study approach, positioning the aviation sector as a representative example of the broader challenges associated with Workforce 4.0 in the global economy. Aviation is a strategic industry with significant economic, social, and environmental impact, employing millions of individuals worldwide and contributing substantially to GDP in Europe and beyond. Its high degree of regulation, technological intensity, and exposure to sustainability imperatives make it an ideal sector through which to explore the evolving alignment between education, workforce competences, and economic transformation.

The research design combines three complementary methods: literature review, expert consultation and survey.

A structured review of academic and policy literature was conducted to establish the conceptual framework of Workforce 4.0, with particular emphasis on digital and green competences. Sources included international organizations such as ICAO, IATA, and the European Commission, as well as broader economic and management studies on human capital development, Industry 4.0, and lifelong learning. This allowed the identification of key trends, emerging occupational profiles, and the role of education systems in supporting sustainable economic growth.

Insights from professionals in higher education institutions and industry partners were gathered to validate trends and refine the analytical framework. These experts were involved through structured discussions and workshops within the AVIONIC project, ensuring that sectoral developments were interpreted in light of broader economic and workforce dynamics.

A tailored survey was administered to three stakeholder groups: aviation employees, managers, and educators/trainers. The questionnaire explored perceptions of initial workplace training, collaboration between industry and higher education institutions, the adequacy of formal education, and the identification of emerging competences. Respondents also reported on their preferred formats for training delivery, offering insights into the potential role of modular learning and microcredentials.

The survey reached participants across multiple subsectors, including airlines, airports, air traffic management, ground handling, training providers, and public authorities, reflecting the diversity of the aviation ecosystem. Responses were analyzed using descriptive statistics and interpreted in relation to both the literature and expert feedback. This triangulated approach strengthens the validity of findings and allows them to be situated within a broader discussion of how global industries can adapt to the twin transition of digitalization and sustainability.

By employing aviation as a sectoral case study and integrating evidence from literature, experts, and practitioners, the methodology ensures that the conclusions are relevant not only to aviation but also to other industries undergoing similar transformations. This design positions the paper to contribute both sector-specific insights and cross-sectoral lessons for policy, education, and business strategy in the era of Workforce 4.0.

4 Results and Discussion

The survey conducted within the AVIONIC project provides insights into how the aviation workforce perceives training effectiveness, education–industry collaboration, and emerging competence requirements. While the data are sector-specific, the results illustrate broader dynamics of Workforce 4.0 that are relevant across industries.

4.1 Effectiveness of Initial Training and Collaboration with Education Providers

Initial workplace training was generally evaluated positively by respondents, with 66% rating it as "very helpful" or "extremely helpful" (Fig. 1). Nevertheless, 34% reported limited or low usefulness, indicating a persistent gap between formal education and workplace requirements. From a business and management perspective, this underscores the importance of bridging the education-to-employment pipeline to ensure that organizations can rely on work-ready employees.

Collaboration between companies and higher education institutions (HEIs) was reported by 62% of respondents, yet 21% stated they were unaware of such partnerships (Fig. 2). This limited visibility points to weak integration and communication channels. For the wider economy, the implication is that HEI–industry collaboration remains fragmented, and without stronger co-design of curricula and joint competence forecasting, mismatches will continue to hinder workforce competitiveness.

4.2 Relevance of Formal Education to Job Requirements

When asked to assess the role of educational institutions in preparing them for their occupations, 26% of respondents rated their education as "very effective" and 8% as "extremely effective," while almost one-fifth expressed doubts about its adequacy (Fig. 3).

This mirrors findings in other sectors, where traditional curricula struggle to keep pace with digital transformation and sustainability imperatives.

From the standpoint of Workforce 4.0, this highlights the incomplete adaptation of education systems to the twin transition. Although most respondents value their education, the perception of limited relevance suggests that universities must integrate more practical, digital, and sustainability-oriented components to align with labor market needs.

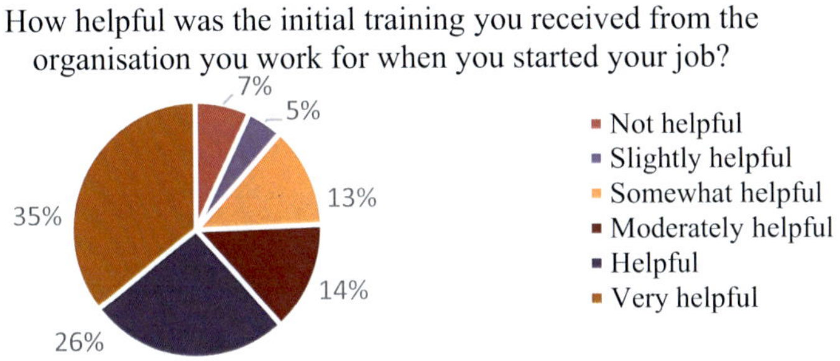

Fig. 1. Effectiveness of initial training received from employer upon job entry

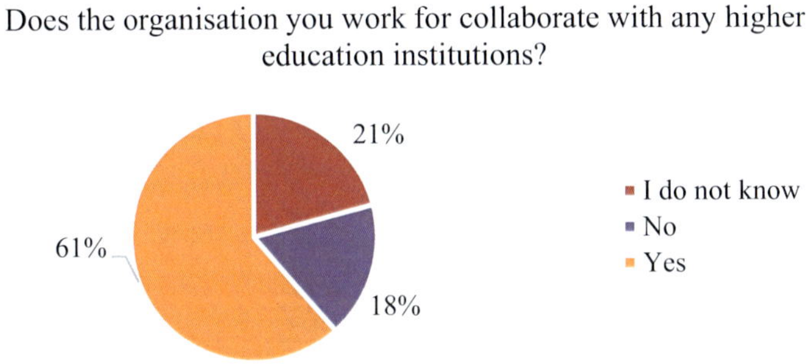

Fig. 2. Collaborate between companies with higher education institutions

4.3 Emerging Training Needs and Competence Gaps

Programming was identified as the most essential training area by 36% of respondents, followed by management (17%) and digital competences (16%) (Fig. 4). Leadership and safety also featured prominently. Strikingly, environmental or green competences were rarely mentioned.

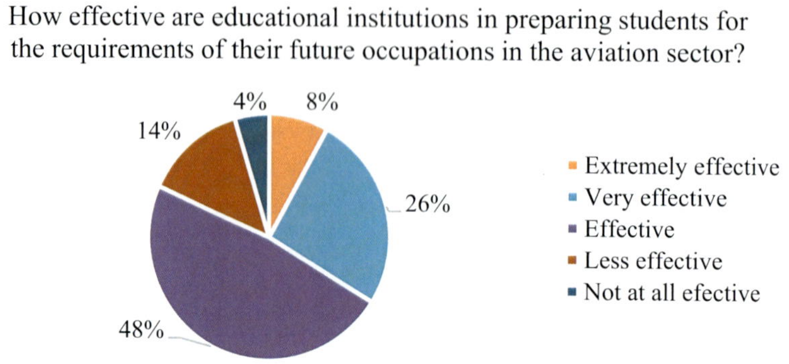

Fig. 3. Effectiveness of educational institutions in preparing students for the requirements of their future occupations in the aviation sector

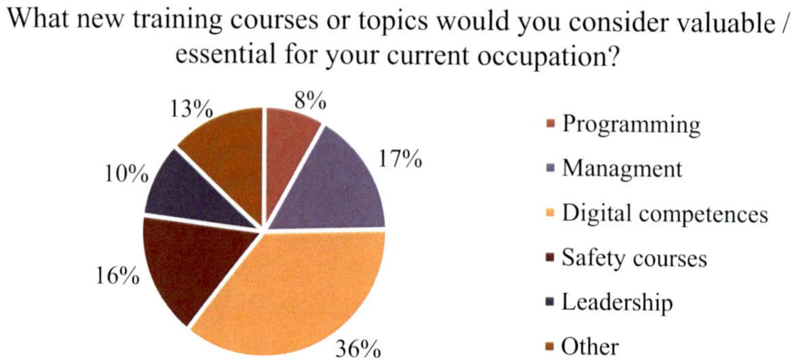

Fig. 4. New training courses or topics essential for current occupation

This distribution reflects a strong emphasis on digital competences and transversal skills, consistent with Workforce 4.0 frameworks that prioritize adaptability, data literacy, and systems integration. However, the low visibility of green competences raises concern. In the context of the European Green Deal and the global push for decarbonization, neglecting environmental skills may compromise both sectoral and broader economic sustainability. This pattern is not unique to aviation and signals a lag in embedding green competences across industries, potentially creating a barrier to achieving climate-neutral economies.

4.4 Preferred Training Formats and Models

Training delivery preferences reveal a strong inclination toward experiential formats: 36% favored face-to-face learning, while 23% preferred interactive practical training (Fig. 5). Online and blended formats were less attractive, though still relevant for theoretical knowledge transfer.

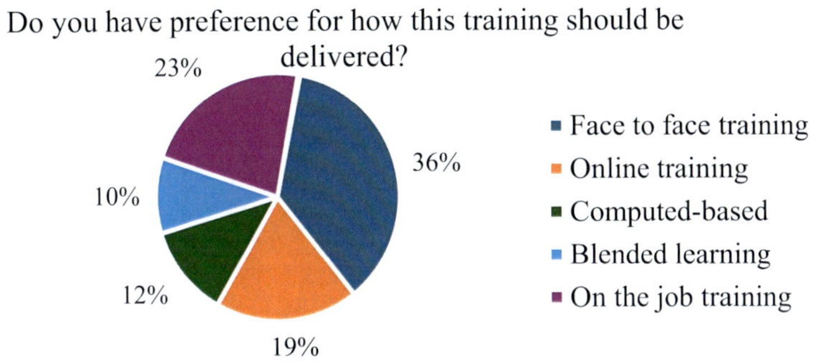

Fig. 5. Training delivery preferences

These preferences highlight the value of work-integrated and hands-on learning in industries where precision, safety, and applied skills are essential. More broadly, they point to the limits of purely digital learning in preparing a resilient Workforce 4.0. For policy and education systems, this suggests the need to combine modular and microcredential approaches with experiential and workplace-based methods, ensuring that digital delivery complements rather than replaces practical competence development.

4.5 Broader Implications for Workforce 4.0 and the Global Economy

Taken together, the results underscore three main lessons with cross-sectoral relevance referring to curriculum alignment, green skills and experiential formats.

Firstly, the curriculum–workplace misalignment persists. Gaps in training effectiveness reflect a broader challenge of adapting education systems to rapid technological and environmental transitions. Secondly, green skills remain underdeveloped. Despite policy emphasis, environmental competences are not yet internalized by employees, signaling an urgent need for systemic integration into curricula and training.

Lastly, experiential formats are essential for workforce readiness. The continued preference for face-to-face and practical training illustrates the enduring importance of human interaction and real-world application in skill development.

These findings reinforce the argument that aviation serves as a microcosm of Workforce 4.0 challenges, where digital and managerial competences dominate but sustainability competences lag behind. The sector's experience offers lessons for the wider economy: industries that fail to anticipate competence shifts risk losing competitiveness, while those that embrace modular, work-integrated, and sustainability-oriented training will be better positioned to thrive in a digital and green global economy.

5 Conclusions

This paper has examined how digitalization and sustainability imperatives are reshaping workforce competences, using aviation as a case study for the broader dynamics of Workforce 4.0. Survey findings from the AVIONIC project highlight both progress and persistent challenges in aligning education and training systems with evolving labor market needs.

Three main conclusions can be drawn. First, while initial training is generally perceived as effective, there remain gaps between formal education and workplace requirements. These gaps illustrate a structural misalignment that extends beyond aviation, underscoring the urgency for stronger collaboration between higher education institutions, industry, and policy-makers in designing curricula and forecasting competences.

Second, the results reveal a strong emphasis on programming, managerial, and transversal competences, consistent with the demands of a digitalized economy. However, the limited focus on environmental sustainability skills suggests that green competences are not yet fully embedded in workforce development strategies. This disconnect may hinder progress toward climate neutrality if not addressed through systemic integration of sustainability principles into both technical and managerial education.

Third, employees continue to value face-to-face and practical training formats, pointing to the enduring importance of experiential and work-integrated learning. While digital platforms and microcredentials can enhance flexibility and access, they cannot fully substitute the value of applied, interactive approaches in preparing a resilient and adaptable workforce.

The implications extend beyond aviation. The sector serves as a microcosm of how industries at the intersection of technological innovation and sustainability must adapt workforce strategies. To remain competitive in the global economy, businesses and education providers must adopt modular, lifelong, and interdisciplinary approaches to skill development, while policy frameworks should incentivize the integration of both digital and green competences across sectors.

In short, the case of aviation illustrates that Workforce 4.0 is not defined solely by technological proficiency but by the capacity to combine digital agility, managerial adaptability, and environmental responsibility. Preparing such a workforce requires coordinated action across education, business, and policy domains. By aligning competences with the twin transition, economies can strengthen resilience, foster sustainable growth, and ensure that the future of work contributes meaningfully to a greener and more digitalized global economy.

Acknowledgements. This report was developed within the framework of the Erasmus+ project Creative Digital Teaching and Learning for Green Air Transport and Logistics Project (AVIONIC) (Project No. 2022–1-RO01-KA220-HED-000086424), co-funded by the European Union. The authors gratefully acknowledge the financial support provided by the Erasmus+ Programme. The work presented here reflects the collaborative efforts of the project consortium. The authors also wish to thank all team members, students, and stakeholders who contributed their expertise, time, and feedback throughout the project activities.

References

1. Sima, V., Gheorghe, I.G., Subić, J., Nancu, D.: Influences of the Industry 4.0 revolution on the human capital development and consumer behavior: a systematic review. Sustainability **12**(10), 4035 (2020)
2. Hecklau, F., Galeitzke, M., Flachs, S., Kohl, H.: Holistic approach for human resource management in Industry 4.0. Procedia CIRP **54**, 1–6 (2016)

3. Becker, G.S.: Human Capital: A Theoretical and Empirical Analysis, with Special Reference to Education, 3rd edn. University of Chicago Press, Chicago (1994)
4. Skills Outlook, O.E.C.D.O.E.C.D.: Learning for Life, p. 2021. OECD Publishing, Paris (2021)
5. European Commission. Digital Education Action Plan 2021–2027. Brussels: European Commission (2021). Available at: https://education.ec.europa.eu/focus-topics/digital-education/action-plan. Accessed: 9 Sept 2025
6. Zaharia, S.E., Pavel, A.P., Munteanu, A.R.: Smart Skills for a Green and Digital Aviation Future: Insights from the AVIONIC Project. Politehnica University of Bucharest, Bucharest (2025)
7. European Training Foundation: Skills for the Green Transition: Evidence from the EU Neighbourhood. ETF, Turin (2023)
8. Koundouri, P., et al.: Twin Skills for the Twin Transition: Defining Green and Digital Skills and Jobs. European Commission, Brussels (2023)
9. International Civil Aviation Organization (ICAO): Innovation for a Green Transition. ICAO, Montreal (2022)
10. International Air Transport Association (IATA): Next Generation Competence Strategies in Aviation. IATA, Montreal (2025)
11. ATHENA Research Centre: Education and Training Needs for a Sustainable Aviation Workforce. ATHENA, Athens (2022)
12. Sursock, A.: Microcredentials and Modular Learning in European Higher Education. European University Association, Brussels (2020)

Modeling and Optimization of Edge Server Location for IoT-Based Urban Traffic Control

Mohammed El-Amrani(✉) and Salma Ibadissen

Faculty of Sciences, Mohammed V University in Rabat, Rabat, Morocco
m-elamrani@um5r.ac.ma, salma-ibadissen@um5.ac.ma

Abstract. Recently, urban traffic management has become a major challenge due to the increasing number of vehicles and the growing demand for mobility, particularly in large cities. Traditional, centralized, and rigid approaches suffer from high latency, significant costs, and poor adaptability to unforeseen events. In this article, we propose an improved strategy based on the deployment of local servers and a hierarchical architecture for processing traffic data. To this end, a mathematical model for the server location problem has been developed to reduce deployment costs, minimize latency, and ensure efficient load balancing between servers. Furthermore, this research also includes evaluating performance against existing solutions and integrating fault-tolerant mechanisms to guarantee service availability. These improvements aim to create a more robust, intelligent, and adaptive urban traffic management system, thereby contributing to sustainable and safe mobility.

Keywords: Facility Location Problem · Traffic Management · Latency · IoT · Edge Computing

1 Introduction

Traffic management has become a major challenge for large modern cities. Without precise and effective regulation, these cities will suffer numerous economic, social, and environmental consequences. Conversely, effective traffic management reduces frequent and unnecessary vehicle stops, thereby decreasing polluting emissions. It also improves emergency services and public transportation and reduces travel times.

The search for road traffic management methods began long ago, with traditional methods based on adapted road signs, and evolved towards dynamic systems that enabled more effective organization. However, none of these methods can adapt to emergency situations or changes in road infrastructure. The emergence of adaptive methods has significantly improved signage management, but their implementation remains very expensive. This has led to the development of modern and highly advanced methods relying on artificial intelligence, the Internet of Things (IoT), autonomous vehicles, and cloud computing. It

Y. Benadada et al. (Eds.): GOL 2026, LNNS 2034, pp. 274–283, 2027.
https://doi.org/10.1007/978-3-032-29677-1_25

should be noted here that high performance is not always sufficient; these methods require significant equipment and high-speed cloud capacity. In this context, some studies, cited in the following section, have shown that centralized processing leads to increased delays in analysis, decision-making, and response times. In other words, it reduces response speed due to network congestion. Edge computing has emerged as a solution to this problem. It allows data processing to be brought closer to its source (cameras, sensors, radars, etc.) in order to improve response time. Furthermore, to optimize this process even further, it is essential to intelligently and optimally choose server locations [12] to ensure efficient resource allocation and complete coverage of the area under study. Several more general variants of facility location problems, which are adaptable to our problem, are discussed in [1,2].

This study has several objectives. First, it analyzes and understands the challenges related to road traffic management [12], highlighting the limitations of traditional methods [4,6–8] and the opportunities offered by new approaches. Second, it optimizes the placement of local servers using mathematical and algorithmic models from operations research [3,10,11], with the aim of reducing costs and improving overall system performance.

The third objective is to offer a real-time road traffic management system capable of collecting and processing sensor data in real time. This system not only guarantees maximum traffic flow with the lowest possible operating costs but also enables immediate intervention in emergencies, such as prioritizing emergency vehicles (ambulances, fire trucks, and police cars). This integrated solution helps reduce response times and improve the overall safety of the road network.

The remainder of this paper is structured as follows: Sect. 2 presents a review of previous work on traffic management, from traditional roads to automated roads, including the Internet of Things and the location of infrastructure (processing servers). Section 3 proposes a methodology combining a hierarchical structure and local optimization of the processing servers for data from the various sensors. Section 4 presents a simulation performed on a random dataset compared to the results of another simulation conducted on two different datasets to test the validity and effectiveness of the mathematical formulation. Finally, Section 5, which concludes the study, summarizes the results of our methodology and their interpretation and outlines future directions.

2 Related Work

Recent advances in traffic management have been largely driven by the integration of artificial intelligence, the Internet of Things (IoT), and distributed computing paradigms such as edge and cloud computing. Traditional machine learning models, including Random Forests and Support Vector Machines (SVM), have demonstrated strong predictive capabilities for traffic flow estimation and congestion detection. For instance, Talbi et al. [19] combined Random Forest with 5G edge computing, achieving 99% accuracy in congestion prediction and enabling adaptive routing in real time [19]. Similarly, Shirsath et al. [20] used an RBF-kernel SVM to forecast traffic density and optimize signal timing dynamically, reducing travel times significantly [20].

Deep learning approaches have further enhanced traffic prediction accuracy by capturing temporal and spatial dependencies. Abdellah et al. [5] compared several recurrent architectures (LSTM, GRU, BiLSTM) using V2X data, highlighting the superiority of BiLSTM models [5]. Chen et al. [13] introduced a spatio-temporal graph neural network with multi-scale attention mechanisms to model complex urban traffic interactions [13]. More recently, transformer-based architectures such as DST-net [16] have combined federated and reinforcement learning, achieving state-of-the-art detection and control performance [16].

Reinforcement learning (RL) and deep reinforcement learning (DRL) techniques have been extensively applied for traffic signal control. Wang et al. [15] proposed an A2C-based framework integrating graph attention for cooperative intersection control [15]. Hybrid models combining RL with multi-objective optimization, such as the Q-learningNSGA-II approach by Al-Heety et al. [18], demonstrated efficiency in prioritizing emergency vehicles [18], and Mutambik et al. [17] introduced a distributed IoT-based system reducing simulated CO_2 emissions by 28% [17].

Computer vision techniques have complemented these methods by providing real-time perception capabilities. Aboah et al. [9] deployed YOLOv5-based embedded systems for anomaly detection in traffic videos [9], while Villa et al. [14] developed an edge-enabled deep learning infrastructure for vehicle detection and tracking with high spatial accuracy [14].

Finally, the convergence of edge and cloud computing has emerged as a key enabler of scalable and responsive traffic management architectures. While cloud computing supports centralized analytics and long-term planning, edge computing ensures low-latency local decision-making. Hybrid edgecloud systems, such as those proposed by Al-Heety et al. [18], effectively combine both paradigms to balance computational efficiency and real-time responsiveness [18].

Despite these advances, few studies address the problem of strategic server placement within such architectures. Efficiently locating edge servers remains essential to minimizing latency, reducing communication costs, and ensuring load balancing across the network. The present work contributes to this area by proposing an optimized IoT-based hierarchical architecture for urban traffic management that integrates mathematical modeling and intelligent decision-making for server placement and data processing.

3 Methodology

The proposed system what is shown in the image below Fig. 1 aims to improve urban traffic management through a hierarchical architecture integrating edge and cloud computing models. This architecture enables the distributed processing of data from a sensor network, ensuring fast and accurate decision-making at the source. By bringing computing power closer to data collection points, the system significantly reduces latency, improves operational responsiveness, and optimizes resource utilization.

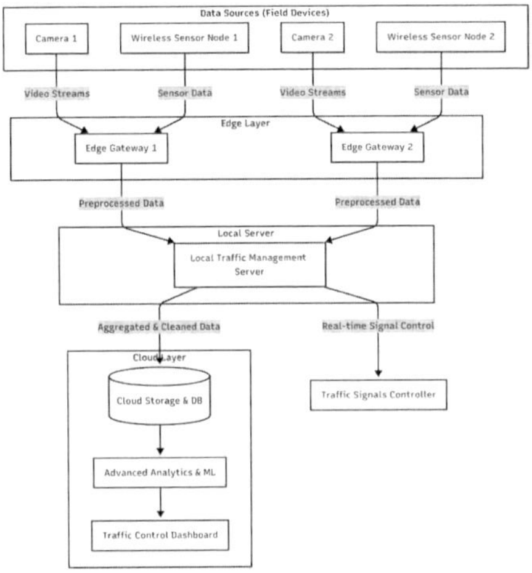

Fig. 1. Proposed hierarchical architecture for urban traffic management system.

3.1 Hierarchical Architecture

The system is structured into four layers: sensing, edge, local server, and cloud.

At the sensing layer, cameras and wireless sensor nodes (WSN) collect traffic data, including vehicle count, speed, and classification. These raw data are transmitted to nearby edge gateways, which form the second layer.

Edge gateways perform local preprocessing and analysis using embedded AI models (e.g., YOLOv5, TensorFlow Lite, OpenCV). They filter and compress the data, reducing bandwidth usage and enabling real-time responses such as dynamic traffic light adjustments or anomaly detection.

The third layer, the local server, acts as the operational core of the system. It aggregates the processed data from multiple edge gateways and runs optimization algorithms for traffic signal control. These algorithms can be rule-based or machine learningdriven, adjusting green times, phase sequences, and offsets dynamically according to real-time conditions. The local server also manages emergency vehicle prioritization and communication with traffic light controllers through standard protocols such as NTCIP.

The cloud layer ensures long-term storage and advanced analytics. It performs predictive modeling, historical trend analysis, and supports strategic planning through dashboards and visualization tools. This hybrid architecture combines low-latency decision-making with high-level coordination and scalability.

3.2 Edge Data Processing Pipeline

The edge data pipeline consists of several stages designed for fast and reliable data processing. Figure 2 illustrates the complete structure of this pipeline at the edge layer. First, data are captured from sensors and cameras via RTSP or wireless links. Preprocessing follows, including denoising, normalization, and signal validation using tools such as OpenCV and NumPy. Next, feature extraction and object detection are carried out using lightweight AI models optimized for embedded hardware such as NVIDIA Jetson devices. Detected objects are classified, and their speeds are estimated.

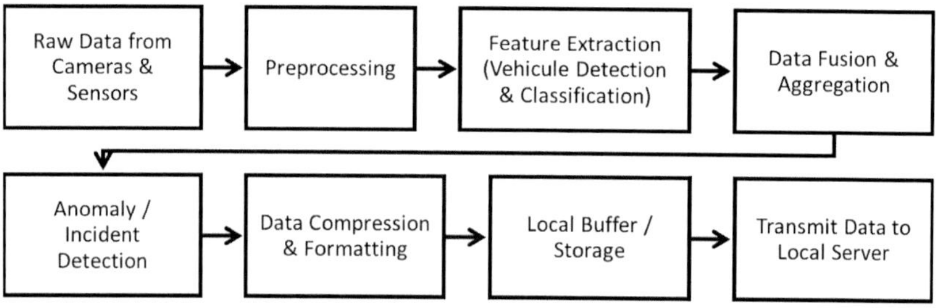

Fig. 2. Data flow diagram of edge processing.

The extracted features are then aggregated and fused from multiple sensors covering the same area, ensuring higher accuracy. Data are structured in lightweight formats (JSON or Protobuf), buffered locally (using SQLite or Redis), and transmitted to the local server through low-latency protocols such as MQTT or REST APIs. This approach ensures fast communication even in case of temporary network interruptions.

3.3 Mathematical Model for Server Placement Optimization

The placement of local servers is formulated as an optimization problem that minimizes deployment cost and communication latency while ensuring balanced computational loads across servers.

Let $I = \{i_1, i_2, \ldots, i_n\}$ represent the set of edge gateways and $J = \{j_1, j_2, \ldots, j_m\}$ the set of potential server locations. Each edge gateway i generates processing, memory, and storage demands (P_i, M_i, S_i), while each server j provides limited capacities (P_j, M_j, S_j). The following Table 1 presents the notations used in this model.

Decision variables:

$$x_{ij} = \begin{cases} 1, & \text{if edge gateway } i \text{ is assigned to server } j \\ 0, & \text{otherwise} \end{cases}$$

$$y_j = \begin{cases} 1, & \text{if a server is installed at location } j \\ 0, & \text{otherwise} \end{cases}$$

Table 1. Notations used in the local server placement model

Symbol	Description
I	Set of edge gateways
J	Set of potential locations for local servers
n	Number of edge gateways
m	Number of potential server locations
p	Number of local servers to deploy
C_j	Deployment cost of a local server at location j
L_{ij}	Latency between gateway i and local server j
$L_{\max}$	Maximum tolerated latency
P_i	Processing demand of gateway i
M_i	Memory demand of gateway i
S_i	Storage demand of gateway i
P_j	Processing capacity of server j
M_j	Memory capacity of server j
S_j	Storage capacity of server j
F	Total deployment cost of local servers

Objective Function The total deployment cost is minimized through the following objective function:

$$\text{Minimize } F = \sum_{j \in J} C_j y_j \tag{1}$$

Constraints

$$\sum_{j \in J} x_{ij} = 1, \quad \forall i \in I \tag{2}$$

$$\sum_{j \in J} L_{ij} x_{ij} \leq L_{\max}, \quad \forall i \in I \tag{3}$$

$$\sum_{i \in I} P_i x_{ij} \leq P_j y_j, \quad \forall j \in J \tag{4}$$

$$\sum_{i \in I} M_i x_{ij} \leq M_j y_j, \quad \forall j \in J \tag{5}$$

$$\sum_{i \in I} S_i x_{ij} \leq S_j y_j, \quad \forall j \in J \tag{6}$$

$$\sum_{j \in J} y_j \leq p \tag{7}$$

$$x_{ij} \in \{0,1\}, \quad \forall i \in I, j \in J \tag{8}$$

$$y_j \in \{0,1\}, \quad \forall j \in J \tag{9}$$

Explanation of the Constraints In the model, the objective function in equation (1) minimizes the total cost of the servers. Constraint (2) ensures that the requests of each edge gateway are entirely satisfied by a single local server.

Constraint (3) ensures that a maximum latency is not exceeded between any edge gateway and the local server it affects.

Constraints (4) to (6) require that the processing, memory, and storage capacities of each local server are not exceeded. This constraint prevents overloading the local servers. Constraint (7) limits the number of local servers to be installed so as not to exceed a predetermined number, p. Finally, the integrity constraints (8) and (9) on the decision variables, by imposing binary values, ensure that each link between a local server and an edge gateway is either active or not, and also guarantee that each local server is either open or not.

3.4 System Summary

The proposed methodology integrates intelligent sensing, real-time edge processing, optimized local control, and strategic cloud analysis. By combining distributed computation and mathematical optimization, the system achieves low-latency decision-making, efficient resource utilization, and scalable management for smart urban traffic systems.

4 Simulation and Results

To validate the mathematical model for optimizing edge server deployment, a simulation was performed using IBM ILOG CPLEX Optimization Studio. The objective of this experiment was to demonstrate the workings of the proposed model and analyze the optimal server configuration for a generated dataset.

4.1 Dataset Description

The dataset used is synthetic and represents an urban traffic management network composed of end gateways and candidate sites for edge server deployment. Each gateway is characterized by its processing, memory, storage, and latency requirements, while each server site is defined by its maximum capacity and installation cost. A latency matrix links the gateways to the candidate servers, and a constraint limits the maximum number of servers deployed.

4.2 Simulation Results

After running the model, the analyzer obtained an optimal solution with a target function value of 6000, which corresponds to the minimum total server deployment cost.

The decision vector y_j indicates the locations where the servers were actually deployed:

$$y = [0, 1, 1, 1, 0, 1, 1, 1, 0, 0]$$

Therefore, six servers were installed at positions j = 2, 3, 4, 6, 7, and 8, according to the constraint $\sum_j y_j \leq p = 6$. The variables x_{ij} represent the assignment of each terminal gateway to the servers actually in use. All assignments satisfy the model constraints, including processing, memory, storage capacities, and maximum allowable latency.

4.3 Results Analysis

The results show that the model is capable of:

- Reducing the overall deployment cost while considering performance constraints;
- Enabling balanced use of available servers, taking into account the capabilities of each;
- Ensuring acceptable latency for all edge gateways.

Thus, this simulation confirms the validity and feasibility of the mathematical model for strategically positioning edge servers in the context of urban traffic management. It also demonstrates the ability of the CPLEX analyzer to provide an optimal solution within a reasonable computational time, even for a medium-sized instance.

4.4 Comparative Analysis on Multiple Datasets

To further evaluate the robustness of the proposed model, three different datasets were tested under various network configurations. Each dataset differed in the number of edge gateways, potential server locations, and maximum deployment limits. Table 2 summarizes the key findings obtained from the CPLEX analyzer for these scenarios.

Table 2. Summary of optimization results for multiple datasets

Dataset	n	m	p	Total cost	Active servers	Observations
Data 1	30	10	6	6000	2, 3, 4, 6, 7, 8	Balanced solution, latency $\leq$ 35 ms.
Data 2	40	12	10	8200	1–8	Scalable solution, all gateways served.
Data 3	30	10	6	6000	1, 2, 3, 4, 6, 8	Stable feasible configuration.

The results confirm the robustness and efficiency of the proposed mathematical model for edge server deployment. In all scenarios, the CPLEX analyst identified practical and optimal solutions that minimized the overall deployment cost while considering latency and capacity constraints.

In Scenario 1, six servers were deployed to cover 30 gateways at the lowest cost. In Scenario 2, the model was effectively adapted to a larger configuration, with eight servers deployed for 40 gateways while maintaining acceptable latency. Scenario 3 deployed the same number of edge gateways and local servers as Scenario 1, except for the capacity differences. In this dataset, CPLEX also delivers optimal results while considering the model's limitations.

In summary, these experiments demonstrate that the proposed approach enables efficient resource allocation in edge computing networks. It ensures minimal infrastructure costs and acceptable performance, making it particularly suitable for real-time applications such as urban traffic management.

5 Conclusion

This work proposes an improved architecture for urban traffic management, based on the strategic deployment of edge servers and a hierarchical system integrating wireless sensor networks and cameras. This approach reduces deployment costs and latency while ensuring balanced server load and prioritization of emergency vehicles. By combining heterogeneous sensors with AI-based predictive models, the system enables reliable real-time traffic management. The results demonstrate the efficiency, robustness, and adaptability of the proposed mathematical model. Future work could focus on performance evaluation and improved fault tolerance to optimize urban mobility and safety.

References

1. Abbal, K., Benadada, Y., Mohammed, E.A.: Bi-level Multi-capacitated Facility location Problem. In: 2020 5th International Conference on Logistics Operations Management (GOL), pp. 1–5 (2020)
2. Abbal, K., El Amrani, M., Benadada, Y.: Multi-level multi-capacitated facility location problem: formulation and resolution. In: 2022 IEEE 6th International Conference on Logistics Operations Management (GOL), pp. 1–6 (2022)
3. Abbal, K., El Amrani, M., Benadada, Y.: Dynamic multi-level multi-capacitated facility location problem: formulation and resolution. In: Benadada, Y., Mhada, F.-Z., Boukachour, J., Ouzayd, F., El Hilali Alaoui, A. (Eds.), Proceeding of the 7th International Conference on Logistics Operations Management, GOL'24. Springer Nature Switzerland, Cham, pp. 259–268 (2024)
4. Abbal, K., El Amrani, M., Benadada, Y.: An accelerated benders decomposition approach for the multi-level multi-capacitated facility location problem. Int. J. Ind. Eng. Prod. Res. **36**(1), 33–44 (2025)
5. Abdellah, A.R., Abdelmoaty, A., Koucheryavy, A.: Accurate v2x Traffic Prediction with Deep Learning Architectures. Frontiers in AI (2025)

6. Amrani, E., Benadada, Y., Gendron, B.: New Exact Method to Solve Multi-Capacitated Location Problem Using Set Partitioning Formulation, pp. 1973–1982 (2017)
7. Amrani, M.E., Benadada, Y.: Multi-stage algorithms for solving a generalized capacitated P-median location problem. Int. J. Adv. Comput. Sci. Appl. (IJACSA) **9**(5) (2018)
8. Amrani, M.E., Benadada, Y.: Solving the multi-capacitated location problem using a Lagrangian relaxation approach. In: 2018 4th International Conference on Logistics Operations Management (GOL), pp.1–5 (2018)
9. Aboah, A., Shoman, M.: A Vision-Based System for Traffic Anomaly Detection using Deep Learning and Decision Trees. IEEE CVPR Workshop (AICity) (2021)
10. El Amrani, M., Benadada, Y., Gendron, B.: Generalization of capacitated p-median location problem: modeling and resolution. In: 2016 3rd International Conference on Logistics Operations Management (GOL), pp. 1–6 (2016)
11. El Amrani, M., Benadada, Y., Gendron, B.: Multi-capacitated location problem: a new resolution method combining exact and heuristic approaches based on set partitioning. In: Talbi, E.-G., Nakib, A. (eds.) Bioinspired Heuristics for Optimization, pp. 109–121. Springer International Publishing, Cham (2019)
12. El Amrani, M., Bouanane, K., Abbal, K., and Benadada, Y.: Solving multi-capacitated facility location problem in traffic control systems using Lagrangian relaxation and decomposition approach. In: Benadada, Y., Mhada, F.-Z., Boukachour, J., Ouzayd, F., El Hilali Alaoui, A. (Eds.),Proceeding of the 7th International Conference on Logistics Operations Management, GOL'24. Springer Nature Switzerland, Cham, pp. 91–101 (2024)
13. Chen, H., Huang, J., Lu, Y., Huang, J.: Multi-scale spatio-Temporal graph neural network for urban traffic flow prediction. Sci. Rep. **15**, 26732 (2025)
14. Villa, J., Garcıa, F.: Intelligent infrastructure for traffic monitoring based on deep learning and edge computing. J. Adv. Transp. (2024)
15. Wang, M., Wu, L.: Traffic signal control with reinforcement learning based on region-aware cooperative strategy. IEEE Trans. Intell. Transp. Syst. **23** (2022)
16. Mohankumar, C.E., Manikandan2, A.: Decentralized Traffic Management with Federated Edge Ai: A Reinforced Transnet Model for Real-Time Vehicle Object Detection and Collaborative Route Optimization (2025)
17. Mutambik, I.: Iot-enabled adaptive traffic management: a multiagent framework for urban mobility optimisation. Sensors (2025)
18. Othman, S., Al-Heety, Z., Zakaria, A.A.: Traffic congestion control with emergency awareness and optimized communication infrastructure using reinforcement learning and non-dominated sorting genetic algorithm. IEEE Access (2022)
19. Chaymae, T., M'hamed, R.: Machine learning and 5g edge computing for intelligent traffic management. Int. J. Adv. Comput. Sci. Appl. **16**(6) (2025)
20. Shirsath, V., Vikas Kaul, R.S.K., Nemade, B.: Intelligent traffic management for vehicular networks using machine learning. ICTACT J. Commun. Technol. (2024)

Supply Chain 5.0 for Small and Medium-Sized Enterprises (SMEs): A Bibliometric Analysis

Malebo Mulaudzi[1(✉)] and Blessing Takawira[2]

[1] Johannesburg Business School, University of Johannesburg, Johannesburg, South Africa
mulaudzim@uj.ac.za

[2] Department of Transport and Supply Chain Management, Institute of Transport and Logistics Studies (Africa), University of Johannesburg, Johannesburg, South Africa
btakawira@uj.ac.za

Abstract. Supply Chain 5.0 has emerged as a new paradigm in supply chain management, combining advanced digital technologies with human-centric and sustainability-driven principles. For small and medium-sized enterprises (SMEs), this approach is particularly significant, as it offers pathways to resilience, competitiveness, and sustainable growth while addressing challenges such as financial constraints, digital readiness gaps, and workforce skills shortages. This article presents the landscape of Supply Chain 5.0 development in the SME context. It presents a systematic bibliometric literature review that describes the evolution of the concept and identifies the dominant themes, influential authors, institutions, and countries contributing to the discourse. The study highlights the rapid increase in publications since 2019, reflecting the growing importance of Supply Chain 5.0 in academic and practical debates. Key findings emphasise the role of technological enablers such as artificial intelligence, blockchain, and IoT, alongside organisational readiness and leadership commitment, in shaping SME adoption. We also propose a conceptual model for SMEs Supply Chain 5.0 and outline a research agenda to address gaps in sectoral adaptation, sustainability integration, and co-innovation ecosystems. Practical implications for managers and policymakers are discussed, underscoring the need for collaborative strategies to empower SMEs in building resilient and sustainable value networks.

Keywords: Supply Chain 5.0 · SMEs · Digital transformation · Sustainability · Resilience

1 Introduction

Supply chains have undergone profound transformations over the past decades, evolving from traditional linear models to digitally integrated and automated systems. The emergence of Industry 4.0 introduced technologies such as artificial intelligence, cloud computing, and advanced analytics, which reshaped supply chain operations through automation and data-driven decision-making [1]. However, the transition toward Supply Chain 5.0 represents a new paradigm that emphasises human-centricity, sustainability, and resilience, integrating technological innovation with ethical and social considerations [2].

Y. Benadada et al. (Eds.): GOL 2026, LNNS 2034, pp. 284–298, 2027.
https://doi.org/10.1007/978-3-032-29677-1_26

The distinctiveness of Supply Chain 5.0 for SMEs lies in its theoretical and practical departure from the capital-intensive, automation-first logic of Industry 4.0 toward a sociotechnical resilience model tailored to resource-constrained environments [3]. Unlike large enterprises that possess the financial slack to pursue systemic digital transformation for efficiency, SMEs operate under significant financial and skill limitations that render generic "one-size-fits-all" automation infeasible [4]. Consequently, Supply Chain 5.0 offers a modular and adaptive paradigm where human-centricity is not merely an ethical add-on but a strategic necessity. It reintroduces the workforce as a flexible asset to navigate sectoral misalignment and drive circular economy practices [5]. By prioritising customised, human-in-the-loop technologies over total automation, this paradigm enables SMEs to convert their inherent agility into a competitive advantage, achieving sustainability and resilience without the prohibitive infrastructure costs associated with traditional Industry 4.0 models [6].

Despite these advances, significant barriers remain. Financial constraints, digital readiness gaps, and resistance to change continue to impede SME adoption of advanced supply chain models [7]. Moreover, empirical studies reveal that while SMEs benefit from enhanced resilience and sustainability through digitalisation, sectoral differences and contextual limitations often restrict the generalizability of existing frameworks [8]. Addressing these challenges requires tailored strategies that integrate technological innovation with organisational readiness, leadership, and collaborative ecosystems [9].

This study aims to contribute to the growing discourse on Supply Chain 5.0 by conducting a bibliometric analysis focused on SMEs. Specifically, it seeks to answer two central research questions:

1. What is the current state of scholarly research on Supply Chain 5.0 in the context of SMEs?
2. Which authors, institutions, and countries have contributed most significantly to the study of Supply Chain 5.0 for SMEs?

By exploring these questions, the paper provides a comprehensive overview of the intellectual structure, thematic evolution, and future directions of Supply Chain 5.0 research. The findings will inform both academics and practitioners, offering insights into how SMEs can leverage human-centric and technology-driven pathways to build resilient, sustainable, and competitive supply chains. The remainder of the paper is organised as follows: Sect. 2 outlines the research methodology, Sect. 3 presents the bibliometric results, Sect. 4 discusses key themes and implications, and Sect. 5 concludes with recommendations for future research.

2 Research Methodology

To ensure methodological rigour and conceptual depth, this study adopts the Bibliometric–Systematic Literature Review (BSLR) framework, distinguishing itself by integrating quantitative network mapping with qualitative theory-building [10]. Unlike purely descriptive bibliometric studies, this approach employs a dual-database validation strategy, utilising Scopus as the principal source while cross-validating with Web of Science to ensure comprehensive coverage and mitigate indexing bias [11]. The review protocol focused on "Supply Chain 5.0" and "SMEs" within business and management domains (2019–2025), yielding a final corpus of 212 peer-reviewed articles following strict inclusion criteria [12]. To guarantee reproducibility and minimise selection bias, independent screening was conducted, achieving high inter-rater reliability (Krippendorff's $\alpha > 0.80$) [13]. Analysis was performed using VOSviewer and Bibliometrix (R-package), utilising bibliographic coupling and co-citation analysis to structurally map the field and empirically ground the proposed conceptual model [14].

Having established rigorous protocols for data selection and the BSLR analytical framework, the subsequent section presents the empirical findings derived from this dual-stage process. By applying these quantitative and structural mapping techniques to the final corpus of 212 documents, we aim to uncover the field's evolutionary trajectory and intellectual geography, providing the evidence base required to address the first research question regarding the current state of Supply Chain 5.0 for SMEs.

3 Results

This section details the descriptive and structural findings derived from the systematic analysis of the identified literature. To map the intellectual landscape of Supply Chain 5.0 for SMEs, we first examine the temporal evolution of publication output and the geographic distribution of research impact to gauge the field's maturity. Subsequently, we analyse the most influential sources, authors, and collaborative networks to identify the core communities and thematic intersections that drive this emerging domain, thereby laying the descriptive foundation for the critical discussion that follows.

Figure 1 Annual Scientific Production illustrates the temporal evolution of scholarly output on Supply Chain 5.0 in SMEs between 2019 and 2025. The data reveal a clear upward trajectory, with only two publications recorded in 2019 and no contributions in 2020–2021. A modest increase of seven articles was observed in 2022, followed by a more substantial rise to 29 in 2023 and 57 in 2024. The most significant surge occurs in 2025, reaching 117 publications. This pattern highlights the rapid acceleration of academic interest in the topic, suggesting that Supply Chain 5.0 has recently emerged as a prominent research domain, particularly in relation to SMEs, and is now consolidating as a distinct scholarly field. This exponential growth addresses RQ1, confirming that Supply Chain 5.0 for SMEs has graduated from a niche topic to a critical research domain driven by post-pandemic resilience and sustainability imperatives.

Figure 2 Most Relevant Sources highlights the journals and outlets that have contributed most significantly to the scholarly discourse on Supply Chain 5.0 in SMEs. The analysis shows that Sustainable Futures leads with 14 publications, followed by the Journal of Cleaner Production with 10 articles. Other prominent sources include Business Strategy and the Environment, Cogent Business and Management, and the Journal of Innovation and Knowledge, each of which has eight contributions. Additional outlets, such as Corporate Social Responsibility and Environmental Management, Logistics, and Technological Forecasting and Social Change, also demonstrate notable engagement, with 7–8 publications each. This distribution indicates that the research field is anchored in journals that emphasise sustainability, innovation, and strategic management, reflecting the interdisciplinary nature of Supply Chain 5.0 scholarship. The concentration of

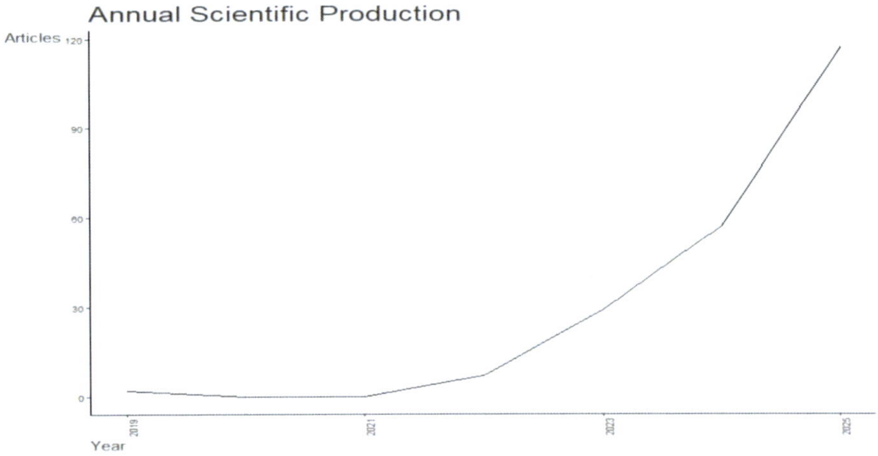

Fig. 1. Annual scientific production

publications in these sources underscores the growing alignment of SME-focused supply chain research with broader debates on sustainability, technological transformation, and organisational responsibility. The dominance of these interdisciplinary outlets suggests that Supply Chain 5.0 is theoretically framed not merely as an operational upgrade, but as a strategic mechanism for achieving environmental and social sustainability goals.

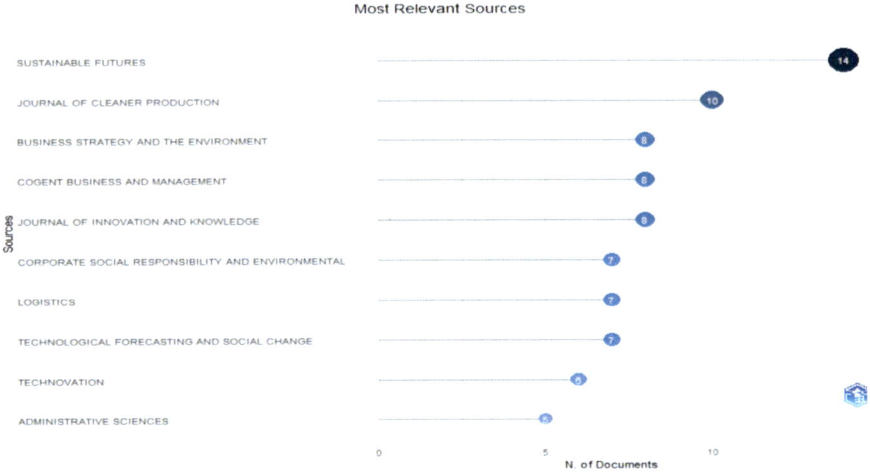

Fig. 2. Most relevant sources

Figure 3 Most Relevant Authors identifies the most prolific contributors to the scholarly discourse on Supply Chain 5.0 in SMEs. The analysis shows that Comite U, D'Adamo I, and Sharma M are the leading authors, each with three publications, underscoring their central role in shaping the field. A second tier of contributors—including

Akbari M, Alazzam FAF, Antony R, Atif S, Costa A, Crupi A, and Fernández Miguel A—have each produced two articles, reflecting a growing body of collaborative and diversified scholarship. The distribution of authorship suggests that while the field is still emerging, it is being driven by a relatively small but active group of researchers whose work spans sustainability, technological innovation, and SME-focused supply chain transformation. This concentration of contributions underscores the significance of these scholars in laying the intellectual foundations of Supply Chain 5.0 research, while also highlighting opportunities for broader participation and cross-disciplinary engagement in future studies. These findings highlight a developing intellectual structure in which foundational concepts are currently being negotiated by a core group of sustainability-focused scholars, thereby directly informing the "Who" component of RQ2.

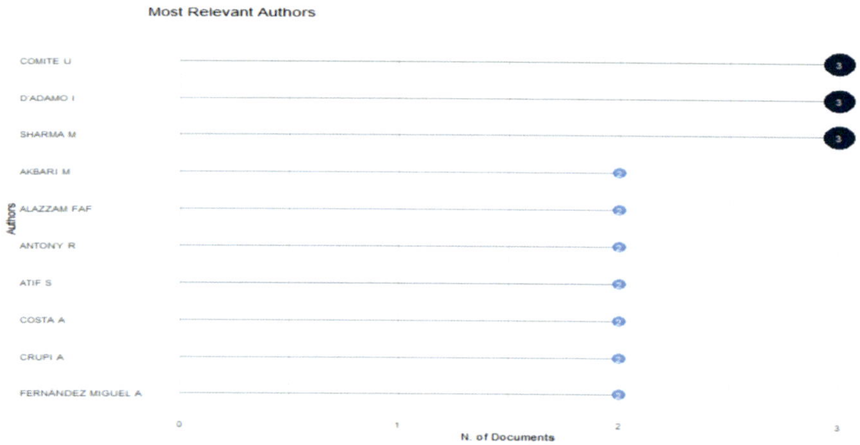

Fig. 3. Most relevant authors

The analysis of the most relevant authors (Fig. 3) reveals a dispersed intellectual structure characteristic of a nascent research domain, in which theoretical leadership is currently being established through the convergence of sustainability and digitalisation. The prominence of scholars such as Comite, D'Adamo, and Sharma suggests that the field's trajectory is being defined by researchers who successfully bridge the "twin transition" of circular economy and Industry 4.0, rather than by pure technologists. The relatively low publication volume among even the top contributors indicates that Supply Chain 5.0 for SMEs lacks a hegemonic core, reflecting a dynamic, exploratory phase where foundational concepts are still being negotiated across fragmented research silos. Consequently, the dominance of these authors indicates that the most impactful scholarship in this area is emerging at the intersection of environmental management and strategic operations, shifting the focus from simple digital adoption to resilience-oriented value creation.

Figure 4 Most Relevant Affiliations illustrates the institutional landscape of research on Supply Chain 5.0 in SMEs. The University of Economics Ho Chi Minh City emerges as the leading contributor with seven publications, reflecting the strong engagement of

Vietnamese scholars in advancing this field. A second cluster of institutions—including Chiang Mai University, Sapienza Università di Roma, Silesian University of Technology, Universidad de Jaén, and the University of Twente—each produced four articles, indicating a geographically diverse set of research hubs spanning Asia and Europe. Other notable contributors, such as Università degli Studi di Pavia, Cranfield University, and Da-Yeh University, each with two publications, further underscore the discourse's interdisciplinary and international character. This distribution demonstrates that Supply Chain 5.0 research is not concentrated in a single region but is instead shaped by a network of institutions across Europe and Asia, underscoring the global relevance of SME-focused supply chain transformation.

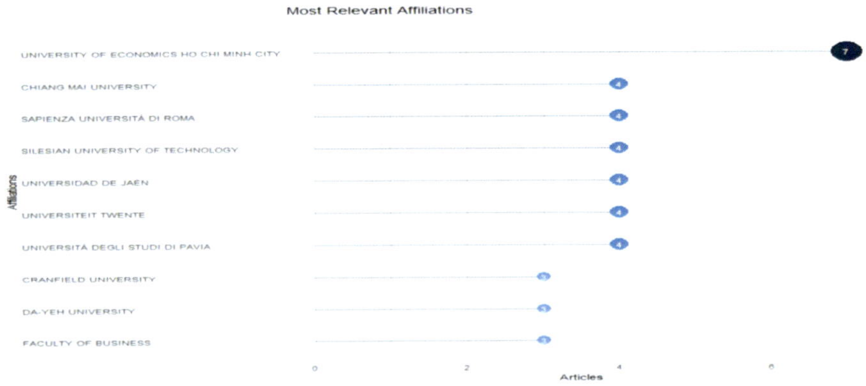

Fig. 4. Most relevant affiliations

Figure 5 Most Cited Countries demonstrates the geographical distribution of scholarly impact in the field of Supply Chain 5.0 for SMEs. The analysis reveals that Italy (383 citations) and the United States (375 citations) are the most influential contributors, reflecting their strong academic engagement and leadership in shaping the discourse. The United Kingdom follows with 249 citations, while China (170 citations), Sweden (151 citations), and Norway also show notable visibility. Other countries, such as Australia, Portugal, Spain, and Poland, contribute meaningfully, albeit with comparatively fewer citations. This distribution highlights the dominance of European and North American scholarship in advancing Supply Chain 5.0 research, while also indicating the growing participation of Asian countries. The citation patterns suggest that the field's intellectual foundations are being established through cross-regional collaboration, with Italy and the USA emerging as central hubs of influence. While the Global North currently leads in citations, the high activity in Vietnam and China indicates that the empirical application of Supply Chain 5.0 is increasingly driven by manufacturing hubs in emerging markets.

Figure 6 The word cloud provides a visual representation of the most frequently occurring keywords in the literature on Supply Chain 5.0 for SMEs. The dominant term is sustainable development, reflecting the centrality of sustainability in current discourse. Closely associated concepts such as "Industry 4.0," "digital transformation," "supply chain management," "circular economy," and "sustainable development goals" highlight

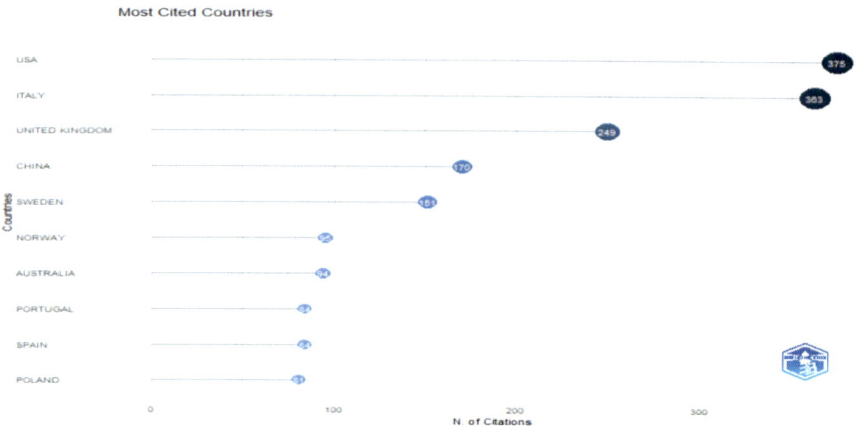

Fig. 5. Most cited countries

the intersection between technological innovation and sustainability imperatives. Other recurrent terms—including "information management," "small and medium-sized enterprise," "technological innovation," "manufacturing," and "decision making"—demonstrate the breadth of research themes, spanning organisational capabilities, environmental economics, and digital readiness. This distribution confirms that the scholarly focus on Supply Chain 5.0 is not limited to technological enablers but is deeply embedded in sustainability-driven agendas, aligning with the rapid growth in publications observed in recent years (see Fig. 1).

Fig. 6. Wordcloud

Figures 7 and 8 collectively illustrate the thematic and relational structure of Supply Chain 5.0 research for SMEs. The keyword co-occurrence map (Fig. 7) identifies five distinct clusters: technological foundations (Industry 4.0/5.0, digital technologies, green innovation), SME adoption and organisational readiness (digital transformation, structural equation modeling), sustainability and circular economy (linked to supply chain management and competitive advantage), industrial transitions, and global policy agendas (sustainable development goals). These clusters position the field at the intersection of technological innovation, sustainability, and organisational transformation. Complementing this, the Multiple Correspondence Analysis (Fig. 8) reveals multidimensional linkages among keywords, sources, and themes, showing three dominant groupings: sustainability-oriented journals (e.g., Journal of Cleaner Production) associated with circular economy and SDGs; technology-driven outlets (e.g., Technological Forecasting and Social Change) linked to innovation and Industry 4.0/5.0; and SME-focused sources emphasising organisational readiness and capabilities. Together, these visualisations confirm the interdisciplinary nature of Supply Chain 5.0 scholarship, integrating technological, environmental, and managerial perspectives within SME contexts. Structurally, these clusters validate the proposed conceptual model by distinguishing between "hard" technological enablers (Cluster 1) and "soft" organisational antecedents (Cluster 2), confirming that technology alone is insufficient for SME resilience.

Figure 9 The co-citation network visualises the intellectual structure of Supply Chain 5.0 for SMEs research by mapping the most frequently co-cited authors and their interconnections. The network reveals several distinct clusters, with Ali, Agrawal, and Bag emerging as central nodes, reflecting their strong influence and frequent co-citation across multiple studies. Other significant contributors, such as Ahmed and Ahmad, form additional clusters, indicating thematic specialisation around technological innovation, sustainability, and SME adoption. The presence of bridging nodes—such as undefined intermediary links—suggests cross-disciplinary integration, connecting sustainability-focused scholarship with technology-driven studies. This distribution aligns with the biblioshiny results, where highly cited countries (Italy, USA, UK) and leading journals (Sustainable Futures, Journal of Cleaner Production) dominate the discourse, reinforcing the idea that Supply Chain 5.0 research is shaped by a small but interconnected set of scholars whose work spans sustainability, digital transformation, and organisational readiness.

Figure 9 The co-citation network visualises the intellectual structure of Supply Chain 5.0 for SMEs research by mapping the most frequently co-cited authors and their interconnections. The network reveals several distinct clusters, with Ali, Agrawal, and Bag emerging as central nodes, reflecting their strong influence and frequent co-citation across multiple studies. Other significant contributors, such as Ahmed and Ahmad, form additional clusters, indicating thematic specialisation around technological innovation, sustainability, and SME adoption. The presence of bridging nodes—such as undefined intermediary links—suggests cross-disciplinary integration, connecting sustainability-focused scholarship with technology-driven studies. This distribution aligns with the biblioshiny results, where highly cited countries (Italy, USA, UK) and leading journals (Sustainable Futures, Journal of Cleaner Production) dominate the discourse, reinforcing the idea that Supply Chain 5.0 research is shaped by a small but interconnected set of scholars whose work spans sustainability, digital transformation, and organisational readiness.

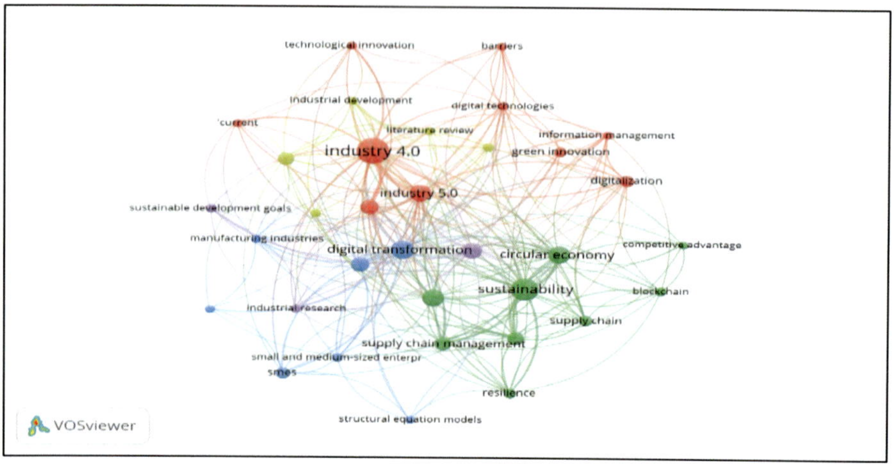

Fig. 7. Keyword co-occurrences

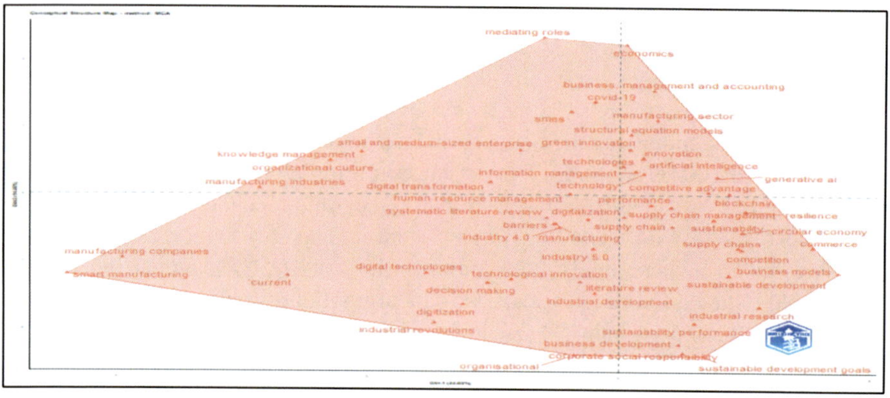

Fig. 8. Multiple correspondence analysis

Figure 10 The Collaboration Network by Countries illustrates the international research collaborations that underpin the field of Supply Chain 5.0 for SMEs. The network reveals several regional clusters, with the USA, Italy, China, Poland, and Mexico forming a dense red cluster, reflecting strong cross-country partnerships and citation linkages. A second grouping, represented in purple, includes the United Kingdom, Canada, Ireland, and Australia, highlighting the Anglo-Saxon research tradition and its emphasis on sustainability and digital transformation. The blue cluster, comprising India, Pakistan, and several European countries, demonstrates emerging contributions from South Asia and continental Europe. In contrast, the green cluster, which includes Malaysia, Indonesia, and Bangladesh, underscores Southeast Asia's growing role in SME-focused supply chain innovation. This distribution aligns with the biblioshiny results, where Italy and the USA lead in citation impact (Fig. 4), and institutions in Vietnam and Europe dominate affiliations (Fig. 4). Collectively, the collaboration network confirms that Supply

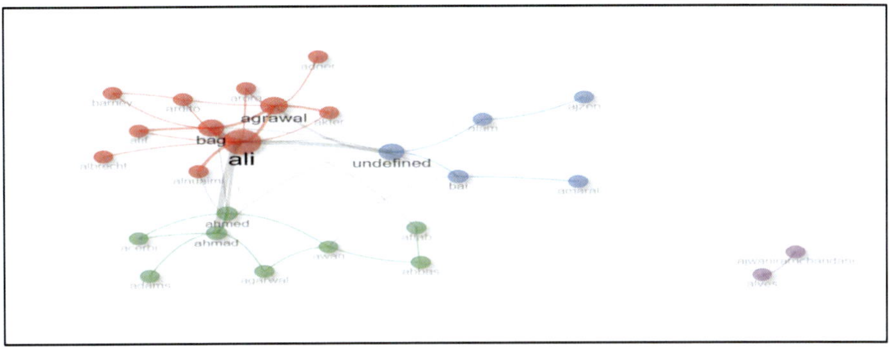

Fig. 9. Co-citation network

Chain 5.0 research is increasingly global, with knowledge production shaped by both established academic hubs and emerging economies.

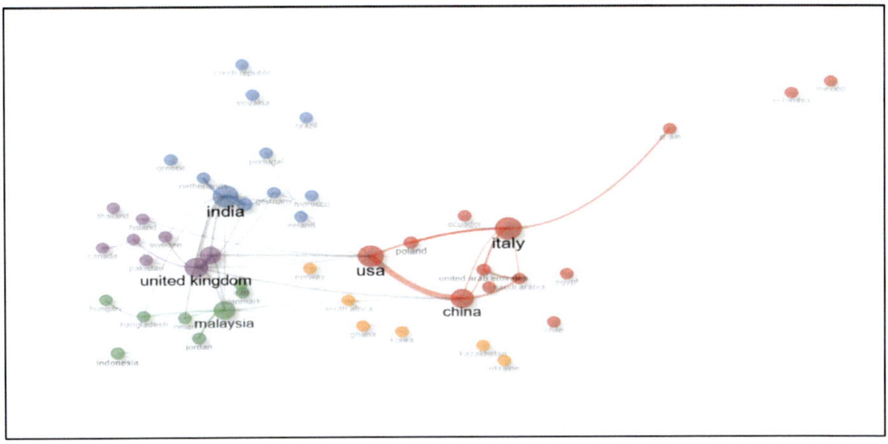

Fig. 10. Collaboration Network by countries

4 Discussion, Research Agenda, and Limitation

While the preceding results provided a structural map of the field, this section synthesises these empirical patterns to generate theoretical insights and address the second research objective. Moving beyond descriptive metrics, we critically interpret the identified bibliographic clusters and network relationships to construct the SMEs Supply Chain 5.0 Model, elucidating the conceptual linkages between technological enablers, organisational readiness, and sustainable outcomes in resource-constrained environments.

The trajectory of scholarly output evidenced in this study signals a fundamental theoretical pivot from the techno-centric determinism of Industry 4.0 toward a sociotechnical systems perspective characteristic of Supply Chain 5.0. While nascent digitalisation literature prioritised operational efficiency through automation, current discourse synthesises resilience, sustainability, and human-centricity into a cohesive survival strategy for volatile markets [2]. This evolution suggests that for Supply Chain 5.0 for SMEs represents a conceptual shift where technology functions not as a standalone objective but as a critical enabler for ethical compliance and circularity, challenging the traditional view that sustainability is a secondary outcome of digital maturity [15].

Theoretically, these findings delineate a distinct "SME Supply Chain 5.0" construct that diverges from large-enterprise models. The analysis reveals that successful transition in resource-constrained environments relies less on capital-intensive infrastructure and more on organisational readiness, leadership commitment, and the agile adoption of modular technologies like AI and blockchain to enhance transparency [16, 17]. By integrating these dimensions, the literature points toward a collective system design approach, positing that human values and workforce empowerment are necessary antecedents for SMEs to leverage digital tools effectively to enhance resilience [18]. Synthesising these patterns addresses RQ2, revealing that the field's intellectual structure is defined by the integration of digital assets with human-centric values, proving that resilience in SMEs is a socio-technical rather than purely technological outcome.

Future research should develop SME-specific, modular frameworks [1]; extend sectoral and regional coverage to services and micro-enterprises, especially in underrepresented regions [19]; integrate and empirically validate circular-economy and ESG principles [20]; examine governance for co-innovation ecosystems among SMEs, technology providers, and policymakers [21]; and conduct longitudinal studies on human-centric technology impacts [9]. Limitations include reliance on Scopus-indexed English-language journals (2019–2025), potential bias toward digitally mature SMEs [22], the inability of bibliometrics to capture contextual nuance across geographies and sectors, and the risk that rapid technological change outpaces SME implementation, rendering findings a snapshot of an evolving domain.

The conceptual architecture of the SMEs Supply Chain 5.0 Model (see Fig. 11) is empirically grounded in the structural relationships identified through the bibliometric analysis, specifically the intersection of keyword co-occurrences (Fig. 7), multiple correspondence analysis (Fig. 8), and co-citation networks (Fig. 9). These quantitative clusters do not merely represent topic frequencies but delineate the four core pillars of the proposed framework: Technological Enablers, Organizational Readiness, Human-Centric Design, and Strategic Benefits.

First, the *Technological Enablers* pillar is directly derived from the dense "technological foundations" cluster identified in Fig. 7, where terms such as "Industry 4.0," "AI," "blockchain," and "IoT" show high centrality [20]. However, the Multiple Correspondence Analysis (Fig. 8) reveals that these technological terms form a distinct grouping (Group 2) separate from SME-specific management constraints. This structural separation empirically validates the need for a distinct *Organisational Readiness* pillar. The bibliometric data suggest that for SMEs, advanced digital tools are not plug-and-play solutions but require specific antecedents—leadership commitment and digital maturity—evident in the distinct "SME adoption" cluster containing terms like "decision making" and "structural equation modelling" [15].

Furthermore, the model's *Human-Centric Design* and *Strategic Benefits* pillars are justified by the co-citation network (Fig. 9) and word cloud (Fig. 6). The dominance of "sustainable development" and "circular economy" as the most frequent keywords indicates that the ultimate output of Supply Chain 5.0 is not merely efficiency, but resilience and sustainability [2]. The co-citation analysis reinforces this by revealing bridging nodes that connect purely technical authors (e.g., Agrawal, Bag) with sustainability-focused scholars (e.g., Ahmad), signalling a theoretical convergence where technology serves human and environmental ends. Consequently, the conceptual model synthesises these disparate clusters into a coherent socio-technical system, positing that SMEs achieve Strategic Benefits only when Technological Enablers are mediated by Organisational *Readiness* and anchored in *Human-Centric Design* [9].

Synthesising the structural relationships identified in the bibliometric analysis, we propose the SMEs Supply Chain 5.0 Model, a theoretical framework comprising four interdependent constructs that govern the transition from digital adoption to socio-technical resilience.

1. Technological Enablers: Defined as the modular digital infrastructure—specifically AI, IoT, and blockchain—that provides the requisite transparency and data velocity for agile decision-making [16, 20].
2. Organisational Readiness: This construct encompasses the managerial antecedents required for technology assimilation, specifically leadership commitment, digital maturity, and change management capabilities, which differentiate successful adopters from those facing implementation failure [7, 15].
3. Human-Centric Design: A mediating construct representing the ethical alignment of technology with workforce empowerment, ensuring that automation augments rather than replaces human labour, thereby fostering creativity and reducing resistance [9, 18].
4. Strategic Benefits: The outcome variable, defined not merely as operational efficiency (Industry 4.0), but as the achievement of dual-purpose resilience and sustainability (circularity) within the supply network [23].

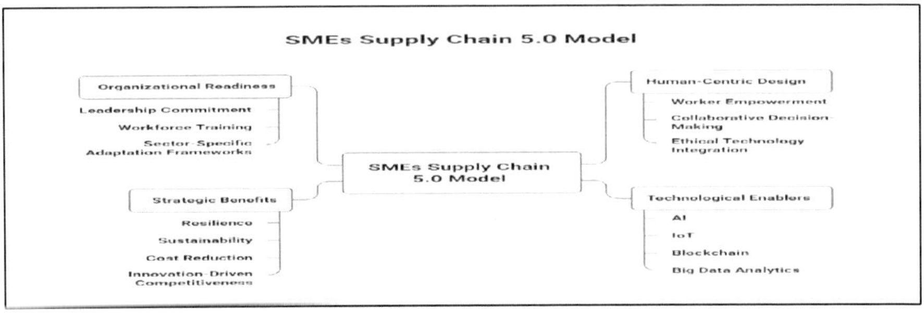

Fig. 11. SMEs supply chain 5.0 model

The explanatory power of the model lies in its causal logic: *Technological Enablers* are posited as necessary but insufficient conditions for Supply Chain 5.0. The model suggests that for SMEs, the path from technology to *Strategic Benefits* is fully mediated by *Human-Centric Design* and moderated by *Organisational Readiness*. Without the "soft" infrastructure of readiness and human-centric alignment, the "hard" infrastructure of digitalisation fails to yield resilience, leading instead to resource waste [24].

However, the applicability of this framework is subject to specific boundary conditions. First, Digital Maturity: the model assumes a baseline level of digitisation; for analogue micro-enterprises, the barrier to entry for *Technological Enablers* may render the Supply Chain 5.0 paradigm premature [3]. Second, Sectoral Context: the conceptual link between human-centricity and resilience is strongest in manufacturing SMEs, where human-machine interaction is physical, whereas service-sector SMEs may require distinct theoretical modifications [8]. Finally, Firm Size: the *Organisational Readiness* construct is highly sensitive to resource slack, suggesting that medium-sized enterprises may leverage this model more effectively than resource-scarce micro-enterprises, which face steeper "sectoral misalignment" risks [25].

5 Conclusion

Supply Chain 5.0 has emerged as a holistic paradigm that integrates advanced digital technologies with human-centric and sustainability-driven principles, offering SMEs a pathway to resilience, competitiveness, and ethical responsibility in volatile global markets. By adopting this model, SMEs can enhance operational efficiency, transparency, and sustainability, positioning themselves as proactive actors in resilient value networks. This study provides a comprehensive overview of Supply Chain 5.0 in the SME context through a bibliometric and systematic literature review, mapping the intellectual structure and synthesising empirical evidence. The analysis identifies four dominant themes: technological enablers (AI, IoT, blockchain, cyber-physical systems), human-centric design, organisational readiness, and sustainability integration, alongside influential authors, institutions, and countries. It highlights rapid growth in publication output since 2019, reflecting both quantitative expansion and qualitative evolution in the discourse on digital transformation and resilience. These findings underscore critical research gaps, including SME-specific frameworks, sectoral adaptation, and longitudinal validation across diverse contexts. In conclusion, Supply Chain 5.0 represents a transformative model balancing technological innovation with human values and sustainability imperatives, requiring collaborative, interdisciplinary efforts to advance theory and practice in SME supply chain management.

References

1. Ahmad, A., Syed, H.F., Joshi, J., Khan, S.A.: (Eds.) Barriers to the Transition from Supply Chain 4.0 (SC4.0) to Supply Chain 5.0 (SC5.0). IEEE International Conference on Industrial Engineering and Engineering Management (2022)
2. Ivanov, D.: The industry 5.0 framework: viability-based integration of the resilience, sustainability, and human-centricity perspectives. Int. J. Prod. Res. **61**(5), 1683–1695 (2023)
3. Matt, D.T., Modrák, V., Zsifkovits, H.: Industry 4.0 for Smes: Challenges, Opportunities and Requirements. 1–401 pp. (2020)

4. Machado, E.A., Scavarda, L.F., Caiado, R.G.G., Santos, R.S.: Industry 4.0 and sustainability integration in the supply chains of micro, small, and medium enterprises through people, process, and technology within the triple bottom line perspective. Sustainability (Switzerland) **16**(3) (2024)
5. Castillo, C., Otero-Romero, T., Alvarez-Palau, E.J.: Navigating the transition to industry 5.0: advancing sustainability, resilience, and human-centricity in Spanish supply chain management. Discov. Sustain. **6**(1) (2025)
6. Basova, Y., Dobrotvorskiy, S., Balog, M., Iakovets, A., Chelabi, M.A., Zinchenko, A.: Increasing sme supply chain resilience in the face of rapidly changing demand with 3d model visualisation. Int. J. Mechatron. Appl. Mechanics. **14**, 35–47 (2023)
7. Kumar, L., Sharma, R.K.: Adapting to Industry 4.0: evaluating SMEs preparedness through a comprehensive digital readiness assessment maturity model by validating stakeholders' perceptions. Bus. Proc. Manage. J. (2025)
8. Christiansen, A.M., Steger-Jensen, K., Vidskjold, K.: (Eds.) Shorter and More Agile Supply Chains in Industry 5.0: Case Studies Provide Insights into Manufacturing Management Theory. IFIP Advances in Information and Communication Technology (2026)
9. Cochran, D.S., Borbieva, N.O.: (Eds.) Collective System Design and Industry 5.0: Building Community, Resilience, and Sustainability at Purdue University Fort Wayne. Lecture Notes in Networks and Systems (2023)
10. Marzi, G., Balzano, M., Caputo, A., Pellegrini, M.M.: Guidelines for Bibliometric-Systematic Literature Reviews: 10 steps to combine analysis, synthesis and theory development. Int. J. Manag. Rev. **27**(1), 81–103 (2025)
11. Mongeon, P., Paul-Hus, A.: The journal coverage of Web of Science and Scopus: a comparative analysis. Scientometrics **106**(1), 213–228 (2016)
12. Tranfield, D., Denyer, D., Smart, P.: Towards a methodology for developing evidence-informed management knowledge by means of systematic review. Br. J. Manag. **14**(3), 207–222 (2003)
13. Hiebl, M.R.: Sample selection in systematic literature reviews of management research. Organ. Res. Methods **26**(2), 229–261 (2023)
14. Donthu, N., Kumar, S., Mukherjee, D., Pandey, N., Lim, W.M.: How to conduct a bibliometric analysis: an overview and guidelines. J. Bus. Res. **133**, 285–296 (2021)
15. Al Rawashdeh, S., Nasaj, M., Ahmad, S.Z.: Driving circular economy adoption through top management commitment and organisational motivation: a quantitative study on small- and medium-sized enterprises. Int. J. Organ. Anal. (2024)
16. Dey, P.K., Chowdhury, S., Abadie, A., Vann Yaroson, E., Sarkar, S.: Artificial intelligence-driven supply chain resilience in Vietnamese manufacturing small- and medium-sized enterprises. Int. J. Prod. Res. **62**(15), 5417–5456 (2024)
17. Asadi, M., Hashemkhani Zolfani, S., Pamucar, D., Salimi, J., Saberi, S.: The appropriation of blockchain implementation in the supply chain of SMES based on fuzzy LMAW. Eng. Appl. Artif. Intell. **123** (2023)
18. Cardamone, M., Cimino, A., Cosma, A., Filice, L., Longo, F., Nicoletti, L., et al.: (Eds.) Ethical Technology Design for Production Systems: A Human-Centered Approach at the LEONARDO Learning Factory. Procedia Computer Science (2025)
19. Chen, C.L.: Value creation by SMEs participating in global value chains under industry 4.0 trend: case study of textile industry in Taiwan. J. Glob. Inf. Technol. Manag. **22**(2), 120–145 (2019)
20. Han, Y., Fang, X.: Systematic review of adopting blockchain in supply chain management: bibliometric analysis and theme discussion. Int. J. Prod. Res. **62**(3), 991–1016 (2024)
21. Benitez, G.B., Ayala, N.F., Frank, A.G.: How can SMEs Participate Successfully in INDUSTRY 4.0 Ecosystems? The Digital Supply Chain, pp. 325–339 (2022)

22. Alam, M.C., Setiawan, B., Toiba, H., Maulidah, S., Rahman, M.S.: Supply chain management practices of SMEs in developing countries: bibliometric and systematic reviews. Agric. Econ. (Czech Republic). **71**(2), 99–112 (2025)
23. Shojaei, P., Rashidi Alavije, A., Askarifar, K.: Supply chain resilience improvement in SMEs: the role of industry 4.0 technologies. Glob. J. Flexible Syst. Manag. **26**(4), 865–81 (2025)
24. Durugbo, C.M.: Uncertainties of supply chain transition and integration for Industry 5.0: a review of operations management literature. Int. J. Prod. Res. **63**(18), 6915–6937 (2025)
25. Bravi, L., Pierli, G., Murmura, F., Musso, F.: Integrating industry 4.0 and industry 5.0: a study of hybrid adoption in SMEs and its impact on sustainability, human-centricity, resilience. Int. J. Prod. Res. (2025)

From Data to Decisions: Adaptive Supply-Chain Management Through Simulation, Data Analytics and Digital Twins

Digital Twin-Driven Real-Time Decision Support in Maritime Transport: A Literature Review and Analytical Study of Emerging Big Data Paradigms

Hamza El Mastik[1(✉)] and Fatima El Khoukhi[2]

[1] IA Laboratory, FS Meknes, Moulay Ismail University of Meknes, Meknes, Morocco
h.elmastik@edu.umi.ac.ma

[2] IA Laboratory, FLSH Meknes, Moulay Ismail University of Meknes, Meknes, Morocco
f.elkhoukhi@flsh.umi.ac.ma

Abstract. Digitalization in maritime operations advances in steps, not in a single leap. We use digital twin (DT) in a narrow sense: a live, asset-level mirror that synchronizes selected variables (position, queues, equipment state) with a computational model. Our review maps Big Data Analytics (BDA) to three families—conventional, learning-based, hybrid—and asks what they deliver in practice. We screened 32 studies from 2019–2025; papers without a concrete task or measurable outcome were excluded. Typical tasks include ETA/trajectory prediction, berth and yard planning, condition-based maintenance, and abnormal-behaviour flags. Findings: learning-based models usually outperform simple rules for prediction/detection, but the effect is uneven and sensitive to data quality; explainability remains a concern for safety-critical use. Hybrid designs keep explicit constraints while letting data capture patterns the rules miss. DTs generally host this combination, although deployments are mostly asset- or subsystem-scoped. Despite better plumbing (IoT feeds, cloud ETL; sometimes blockchain for auditability), fragmentation, weak standardization, and limited interpretability still slow routine adoption.

Keywords: Digital Twin · Big Data Analytics · Real-Time Decision Support · Maritime Logistics · Literature Review · Artificial Intelligence · Hybrid Approaches

1 Introduction

Maritime transport still carries roughly four-fifths of global merchandise trade [15]. The pressure points are well known: quicker turnarounds, tighter safety and security, and lower emissions—all at once. In this setting, BDA is used less as a dashboard and more as a decision aid. Typical inputs are heterogeneous: AIS tracks, engine and hull sensors, numerical weather forecasts, port-call and yard activity logs. Combined, they support tasks that practitioners care about on a daily basis: ETA and delay anticipation,

Y. Benadada et al. (Eds.): GOL 2026, LNNS 2034, pp. 301–311, 2027.
https://doi.org/10.1007/978-3-032-29677-1_27

berth and tug scheduling, routing under weather and traffic, and early warnings for unusual behaviour [3, 4, 13]. Progress, however, is uneven. Data remain scattered across legacy and vendor-locked systems; interfaces are inconsistent; and analytical capacity is not the same from one operator to another [10, 30]. Classical descriptive or rule-based methods are transparent and cheap to implement, but they struggle when the data arrive faster, from more sources, and with shifting patterns [9]. Learning-based models help with prediction and anomaly detection, yet they come with their own issues: labels are often limited, data quality varies by port or fleet, and model decisions can be difficult to justify in safety-critical contexts [14]. For this reason, hybrids—plain rules plus learned components—are attracting interest because they retain domain constraints while improving accuracy. DTs tie these layers together. In practice, a maritime DT is rarely the whole system; it is usually an asset- or subsystem-level mirror (a vessel, a berth, a crane, sometimes a small terminal slice). Even so, synchronizing a few state variables (position, machinery health, queue and resource status) with a computational model enables what-if checks, maintenance scheduling windows, and smoother coordination among actors [1, 6]. This article reviews 32 studies (2019–2025) on DT-enabled and data-intensive decision support for maritime freight. We compare conventional, learning-based, and hybrid approaches across route/ETA prediction, port and terminal operations, maintenance, environmental performance, and resilience. We also examine how DT architectures act as the common scaffold that hosts these methods and their data flows. The study follows a descriptive, conceptual mapping logic rather than an empirical or systematic methodology. The remainder of the article is organized as follows. Section 2 sets the conceptual background for the three paradigms and situates DTs within it. Section 3 synthesizes the main application areas and data sources. Section 4 compares performance and limitations while noting recent shifts in practice. Section 5 closes with implications for deployment and next steps. The remainder of this article is organized as follows. Section 2 explains the conceptual context of the three paradigms and situates the concept of the digital twin within this framework. Section 3 presents the main application areas. Section 4 compares their performance and limitations while highlighting current trends. Finally, Sect. 5 concludes with the main findings and some directions for future research.

2 Conceptual Framework: Traditional, AI-Based, and Hybrid Approaches

BDA in the maritime domain can be organized into three categories—traditional, AI-based, and hybrid—whose complementary strengths underpin DT decision-support systems. Traditional methods rely on descriptive statistics, regression techniques, and predefined rules; they are transparent, easy to audit, and inexpensive, which makes them effective for compliance monitoring and for well-structured datasets [9, 10]. Their limits become clear when inputs arrive as heterogeneous, high-velocity streams from ships, ports, and sensors—often in (near) real time—where predictive performance tends to deteriorate quickly [29, 30]. AI-based techniques (machine learning, deep learning, and at times reinforcement learning) recover part of this loss by modeling non-linear relations and coping with unstructured or very large data; in the maritime literature they are used for fuel-consumption estimation, anomaly detection, and ship-trajectory/ETA prediction [3, 16]. Even so, these models often require large labeled datasets and may function as "black boxes," which makes explanation and validation difficult in safety-critical contexts [14]. A practical response is to pair physical or rule-based components with learned modules so that accuracy gains do not come at the expense of interpretability [19]. This aligns with DT practice: a live process—such as a vessel leg, a berth queue, or a crane cycle—is mirrored by a virtual model that refreshes continuously [1, 6]. Inside that mirror, expert rules can enforce navigation and safety constraints while AI predictors refine ETA, flag off-nominal behaviour, or anticipate equipment condition; together they support routine decisions in route planning and port management [7, 24]. In short, the DT provides the scaffold that keeps the transparency of traditional tools while adding the adaptability of AI, within a single workflow aimed at real-time, data-driven maritime operations. It should be noted that this study does not rely on a systematic method for identifying fragmentation points; rather, it is based on a targeted qualitative review of peer-reviewed publications.

3 Literature Review

Digitalization in maritime operations advances stepwise. In this review, we map BDA to three families—conventional, learning-based, and hybrid—and focus on what they deliver in practice. We queried ScienceDirect, Scopus, Web of Science, and Google Scholar for 2019–2025 and retained only studies reporting a concrete operational task with a measurable outcome (ETA/trajectory prediction, berth or yard planning, condition-based maintenance, abnormal-behaviour alerts, sustainability metrics). Papers without an evaluable result were excluded. Table 1 summarizes the set by domain, data source, and method.

Across navigation and traffic safety, AIS remains central: Ma et al. assess navigation risk in congested waters using machine learning [4]; Spyrou-Sioula et al. couple routing with weather to reduce cargo losses [16]; Stach et al. review anomaly-detection tools reflecting a broader shift toward AI for situational awareness [9]. More recent pipelines extract traffic networks directly from AIS using unsupervised steps—early DT-type monitoring of vessel flows—by Guo et al. [23] and by Liu et al. [24]. For ports and terminals, Long et al. blend time-series forecasting with LSTM to predict throughput under disruption and beat traditional baselines [8]; Rao et al. examine predictive berth planning and congestion; Madusanka et al. describe a metaverse-like, digital twin–based logistics environment designed to improve safety and efficiency in maritime operations [15]. Surucu-Balci et al. underline the "plumbing" (cloud plus blockchain) needed to break silos and secure exchanges [18]. On asset health, Simion et al. mine machine-sensor streams to flag early failures and schedule interventions [5]; Bertagna et al. mix IoT sensing, analytics, and expert knowledge in a hybrid design [28]; Chen et al. show that data-driven voyage and routing optimization can significantly improve operational efficiency and reduce fuel consumption, linking predictive analytics to sustainability goals [16]. Sustainability and governance threads include DT-based routing aligned with decarbonization policies by Wei et al. [10] and DT evaluation of the Carbon Intensity Indicator (CII) by Vasilikis et al. [11]; Zhou et al. pair big-data and geospatial workflows for real-time risk maps [29]; Yuan et al. discuss data monetization and the rising strategic value of maritime data assets [19]. At the transformation layer, several strands recur. Tijan et al. argue for tighter integration of IoT, AI, and blockchain within "smart" maritime systems [20]. Tang et al. chart the shift toward real-time decision analytics and streaming pipelines [22]. Zhang and Lam document adoption barriers—management resistance, cybersecurity exposure, and fragmented ownership—inside maritime organizations [10]. Zhao et al. corroborate these obstacles with additional industry cases [30]. Madusanka et al. review practical roll-outs across multiple functions [31]. Kautsar et al. report concrete implementations for maintenance, energy management, and terminal operations [32].

To structure the findings more precisely, the analysis can be aligned with the Supply Chain Operations Reference (SCOR) model processes to illustrate concrete fragmentation points throughout the maritime logistics chain.

- Plan: Port scheduling forecasts are often not synchronized with shipping line schedules, leading to berth allocation conflicts and inefficient sequencing.
- Source: Procurement systems (for fuel, spare parts, services) operate in isolation, with little cross-actor visibility, making joint planning or bundling difficult.
- Make (Operations): Real-time equipment and engine data are rarely integrated into port planning systems, limiting dynamic coordination between ships and terminals.
- Deliver: Tracking platforms remain fragmented across modes (sea, road, rail), preventing end-to-end visibility and delaying decision-making during intermodal transitions.
- Return: Empty container repositioning is often inefficient due to the lack of shared information between carriers, leading to redundant moves or underutilized stock.

A formal mapping table connecting SCOR processes to these fragmentation issues could enhance clarity and operational relevance—this will be considered in future work.

In terms of integration levers, several actionable solutions emerge: adopting open data exchange standards (e.g., EDIFACT, APIs), deploying collaborative platforms like Port Community Systems, and establishing real-time system interconnectivity across stakeholders. These levers directly address the specific disconnections identified at each SCOR stage.

Overall, the literature shows a sector in transition: isolated BDA pilots are giving way to more integrated, near-real-time ecosystems. Traditional methods remain useful where clarity and auditability dominate; learning-based techniques lead on data-rich prediction and detection; hybrids keep constraints explicit while exploiting patterns; and DTs act as the integration layer that ties these elements and data flows into operations-grade decision support.

The results in Table 1 sketch how BDA is being applied across maritime domains and how the toolkit has shifted in recent years. The mix of methods is broad. Earlier work tends to set out conceptual frameworks and policy-oriented analyses that paved the way for data-driven decision support. More recent AI-based studies push predictive modelling and near-real-time analysis, with tangible gains for navigation, port operations, and maintenance. Hybrid designs sit between the two: they keep expert rules visible while letting machine-learning components capture residual patterns, which improves adaptability without losing interpretability. These patterns also appear in the figures: Fig. 1 compares the share of each methodological family, and Fig. 2 tracks their use from 2019 to 2025. The post-2020 rise of AI-based and hybrid approaches aligns with

Table 1. Comparison of a set of representative studies applying digital twin technologies in intelligent maritime transport and logistics.

Ref. (Year)	Focus/application domain	Data sources & methods	Method type	Key findings/contributions
[1] (2024)	BDA for maritime shipping (challenges & opportunities)	Conceptual analysis (literature-based)	Traditional	Identified key challenges and future directions for BDA adoption in shipping
[12] (2020)	Digitalization in shipping—industry outlook (ICS)	Expert insights; industry case highlights (report)	Traditional	Outlined emerging digital technologies in shipping and the need to navigate digital transformation for future competitiveness in maritime freight
[25] (2020)	ML for AIS data-driven maritime research (comprehensive review)	Review of ML applications on AIS datasets	Traditional	Provided a comprehensive review of machine-learning techniques for AIS data in maritime research, summarizing key applications (routing, anomaly detection, etc.) and identifying research gaps
[11] (2019)	Global maritime transport—annual review (UNCTAD)	Industry data and trend analysis (report)	Traditional	Provided a comprehensive overview of maritime trade and highlighted the impact of digitalization on shipping operations (context for BDA adoption)
[10] (2019)	Barriers to BDA adoption in maritime organizations	Delphi-AHP-TOPSIS multi-criteria framework	Traditional	Prioritized managerial, cultural, and technical challenges to BDA adoption in smart maritime industries

(*continued*)

Table 1. *(continued)*

Ref. (Year)	Focus/application domain	Data sources & methods	Method type	Key findings/contributions
[6] (2024)	Predictive optimization of port operations	Data-driven predictive modeling	AI	Enhanced berth scheduling and resource allocation through predictive analytics
[5] (2024)	Predictive maintenance in maritime transport	AI-driven predictive analytics	AI	Improved ship maintenance reliability and operational efficiency via AI
[8] (2023)	Port throughput forecasting under disruption	Time-series data analytics; forecasting (LSTM)	AI	Modeled port performance under unexpected disruption scenarios (hybrid model outperformed benchmarks)
[24] (2023)	Data mining for maritime traffic network extraction	AIS data mining method; graph construction	AI	Developed a method to construct maritime traffic networks from AIS data, facilitating improved transport management through data-driven routing insights
[7] (2019)	Fuel consumption prediction for ships	Voyage data; machine learning models	AI	Achieved accurate en-route fuel consumption predictions based on operational data
[15] (2023)	Smart port logistics	Big data integration; AI + virtual environment	Hybrid	Proposed an AI-based logistics framework for enhanced port productivity and safety, combining domain knowledge with AI
[26] (2020)	Ship-ship collision risk evaluation under environmental conditions	AIS and hydrometeorological data; hybrid risk modeling	Hybrid	Developed a DBA method to evaluate ship collision risk under complex weather/sea conditions, combining data-driven analysis with expert knowledge for better risk assessment

the spread of DT-enabled analytics, where simulation and integrated data streams support continuous, operational decision-making.

4 Analysis and Discussion

This discussion draws on the studies in Table 1 and the distribution in Fig. 1. Together, they show how the three methodological families—traditional, AI-based, and hybrid—cover the main maritime application areas. Figure 1 makes the shift clear: after 2020, AI-based and hybrid work take a larger share. That timing aligns with the spread of DT-enabled analytics, where live data streams are coupled with simulation to guide decisions as operations unfold. In this context, digital twins serve both as a common scaffold that links the analytical paradigms and as a practical engine for moving from static, one-off analyses to adaptive, operations-ready maritime systems.

4.1 Emerging Trends in Smart Maritime Analytics

Recent work shows a sharp tilt toward AI in maritime analytics. Traditional statistics and rule-based tools lose ground to learning models that absorb large, messy inputs, with fuel-consumption prediction as one clear case [7] and port-throughput estimation under disruption as another [8]. The focus shifts from descriptive reporting to near real time decisions for crane scheduling, route planning, and just-in-time arrivals. Stream processing replaces batch runs and triggers immediate actions. This pairs with the rise of digital twins that keep live views of ships, ports, or logistics subsystems, from a metaverse-style port logistics setup [15] to predictive maintenance fed by machine sensors [5]. A second theme keeps growing: hybrids that mix domain knowledge with learned parts. They embed physical laws or expert constraints inside data-driven models, lift accuracy, and keep explanations intact, which answers black-box worries; see energy optimization [16] and collision-risk analysis [26]. Many DT stacks follow the same logic, with simulation models of vessels or port processes updated by operational data to produce forecasts people can audit. The tech around it holds the line: IoT for continuous sensing, cloud for scalable compute and storage, blockchain for transparent and secure exchange across stakeholders [18]. Figures 1and2 pull the threads together: AI-based work sits near half of recent studies, traditional near one third, hybrids around one fifth, and after 2020 the AI plus hybrid share climbs fast; by 2024 these two camps dominate digital transformation efforts, which signals a lasting move to AI-driven and hybrid analytics built on broader data sharing.

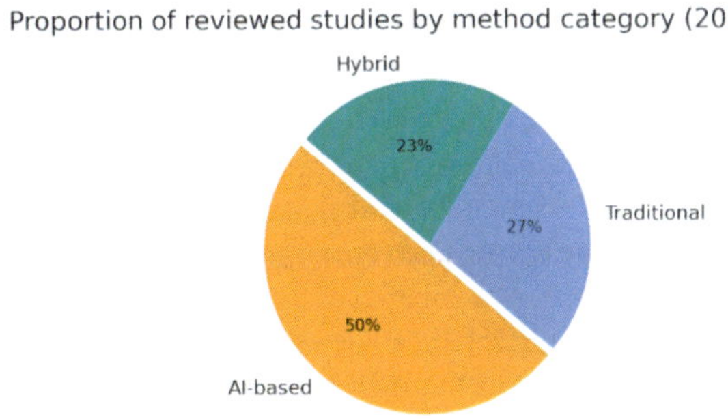

Fig. 1. Proportion of reviewed studies by method category (2019–2025): AI-based, traditional, and hybrid approaches

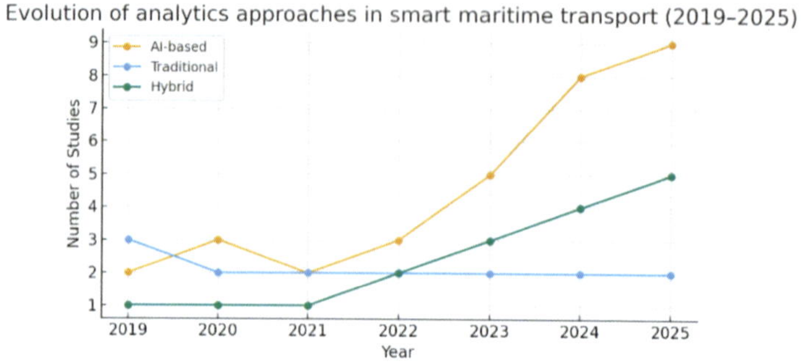

Fig. 2. Evolution of analytics approaches in smart maritime transport from 2019 to 2025.

4.2 Strengths and Limitations by Approach

Traditional methods such as structured databases, statistical tools, and expert rules give clarity, full audit trails, and low compute. They fit stable contexts and decisions that must stay traceable, for example compliance checks and simple ETA calculations [10]. They slip once inputs arrive fast and messy, like AIS streams or maintenance logs, or when behavior turns nonlinear, which shows up often in current operations [9, 30]. AI based approaches lift predictive power and adaptability; they beat classical baselines in pattern and anomaly detection and they can ingest mixed inputs, AIS plus weather plus engine signals, for routing, berth allocation, and cargo planning [9, 25]. The price is familiar: models that act like black boxes, large labeled datasets, and higher runtime costs on constrained platforms, all of which slows broad industrial uptake [14]. Hybrid designs step in by pairing physics or rule based parts with learned modules, which improves the accuracy–interpretability balance and often trims data and compute needs because domain structure guides learning. Typical cases include digital twins that fuse a physics fuel use model with AI corrections, or voyage planners where AI suggested routes pass safety checks before adoption [16, 26]. In practice, hybrids look like the most pragmatic path to reliable, scalable analytics in real maritime settings, joining transparency with performance and forming a solid base for DT centric decision support.

4.3 Visual Comparison Based on Performance Criteria

Figure 3 compares the three families—traditional, AI-based, and hybrid—on six axes: predictive accuracy, interpretability, data needs, adaptability, computational load, and ease of adoption. Traditional methods score well on interpretability and low compute, so they are easy to deploy and acceptable for traceable tasks (compliance checks, routine analyses with limited data) [10]; they fare worse on accuracy and adaptability, and they struggle with continuous AIS streams and the non-linear behaviour common in modern operations [9, 30]. AI-based approaches lead on accuracy and adaptability, spotting patterns and forecasting events such as anomalies or port-traffic shifts while scaling to heterogeneous, high-volume inputs [9, 25]; the trade-offs are reduced interpretability and trust, heavy labeled-data needs, and higher compute/infra requirements—constraints that complicate use in safety-critical or resource-constrained settings [14]. Hybrids are the most balanced: by embedding domain rules or physics inside learned components, they keep explanations intact and data/compute manageable while capturing residual patterns (e.g., fuel-use models or voyage planners that fuse physics with AI corrections) [16, 26]. In short, the radar chart suggests AI-based models lead in raw analytics, traditional models remain strong baselines for transparent auditing, and hybrids offer the most practical trade-off for operations, especially within DT-centric workflows.

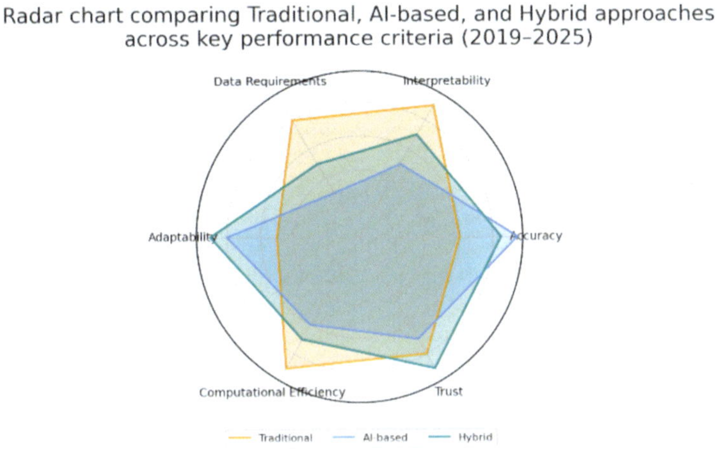

Fig.3. Radar chart comparing optimization approaches across multiple criteria

5 Conclusion and Future Directions

Based on 32 studies from 2019–2025, the field is clearly moving from small pilots to working, DT-backed decision tools across navigation (ETA windows, weather routing), port operations (berth/tug scheduling), maintenance, and sustainability. Three signals came up repeatedly in our screening: wider use of AI for prediction; growing reliance on hybrids that keep rules/physics visible while letting models learn the rest; and sturdier data plumbing—IoT feeds, cloud pipelines, and, at times, blockchain for audit trails. The reported gains are concrete: dynamic routing and just-in-time arrivals, sharper resource allocation, and predictive maintenance that cuts failures and fuel burn. Two classes of limits explain the slower industrial uptake: classical methods don't scale to today's data volumes, and many AI systems are still hard to justify in safety-critical settings [14]; governance frictions add to this—fragmented responsibilities, uneven standards, and persistent silos [10, 30]. These limitations directly justify the need for the perspectives discussed earlier (digital twins, blockchain interoperability, etc.), as they target the core issues identified in this study: lack of real-time integration, data silos, and system heterogeneity. The priorities follow directly: interoperable data standards and secure sharing for connected DT ecosystems; explainable AI with human-in-the-loop controls; large-scale validation and simulation with regulators and operators; stronger data governance and incentive models (use blockchain only where it helps); and transfer learning plus lessons borrowed from aviation, logistics, and smart-city systems. With steady investment in both technology and coordination—skills, standards, and joint trials—DT-based, hybrid analytics can mature into reliable, resilient, lower-emission maritime operations over the next decade.

References

1. Rao, B.V.R., Singh, A., Kumar, M.: Challenges and opportunities of big data analytics for maritime and shipping industry. Int. J. Emerg. Technol. Maritime Syst. **5**(1), 15–27 (2024)
2. Jović, M., Tijan, E., Marx, R., Gebhard, B.: Big data management in Maritime transport. Pomorski Zbornik: J. Maritime Transp. Sci. **57**(1), 123–141 (2019)
3. Ma, Q., Tang, H., Liu, C., Zhang, M., Zhang, L.: A big data analytics method for the evaluation of maritime traffic safety using automatic identification system data. Ocean Coast. Manag. **251**, 107077 (2024)
4. Spyrou-Sioula, K., Kontopoulos, I., Kaklis, D., Makris, A., Tserpes, K.: AIS-enabled weather routing for cargo loss prevention. J. Marine Sci. Eng. **10**(11), 1755 (2022)
5. Simion, D., Postolache, F., Fleacă, B.: AI-driven predictive maintenance in modern maritime transport: Enhancing operational efficiency and reliability. Sci. Bull. Naval Acad. **27**, 45–57 (2024)
6. Rao, A.R., Wang, H., Gupta, C.: Predictive Analysis for Optimizing Port Operations. arXiv preprint arXiv:2401.14498 (submitted to IEEE BigData 2024) (2024)
7. Hu, Y., Li, Z., Wang, X., Chen, G.: Fuel consumption prediction for ships based on machine learning and voyage data. In: Proceedings of the IEEE International Conference on Data Science, pp. 185–192 (2019)
8. Long, T.-L. (Cuong), You, S.-S., Lee, P.T.-W.: Data analytics and throughput forecasting in port management systems under disruptions: a hybrid approach. Res. Transp. Bus. Manag. **52**, 101066 (2023)
9. Stach, T., Kinkel, Y., Constapel, M., Burmeister, H.: Maritime anomaly detection for vessel traffic services: a survey. J. Marine Sci. Eng. **11**(6), 1174 (2023)
10. Zhang, X., Lam, J.S.L.: A fuzzy Delphi–AHP–TOPSIS framework to identify barriers in big data analytics adoption: case of maritime organizations. Marit. Policy Manag. **46**(8), 977–994 (2019)
11. UNCTAD: Review of Maritime Transport 2019. United Nations Conference on Trade and Development, New York and Geneva (2019)
12. International Chamber of Shipping: Digitalization in Shipping: Navigating the Future—Annual Review 2020. International Chamber of Shipping, London (2020)
13. Lu, L., Wang, Y., Cheng, J.: Big data and artificial intelligence in the maritime industry: a bibliometric review. Marit. Policy Manag. **48**(8), 1159–1176 (2021)
14. Di Vaio, A., Varriale, L., Alvino, F.: Key determinants of big data analytics adoption in sustainable maritime transport management. J. Clean. Prod. **253**, 119926 (2020)
15. Madusanka, N.S., Fan, Y., Yang, S., Xiang, X.: Digital twin in the maritime domain: a review and emerging trends. J. Marine Sci. Eng. **11**(5), 1021 (2023)
16. Chen, Y., Mao, W.: An isochrone-based predictive optimization for efficient ship voyage planning and execution. In: IEEE Transactions on Intelligent Transportation Systems (2024)
17. Balci, G., Surucu-Balci, E.: Blockchain adoption in the maritime supply chain: Examining barriers and salient stakeholders in containerized international trade. Transp. Res. Part E: Logistics Transp. Rev. **156**, Art. no. 102539 (2021). https://doi.org/10.1016/j.tre.2021.102539
18. Surucu-Balci, E., İris, Ç., Balci, G.: Digital information systems in maritime supply chains: a literature review on blockchain and cloud-enabled platforms. Maritime Transp. Res. **1**(1), 100001 (2024)

19. Yuan, Z., Yu, X., Jiang, Y., Sun, J., Liu, Z., Li, B.: Current Status and Governance of Data Asset Monetization in the Global Maritime Industry: A Comparative Study of the US, Europe, and China. SSRN Working Paper No. 4690911 (2024)
20. Tang, Y., Xie, Y., Vo, A., Zhang, D.: Maritime logistics and digital transformation with big data: a review and research trend. Maritime Bus. Rev. **8**(4), 350–368 (2023)
21. Mohandu, A., Kubendiran, M.: Survey on big data techniques in intelligent transportation systems (ITS). Int. J. Adv. Sci. Technol. **29**(7), 6026–6034 (2020)
22. Tijan, E., Jović, M., Aksentijević, S., Pucihar, A.: Digital transformation in the maritime transport sector: a literature review. Technol. Forecast. Soc. Chang. **170**, 120879 (2021)
23. Guo, Z., Qiang, H., Xie, S., Peng, X.: Unsupervised knowledge discovery framework: from AIS data processing to maritime traffic networks generation. Appl. Ocean Res. **146**, 103924 (2024)
24. Liu, Z., Gao, H., Zhang, M., Yan, R., Liu, J.: A data mining method to extract traffic network for maritime transport management. Ocean Coast. Manag. **239**, 106622 (2023)
25. Yang, Y., Liu, Y., Li, G., Zhang, Z., Liu, Y.: Harnessing the power of machine learning for AIS data-driven maritime research: a comprehensive review. In: Proceedings of the IEEE International Conference on Big Data, pp. 3646–3655 (2020)
26. Zhang, M., Montewka, J., Manderbacka, T., Kujala, P., Hirdaris, S.: A big data analytics method for the evaluation of ship-ship collision risk under hydrometeorological conditions. Saf. Sci. **130**, 104876 (2020)
27. Feng, H., Lin, Q., Zhang, X., Lam, J.S.L., Yap, W.Y.: Port selection by container ships: a big AIS data analytics approach. Res. Transp. Bus. Manag. **52**, 101066 (2024)
28. Bertagna, S., Braidotti, L., Bucci, V., Marinò, A.: An innovative methodology for enabling predictive maintenance of ship systems based on Industry 4.0 technologies. In: Proceedings on 6th International Confererrnce on Industry 4.0 and Smart Manufacturing (ISM), pp. 285–292. Procedia Manufacturing, vol. 59, pp. 221–228 (2023)
29. Zhou, X., et al.: Integrated application of big data and geospatial analysis in maritime transportation safety management: a comprehensive review. Int. J. Appl. Earth Obs. Geoinf. **117**, 103168 (2024)
30. Zhao, G., Xie, X., Wang, Y., Liu, S., Lopez, C.: Barrier analysis to improve big data analytics capability of the maritime industry: a mixed-method approach. Technol. Forecast. Soc. Chang. **191**, 122481 (2024)
31. Madusanka, N.S., Fan, Y., Yang, S., Xiang, X.: Digital twin in the maritime domain: a review and emerging trends. J. Marine Sci. Eng. **11**(5), Article 1021 (2023). https://doi.org/10.3390/jmse11051021
32. Kautsar, S., Aisjah, A.S., Arifin, S., Syai'in, M.: Digital twin applications in maritime systems. In: Maritime Infrastructure for Energy Management and Emission Reduction Using Digital Transformation. Springer, pp. 113–125 (2025). https://doi.org/10.1007/978-981-96-4438-4_5

Exploring Digital Twin Applications and Maturity in Supply Chain Management: Systematic Review

Hanane Ouamer[1(✉)], Fayçal Fedouaki[1], and Mouhsene Fri[2]

[1] Laboratory of Intelligent Systems Industrial and Mechanical Engineering, École Nationale Supérieure d'Arts et Métiers, Casablanca, Morocco
ouamerhanane9@gmail.com

[2] Euromed University of Fes, UEMF, Fes, Morocco

Abstract. Digital Twins (DTs) developments are still in the pilot stages of deployment in supply chain management (SCM), and their full integration with real-time synchronization and autonomous decision-making poses many challenges. While sectors such as manufacturing, aerospace, and automotive have been pioneers in adopting DTs to enhance operational efficiency, enable predictive maintenance, and support data-driven decision-making, the process industry. This paper seeks to address this gap by systematically reviewing existing DT applications across various industries and exploring how these insights can be transferred and scaled within SCM. A systematic literature review (SLR) was conducted to define DT functions and analyze their characteristics from an organizational perspective, encompassing supply chain phases. Based on the review findings, a four-layered DT maturity framework is proposed to enhance SCM performance. This framework provides a holistic and integrated approach to DT integration, outlines common enabling technologies, and offers clear directions for future research on leveraging DTs for improved supply chain performance.

Keywords: Digital Twins · Supply Chain Management · Maturity Framework · Systematic Literature Review

1 Introduction

The rapid evolution of digital infrastructure has revolutionized the way organizations collect, analyze, and utilize real-time data to evaluate system performance and operational efficiency. For instance, modern machines and systems have advanced with technology, leading to better control with cyber-physical systems (CPSs) and less human interaction [1]. One of the most transformative innovations in recent years is the digital twins (DTs) concept. This emerging concept enables continuous synchronization between the physical and digital domains, implying the extensive utilization of industrial data-driven optimization and clustering techniques have demonstrated strong effectiveness in handling complex decision environments [2], to solve complex problems related to

Y. Benadada et al. (Eds.): GOL 2026, LNNS 2034, pp. 312–321, 2027.
https://doi.org/10.1007/978-3-032-29677-1_28

planning, scheduling, and control of production and logistics requirements. The utilization of industrial data became possible due to the substantial advancements of Data Analytics, computational intelligence, Artificial Intelligence AI, and operations research. Data-driven predictive models, particularly deep learning approaches, have shown strong capability in forecasting complex operational variables such as lead times [3] In the DTs context, these approaches could be significantly leveraged thanks to technologies like the Industrial Internet of Things (IIoT) and Cloud computing [4].

In today's complex global supply chains, consistency and predictability have emerged as critical factors. The importance of structured performance evaluation has been widely emphasized in complex logistics systems, particularly through hybrid assessment frameworks and consistency-oriented analytical models [5, 6]. According to [7], organizations adopting DT technologies have achieved significant performance gains and improved operational efficiency. Despite these promising outcomes, the process industry and several other sectors remain in the early stages of DT adoption.

While DT research has rapidly evolved, it is still developing, leaving substantial opportunities for further exploration. Interestingly, the number of studies focusing on DT applications in manufacturing and production far exceed those in SCM, where implementation remains fragmented and largely conceptual [8]. Much of the existing literature reports partial DT integration or framework-based models that have yet to be validated through large-scale industrial applications [9, 10].

The maturity of DT integration reflects the degree to which organizations can align digital models with physical processes [11]. Evaluating DT maturity is therefore essential for identifying technological bottlenecks, organizational readiness, and strategic pathways for digital transformation.

Consequently, this study aims to address the following research questions:

1. How have Digital Twin technologies been implemented across various SC phases?
2. What are the key functional roles and integration levels of DTs in current SCM research and practice?
3. How can DT maturity be systematically assessed to guide future adoption and performance improvement?

To answer these questions, this study conducts a systematic literature review (SLR), classifying DT applications according to their technological, functional, and organizational dimensions. The findings contribute to developing a conceptual matrix that provides structured guidance for advancing DT integration in supply chain management. Section 2 discusses the state of the art of DTs from a supply chain simulation perspective. Section 3 details the review methodology. Section 4 presents the results and analysis. Discusses the main challenges. Section 5 concludes.

2 Review Methodology

The research approach adopted in this study was a SLR to provide a comprehensive synthesis of existing literature. The review strategy of this study is in line with the guidelines provided by [1] which was also followed by [13]. The SLR procedure consists of multiple steps presented in Fig. 2 (Fig. 1).

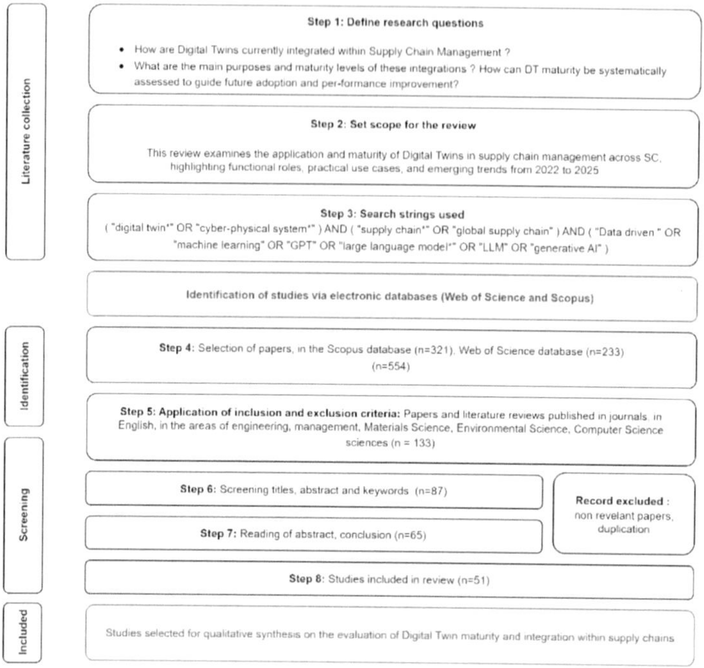

Fig. 1. PRISMA flowchart showing the research process.

The review planning phase began with the objective of systematically analyzing the current state of DT adoption and integration within SCM. This study aims to evaluate the maturity level of DT integration within SCM, aiming to identify the current challenges in developing a fully integrated ecosystem.

The search execution was conducted in September 2025 on Elsevier's Scopus and Web of Science (WoS), the largest abstract and citation databases of peer-reviewed literature. The first step was to implement the advanced search string, detecting only the title, abstract, and keywords fields through the following query:

TITLE-ABS-KEY (("digital twin*" OR "cyber-physical system*") AND ("supply chain*" OR "global supply chain") AND ("Data driven " OR "machine learning" OR "GPT" OR "large language model* OR "LLM" OR "generative AI")).

In this process, we used a combination of keywords to assess the significant research conducted in DTs applied to or related to SCM and logistics streams. To ensure a transparent of the literature selection process, this study follows the PRISMA (Preferred Reporting Items for Systematic Reviews and Meta-Analyses) guidelines to provide a structured framework and manage various stages. Starting from an initial sample of 554 papers identified, a process of filtering was carried out by defining four inclusion criteria. The initial search process excludes non-refereed work, maintaining our focus on validated contributions. Since papers were retrieved from two electronic databases, duplication was checked to ensure that each study was included only once in the review. The literature search covered the period from 2022 to capture the most recent developments in the application of DT after the COVID-19 pandemic, which is particularly relevant in fast-evolving domains such as AI and scheduling optimization. The articles were filtered out in the next step by screening the titles, abstracts, and keywords, and then any non-relevant papers were excluded. The final stage involves reporting the results of the full-text analysis, aimed at addressing the defined research objectives.

3 Observations and Discussion

3.1 Overview of the Literature Sample

This section provides the answer to our first research question, regarding the current state of research in DT integration in supply chain management. The answer to this question is based on Fig. 3. The most prominent application area within the sample, with more than 32%, is manufacturing. The applications are as diverse as the goods produced and range from classic decision support in production logistics [14] to virtual representation within human-based production in factory floors [15]. The logistics sector follows with around (10%), while both the automotive and agriculture sectors represent about 8% each. About (27%) of the identified DT applications are distributed across diverse sectors such as healthcare and maritime, while (15%) correspond to domain-neutral approaches applicable across multiple industries.

This predominance of manufacturing sector reflects the historical development of DT in this process, where machines and production lines are already instrumented with sensors and real-time data acquisition technologies.

After conducting a thorough analysis of this finding, manufacturing environments offer a high level of technical maturity and data availability, facilitating effective DT implementation and extensive investigation. Moreover, manufacturing can be considered a foundational phase within the supply chain framework. In contrast, as observed through the review, domains such as inventory management remain underrepresented, not due to limited strategic importance, but because of the increased complexity associated with multi-actor coordination, heterogeneous information systems, and fragmented data ownership across organizational boundaries. Overall, these findings reveal an imbalance in the current body of knowledge and highlight manufacturing-oriented DT as a critical enabler for the progressive generalization of DT applications across end-to-end supply chain processes.

Additionally, the classification of research types shown in Fig. 4. Four categories were identified: empirical. Most papers deal with implementation in the real world (35%), demonstrating a clear shift from theory to application. Conceptual research (28%) continues to establish theoretical foundations, (23%) of the papers handle both applied and conceptual studies, while review papers constitute a minor portion of the sample (14%). Understanding the evolution of DT adoption is crucial for assessing the technological readiness, scalability, and strategic alignment across supply chain processes. The comprehensive analysis of current literature underscores pressing need for case-based studies. Such studies are essential to validate integration of mechanisms in operational environments, assess feasibility under real-world constraints, and translate DT capabilities into measurable improvements in supply chain performance.

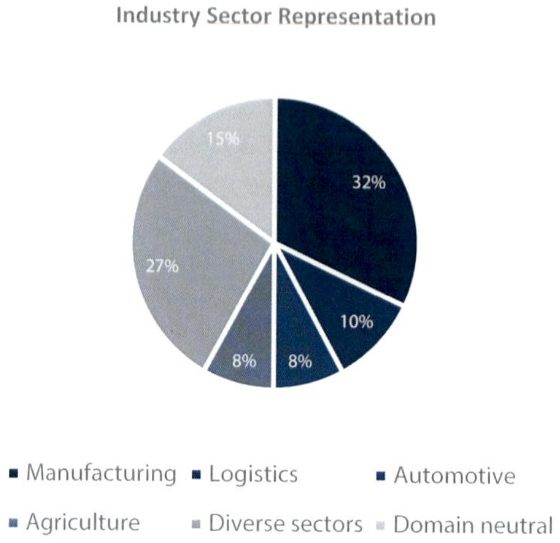

Fig. 2. Distribution of research papers according to Industry sector.

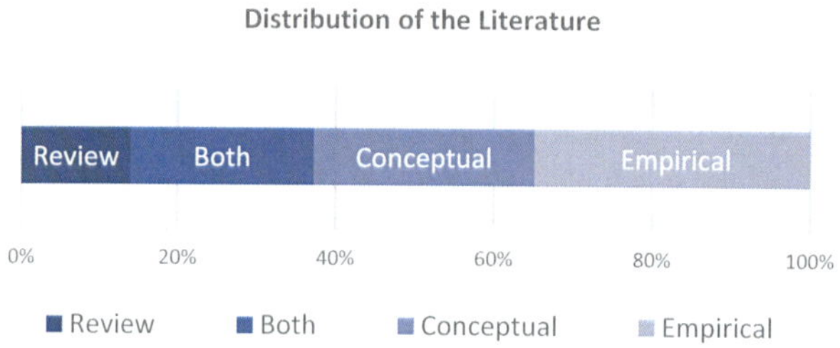

Fig. 3. Distribution of literature.

3.2 Proposed Digital Twin Maturity Model

The maturity model is a structured framework that helps assess and improve a target's level of maturity in a particular concern. It provides a roadmap to move from an initial state to a more advanced, optimized state in terms of processes, practices, capabilities, and performance. The digital twin maturity model (DTMM) is a theoretical framework designed to assess the gap between DT's current development state and its target state. In our study we followed the model presented by [16] that follow a two-step process: (1) Conceptual synthesis: Analyzing various definitions and interpretations of DT to identify its core components, functional competencies, and distinguishing characteristics. (2) Maturity level definition: Building upon the functional and capability requirements of DT, along with established maturity model frameworks from related disciplines and relevant standards.

From the literature, we initially identified over 18 distinct dimensions and components of DT, which, after removing duplicates and grouping similar ideas, were refined into 4 main dimensions providing a clear and structured framework for understanding DTs in supply chains. Context: Defined the scope and intended outcomes of DT such as cost and risk reduction and resilience improvement. Control: the extent to which the DT supports analysis, decision-making, and intervention within the physical system

[17]. Integration: addresses connectivity across digital systems, data sources, and organizational boundaries environments. Architecture: constitutes the technical backbone of supply chain DT containing data management, computing infrastructure and Lifecycle management of the system [18].

During this review, comparing existing frameworks and structured DT components laid the foundation for developing a maturity model predicated on the synergy between DT dimensions and their evolving capabilities as listed in Table 1. The model serves as a diagnostic and roadmap tool, enabling organizations to assess their current DT functions and chart a path toward more advanced, value-generating implementations.

A critical design decision was to use capabilities as the primary indicator for maturity levels. This approach was chosen because capabilities represent the tangible, functional outcomes and value delivered by the DT system, it answers the question "What can the DT do?", moving from descriptive "we can see" to prescriptive "we can autonomously optimize" functionalities.

The relationship between the four foundational dimensions and these capabilities is not merely correlative but foundational and proportional. The dimensions represent the enabling conditions or the "how," while the capabilities represent the resulting functions or the "what." Progression to a higher maturity level is only possible through commensurate and synergistic advancements across all dimensions.

The resulting papers are decomposed in Table 1. Approximately more than one-third of those works remain at the monitoring and visibility level.

Other studies reside between level 2 and 3 have progressed to the predictive or analysis stage, showcasing DT's validation through controlled trials or pilot implementations that bridge theoretical design and practical feasibility. A smaller but significant group of works demonstrate autonomous simulation cases, particularly within manufacturing fields remaining an aspirational goal for most entities of the global supply chains [19].

Table 1. Digital twin maturity model from the capability perspective.

Maturity level	Capabilitises	Technologies	Integration
Level 4 simulation	Automatically optimized operations in real-time Execute decisions without human intervention Adapt to changing conditions autonomously	Reinforcement learning and deep reinforcement learning Autonomous control systems Edge AI and distributed intelligence	Full ecosystem integration (supply chain, partners, customers) Autonomous data exchange Self-configuring interfaces

(*continued*)

Table 1. (*continued*)

Maturity level	Capabilitises	Technologies	Integration
Level 3 predictive	Predict future states and behaviors Forecast failures and maintenance needs Optimize operations proactively Assess risks before they materialize	Advanced machine learning (regression, time series, neural networks) Predictive analytics platforms Digital twin platforms	Deep integration with enterprise systems Real-time data synchronization Automated data pipelines Standardized interfaces
Level 2 analysis	Identify deviations from normal operations Compare scenarios and configurations Analyze trends and patterns Provide diagnostic insights	Statistical process control tools Machine learning Data mining and pattern recognition algorithms Business intelligence tools	Integrated with operational systems (ERP, MES, WMS) Bidirectional data flow API-based integration Shared data models
Level 1 monitoring and visibility	Monitor current state of assets and processes Track historical performance Generate status reports Visualize operations in real-time Basic KPI tracking	IoT sensors and RFID tags Basic data acquisition systems (SCADA, MES) Cloud/edge storage platforms Visualization tools	Standalone or loosely coupled systems Manual data export/import Limited API integration Point-to-point connections

3.3 Challenges and Future Directions

The complexity of implementing DT depends largely on the application scale and structural complexity, often leading to challenges related to engineering methods, project coordination, and organizational or commercial constraints. Based on insights from the reviewed literature, several barriers affect DT improvement as illustrated in Table 5.

The concentration of supply chain DT studies at low to mid maturity levels primarily reflects two structural limitations. First, the absence of mature data ecosystems restricts DT usage to descriptive and exploratory applications. Second, the inherently multi-objective nature of supply chain decision-making, involving competing trade-offs between cost, service level, resilience, and sustainability, significantly increases modeling and control complexity (Table 2).

Table 2. Research Gap and challenges across Supply chain dimensions.

DT dimensions	Barriers	Studies
Data management	Lack of Data standardization Limited availability of Real-time data Distrust in sharing data/Data Privacy and Security	[20]; [21]; [1]
System integration	Lack of adequate infrastructure Limited integration with enterprise systems	[22]; [1]; [23]
Control	Complexities in the supply chain Management for large fleets and networks Lack of collaboration Lack of clear objectives of DT	[24]; [14]; [17]

The analysis of DT challenges indicates that the higher DT maturity does not inherently imply superior suitability or value; in many supply chain contexts, lower-maturity DTs may better align with functional requirements, data availability, and organizational readiness. This highlights the need for fit-for-purpose DT design, where maturity reflects contextual adequacy rather than technological ambition. Addressing this gap requires coordinated efforts in standardization, interoperability, and validation to enable scalable, sustainable, and trustworthy DT evolution.

Progressive and controlled maturity advancement: Future DT research should emphasize stepwise progression between maturity levels, ensuring that each level is fully mastered before transitioning to more advanced functionalities.
Data management prerequisite: Robust data management practices should be treated as mandatory enablers for DT maturity advancement.

4 Conclusion, Contributions, and Further Research

This study presents a systematic literature review on the application of DT to evaluate the research methods employed by previous studies and to identify future research directions. A distinct feature of this paper is its detailed categorization of DT integration levels into SCM, with examples and contributions for each level, offering a unique understanding compared with other surveys.

Regarding the research questions, the analysis indicates that DTs are primarily integrated in SCM through staged progression, from early shadowing models and prototypes to autonomous implementations. Most studies remain in early or intermediate maturity levels, with a smaller subset demonstrating fully operational DT in real-world contexts. Furthermore, the review highlights that DT dimensions are closely linked to capabilities-oriented commonly achieved across various components of supply chain, while conceptual frameworks remain an unevenly adapted to use cases rather than fully standardized models.

Our work is subject to certain limitations. Each study included in the review inherently presents its own perspective on DT maturity. As a result, the classification and conclusions drawn here may not capture the full diversity of approaches in the field. Future research should therefore focus on enriching these dimensions by incorporating additional attributes, such as guiding questions or scoring criteria for each dimension, to support standardization and self-assessment of DT integration.

References

1. Zaidi, S.A.H., Khan, S.A., Chaabane, B.A.: Unlocking the potential of digital twins in supply chains: A systematic review. Supply Chain Anal. **7** (2024)
2. Oucheikh, R., Touil, A., Fri, M.: Data clustering using two-stage eagle strategy based on slime mould algorithm. J. Comput. Sci. **18**(11), 1062–1084 (2022)
3. Euromed University of Fez (UEMF), Fès, Morocco, Fri, M.: Lead time prediction using advanced deep learning approaches: a case study in the textile industry. Logforum **20**(2), 149–159 (2024)
4. Freese, F., Ludwig, A.: A conceptual framework for supply chain digital twins—development and evaluation. Int. J. Log. Res. Appl. **28**(6), 676–698 (2025)
5. Fri, M., Douaioui, K., Lamii, N., Mabrouki, C., Semma, E.A.: A hybrid framework for evaluating the performance of port container terminal operations: Moroccan case study. Pomorstvo (Online) **34**(2), 261–269 (2020). https://doi.org/10.31217/p.34.2.7
6. Fri, M., Douaioui, K., Mabrouki, C., El Alami, S.: Reducing inconsistency in performance analysis for container terminals. IJSOM **8**(3) (2021)
7. Attaran, M., Celik, B.G.: Digital twin: benefits, use cases, challenges, and opportunities. Decision Anal. J. **6**, 100165 (2023)
8. Negri, E., Fumagalli, L., Macchi, M.: A review of the roles of digital twin in cps-based production systems. Procedia Manuf. **11**, 939–948 (2017)
9. Qi, Q., Tao, F., Zuo, Y., Zhao, D.: Digital twin service towards smart manufacturing. Procedia CIRP **72**, 237–242 (2018)
10. Sjarov et al.: The digital twin concept in industry—a review and systematization. In: 2020 25th IEEE International Conference on Emerging Technologies and Factory Automation (ETFA). IEEE, Vienna, Austria, Sep. 2020, pp. 1789–1796 (2020)
11. Kamble, S.S., Gunasekaran, A., Parekh, H., Mani, V., Belhadi, A., Sharma, R.: Digital twin for sustainable manufacturing supply chains: current trends, future perspectives, and an implementation framework. Technol. Forecast. Soc. Chang. **176**, 121448 (2022)
12. van der Valk, H., Haße, H., Möller, F., Otto, B.: Archetypes of digital twins. Bus. Inf. Syst. Eng. **64**(3), 375–391 (2021). https://doi.org/10.1007/s12599-021-00727-7
13. Delgado, F.C.M., Garrido, S.G., Bezerra, B.S.: Barriers to visibility in supply chains: challenges and opportunities of artificial intelligence driven by industry 4.0 technologies. Sustainability (Switzerland) **17**(7) (2025). https://doi.org/10.3390/su17072998
14. Digital Learning. In: Proceedings of the 33rd European Modeling & Simulation Symposium, CAL-TEK srl (2021). https://doi.org/10.46354/i3m.2021.emss.062
15. Jeong, D., et al.: Digital twin: technology evolution stages and implementation layers with technology elements. IEEE Access **10**, 52609–52620 (2022)
16. Li, Q., Gao, X., Peng, C., Zhang, Y., Yi, J.: Data-driven change control design for product and supply chain synchronous evolution systems under cyber-attacks. Scientia Sinica Informationis **53**(2), 325–343 (2023)
17. De Giovanni, M., Lazoi, M., Bandinelli, R., Fani, V.: Advanced planning systems in production planning control: an ethical and sustainable perspective in fashion sector. Applied Sciences (Switzerland) **15**(13) (2025)

18. Uhlenkamp, J.-F., Hauge, J.B., Broda, E., Lutjen, M., Freitag, M., Thoben, K.-D.: Digital twins: a maturity model for their classification and evaluation. IEEE Access **10**, 69605–69635 (2022)
19. Ziari, M., Taleizadeh, A.A.: Integrated data-driven and artificial intelligence framework to develop digital twins in distribution system of supply chains: a real industrial case. Int. J. Prod. Econ. **289** (2025)
20. Zhang, Y., Hong, Z., Wei, Y., Qu, T., Huang, G.: A digital twin and big data-driven opti-state control framework for production logistics synchronisation system. IET Collab. Intell. Manuf. **7**(1) (2025)
21. Biller, B., Biller, S.: Implementing digital twins that learn: AI and simulation are at the core. Machines **11**(4) (2023)
22. Mabkhot, M.M., Kalawsky, R.S., Liaqat, A.: Introducing the manufacturing digital passport (MDP): a new concept for realising digital thread data sharing in aerospace and complex manufacturing. Systems **13**(8) (2025)
23. Badakhshan, E., Mustafee, N., Bahadori, R.: Application of simulation and machine learning in supply chain management: a synthesis of the literature using the Sim-ML literature classification framework. Comput. Ind. Eng. **198** (2024)
24. Mhiri, M., Mkacher, H., Al-Khatib, M., Kharbeche, M.B.M., AlNouss, A.M.N., Haouari, M.: Enhancing liquefied natural gas supply chain robustness through digital twin-driven machine learning models: a special case of cryogenic heat exchanger. Gas Sci. Eng. **143** (2025)

Personalized Assignment of Operators to Workstations in Industry 5.0: A TOPSIS-Based Approach—Case Study in an Automotive Assembly Line

Houda Lahyani[1(✉)], Mariem Bouri[2], and Abdelfettah Sedqui[1]

[1] Department of Industrial and Electrical Engineering, Innovative Technologies Laboratory, National School of Applied Sciences of Tangier, Abdelmalek Essaâdi University, 2117 Tetouan, Morocco
lahyani.houda@etu.uae.ac.ma

[2] Laboratory of Intelligent Systems and Applications (lsia), Moroccan School of Engineering Sciences, Tangier, Morocco

Abstract. In the evolving landscape of smart manufacturing and Industry 5.0, aligning human operators with production workstations has become a critical and complex challenge. The growing diversity of the workforce—in terms of anthropometric dimensions, physical strength, endurance, and experience—makes it increasingly difficult to ensure that each operator is assigned to a workstation that suits their capabilities and limitations. Mismatches between human profiles and workstation requirements can lead to increased fatigue, postural discomfort, and a higher risk of musculoskeletal disorders, ultimately affecting both well-being and productivity. To address this challenge, this study proposes a multicriteria decision-making framework to support ergonomic operator–station assignments. The approach integrates physical, ergonomic, and organizational criteria and applies the TOPSIS (Technique for Order Preference by Similarity to Ideal Solution) method to rank and select the most suitable operator–workstation combinations. A case study in the automotive assembly industry is conducted to validate the applicability and robustness of the proposed model. The results demonstrate that this method can help reduce ergonomic risks, enhance operator comfort, and improve production efficiency.

Keywords: Ergonomics · Industry 5.0 · Operator assignment · Workstation · Multicriteria decision analysis (MCDA) · TOPSIS · Automotive assembly

1 Introduction

Industry 5.0 represents a new evolution in manufacturing where technological performance is pursued together with human well-being. Unlike Industry 4.0, which mainly focused on automation and cyber-physical integration, Industry 5.0 repositions the operator as a central, intelligent, and adaptive actor within production systems [1–2]. In this

Y. Benadada et al. (Eds.): GOL 2026, LNNS 2034, pp. 322–332, 2027.
https://doi.org/10.1007/978-3-032-29677-1_29

paradigm, performance depends not only on advanced digital capabilities but also on the ability to assign each operator to a workstation that is compatible with their physical and cognitive capacities.

Human variability remains a critical factor on the shop floor. Anthropometric characteristics such as height or reach, combined with physiological attributes (strength, fatigue resistance, age) and professional experience, strongly influence productivity, comfort, and safety. When these differences are ignored, workstation assignments may generate excessive postural constraints and increase the risk of work-related musculoskeletal disorders (WMSDs), which continue to represent one of the main causes of productivity loss in industry [3].

Recent ergonomic studies emphasize the importance of personalized task allocation and the integration of quantitative decision-support methods. For example, Realyvásquez-Vargas et al. demonstrated that anthropometric-based workstation adaptation can reduce operator discomfort and error rates [4]. Dul and Neumann, Liu et al. and Fan et al. further highlighted the strategic value of integrating ergonomics and data-driven insights in human-centered manufacturing [5–7].

To support such decision-making, Multi-Criteria Decision-Making (MCDM) methods such as AHP (Analytic Hierarchy Process), MAUT (Multi-Attribute Utility Theory), VIKOR, MOORA, and TOPSIS (Technique for Order Preference by Similarity to Ideal Solution)—have been widely employed to combine quantitative and qualitative indicators into rational and transparent decisions [8–9]. Rathore emphasized the importance of MCDM techniques in supporting the implementation of ergonomic interventions in industrial environments by helping organizations prioritize the most influential factors that improve worker well-being and productivity [10].

For instance, AHP has been applied to prioritize ergonomic criteria in manufacturing workstations [11], while MOORA and VIKOR have been used to evaluate operator fatigue or safety conditions in dynamic production systems [12–13]. Among these techniques, TOPSIS stands out for its conceptual simplicity and intuitive interpretation. Introduced by Hwang and Yoon [14], TOPSIS determines the best alternative by simultaneously minimizing the distance from the negative ideal and maximizing proximity to the positive ideal solution. It is particularly suitable for ergonomic evaluation because it accommodates both benefit-type and cost-type criteria, integrates objective or subjective weighting, and yields a ranked list of alternatives. Recent works confirmed its effectiveness in industrial decision-making contexts where human, technical, and organizational dimensions must be balanced [15–16]. In the field of ergonomics and human factors, several studies have applied TOPSIS and its fuzzy extensions to support ergonomic evaluation and risk prioritization. TOPSIS has been used to prioritize manufacturing halls based on ergonomic risk factors derived from repetitive task assessments, while Fuzzy TOPSIS has been integrated with methods such as ART to rank corrective actions related to musculoskeletal disorder risks. Hybrid MCDM models combining AHP and fuzzy TOPSIS have also been proposed to assess ergonomic considerations in setup processes such as SMED, demonstrating improved ergonomic outcomes and reduced setup times [17–19].

Although these works confirm the relevance of TOPSIS in ergonomic assessment and risk prioritization, few studies focus specifically on the personalized assignment of operators to workstations in production lines, particularly through the simultaneous integration of anthropometric, physiological, and organizational criteria within a human-centric Industry 5.0 framework.

Based on this background, this work introduces a TOPSIS-based decision framework for operator–workstation assignment in the automotive harness industry. The proposed approach integrates ergonomic, physical, and organizational indicators to determine the most appropriate operator–station combinations. To assess the practicality of the proposed approach, the methodology is applied in an industrial context, enabling the evaluation of its performance and added value. The results show that this framework supports improved ergonomic safety, enhanced operator comfort, and increased operational efficiency, contributing to the transition toward human-centric manufacturing promoted by Industry 5.0.

This contribution also constitutes a step toward integrating ergonomic decision models into broader intelligent support systems, such as computerized decision-support systems (CDSS) and digital twin platforms, which are emerging trends in human-centric manufacturing [20–21].

2 Methodology

The methodological framework of this study is structured around a multicriteria decision-making (MCDM) logic designed to support the ergonomic assignment of operators to workstations. The process integrates quantitative and qualitative factors that reflect human variability and workstation constraints. The approach follows three main stages:

1. Identification of relevant criteria, combining anthropometric, ergonomic, and organizational dimensions;

The selection of evaluation criteria was based on an extensive literature review of ergonomic and direct observations from the automotive assembly line.

Three categories of indicators were retained:

Anthropometric criteria: body height, arm reach, and strength. These parameters determine whether an operator can perform manual tasks comfortably and safely within an acceptable range of motion, minimizing excessive trunk bending or arm extension.

Physiological criteria: fatigue resistance and age. These factors influence task sustainability and recovery capacity, particularly in repetitive or constrained work.

Organizational and experiential criteria: experience level, task familiarity, and precision requirement. These reflect the cognitive and procedural aspects influencing performance consistency and quality.

Each criterion was quantified using measurable or scored scales (e.g., height in cm, strength on a 1–5 scale, experience in years) to allow consistent integration into the TOPSIS framework.

These criteria are used to evaluate the compatibility between the operator and the workstation (see Table 1).

2. Evaluation and weighting of criteria according to their relative importance in the assignment process;

Table 1. Selected evaluation criteria for ergonomic assignment of operators to workstations.

Code	Criterion	Type	Justification
C1	Height (cm)	Maximize	A height suitable for the workstation reduces trunk flexion and extension.
C2	Arm span (cm)	Maximize	Ensures access to work areas without excessive arm extension.
C3	Average physical strength (score 1–5)	Maximize	Tasks requiring moderate manual effort demand adequate muscle strength.
C4	Experience on the line (years)	Maximize	Experience improves speed, accuracy, and safety in task execution.
C5	Fatigability or endurance (score 1–5)	Minimize	Lower fatigability helps maintain performance throughout the work shift.

3. Ranking of operator–workstation combinations using the TOPSIS (Technique for Order Preference by Similarity to Ideal Solution) method to determine the most suitable match.

The TOPSIS (Technique for Order Preference by Similarity to Ideal Solution) method is a popular multicriteria decision-making (MCDM) method introduced by Hwang and Yoon [17]. The concept is to select the best alternative which has the longuest distance from the positive ideal solution and the shortest distance from the negative ideal solution. The stepwise procedure for implementing TOPSIS method is as follows:

Step 1. Construct the normalized decision matrix. The normalized value n_{ij} is obtained as

$$n_{ij} = x_{ij} / \sqrt{\left(\sum_{i=1}^{n} x_{ij}^2\right)}, i = 1, 2, \ldots, n; j = 1, 2, \ldots, m \quad (1)$$

Where, x_{ij} represents the rating of the alternative A_i with respect to the criteria C_j

Step 2 Construct the weighted normalized decision matrix, the weighted normalized value can be written as

$$v_{ij} = \omega_j \; n_{ij}, \quad i = 1, 2, \ldots, n; j = 1, 2, \ldots, m \quad (2)$$

Where, ω_j is the weight of the criteria C_j, $\sum_{j=1}^{m} \omega_j = 1$

Step 3 Determine the positive and negative ideal solution. The positive ideal solution A^+ can be expressed as follows:

$$A^+ = \{v_1^+, v_2^+, \ldots, v_m^+\} = \left\{\left(\max_i v_{ij} \backslash j \in J\right), \left(\min_i v_{ij} \backslash j \in J'\right)\right\} \quad (3)$$

The negative ideal solution A^- is represented as follows:

$$A^- = \{v_1^-, v_2^-, \ldots, v_n^-\} = \left\{\left(\min_i v_{ij} \backslash j \in J\right), \left(\max_i v_{ij} \backslash j \in J'\right)\right\} \quad (4)$$

Where, J is the benefit criteria and J' represents the cost criteria, $i = 1, 2, \ldots, n; j = 1, 2, \ldots, m$

Step 4 Compute the separation measures for each alternative A_i by using the Euclidean distance.

The separation of each alternative from the positive ideal solution A^+ is formulated as follows:

$$d_i^+ = \left\{\sum_{j=1}^{m} \left(v_{ij} - v_j^+\right)^2\right\}^{1/2}, i = 1, 2, \ldots, n \quad (5)$$

The separation of each alternative from the negative ideal solution A^- can be calculated as follows:

$$d_i^- = \left\{\sum_{j=1}^{m} \left(v_{ij} - v_j^-\right)^2\right\}^{1/2}, i = 1, 2, \ldots, n \quad (6)$$

Step 5 Compute the relative closeness coefficient R_i of each alternative A_i with respect to the positive ideal solution A^+.

$$R_i = \frac{d_i^-}{d_i^- + d_i^+}, \quad i = 1, 2, \ldots, n \quad (7)$$

Where$0 \leq R_i \leq 1$

Step 6 Rank the alternatives according to the descending order of R_i

This structure enables decision-makers to capture the complexity of human–system interactions while ensuring that the allocation process remains transparent, traceable, and reproducible.

3 Results and Discussion

Data were collected from ten operators across multiple assembly workstations within an automotive production line. For each workstation, ergonomic analyses identified task type, posture constraints, and primary risks (e.g., trunk flexion, overhead work, prolonged squatting).

Operators were assessed using both direct measurements (anthropometry) and expert evaluations (ergonomic risk and skill level). The resulting dataset provided a matrix linking operators (alternatives) to workstations (criteria). (See Table 2)

Table 2. Anthropometric and performance data matrix collected from assembly line operators.

Operator	Height (cm)	Arm span (cm)	Strength (1–5)	Experience (Years)	Fatigue resistance (1–5)
O1	160	155	3	2	4
O2	175	178	4	5	2
O3	182	185	5	7	1
O4	165	160	2	1	5
O5	172	175	3	3	3
O6	180	182	5	6	2
O7	168	165	2	2	4
O8	177	180	4	4	3
O9	170	172	3	2	4
O10	185	190	5	8	1

The relative importance of each criterion was established through brainstorming and expert consultation collaboratively with ergonomists, assembly supervisor and production supervisors. Weights were normalized so that their sum equals 1 (see Table 3).

This weight reflects a balance between physical feasibility and operational performance, ensuring that both ergonomic safety and productivity are taken into account.

Table 3. Assigned weights of evaluation criteria.

C1	C2	C3	C4	C5
0,35	0,25	0,1	0,2	0,1

The weighted normalized matrix was then computed using Eq. (2), as shown in Table 4.

The positive and negative (nadir) ideal solution were computed using Eqs. (3) and (4), as described in Table 5.

The collected data (anthropometric, physiological, and ergonomic) were integrated into the TOPSIS decision matrix. After normalization and weighting, the relative closeness coefficients (Ci) were calculated for each operator, as described in Table 6.

The final ranking derived from the TOPSIS analysis reveals that Operator O10 achieved the highest relative closeness coefficient ($R_i = 1.000$), followed by Operator O3 ($R_i = 0.863$) and Operator O6 ($R_i = 0.777$). These three operators exhibit the most balanced profiles across anthropometric, physiological, and ergonomic dimensions. In practical terms, they represent the most compatible candidates for the evaluated workstation, combining favorable reach, height, endurance, and posture safety.

Table 4. Weighted normalized matrix

	C1	C2	C3	C4	C5
O1	16,32505201	10,88288798	0,075526322	0,054944226	0,15920595
O2	19,52948117	14,35227566	0,134269017	0,34340141	0,039801488
O3	21,12308683	15,50330243	0,20979534	0,673066763	0,009950372
O4	17,3613102	11,59633433	0,033567254	0,013736056	0,248759298
O5	18,86563823	13,87256792	0,075526322	0,123624508	0,089553347
O6	20,66139395	15,00456946	0,20979534	0,49449803	0,039801488
O7	17,99836984	12,33242977	0,033567254	0,054944226	0,15920595
O8	19,9784201	14,67661063	0,134269017	0,219776902	0,089553347
O9	18,42945325	13,40101386	0,075526322	0,054944226	0,15920595
O10	21,8251916	16,35264333	0,20979534	0,879107609	0,009950372

Table 5. Determination of positive and negative ideal solutions (A + and A−) in the TOPSIS model.

	C1	C2	C3	C4	C5
A+	21,8251916	16,3526433	0,20979534	0,87910761	0,00995037
A−	16,325052	10,882888	0,03356725	0,01373606	0,2487593

Table 6. Computed TOPSIS results and ranking of operators

Opérateur	di+	di−	Ri
O1	7,80316025	0,10713766	0,01354407
O2	3,09278737	4,73998951	0,60514803
O3	1,12106374	6,70016201	0,85666393
O4	6,58677968	1,25810839	0,16037302
O5	3,9376198	3,9283517	0,49941087
O6	1,82223592	6,00815874	0,76728684
O7	5,61598261	2,21605097	0,28294707
O8	2,58193889	5,27421749	0,67134833
O9	4,5785048	3,28343433	0,4176367
O10	0	7,81067943	1

Conversely, operators with lower R_i values (e.g., O1, O4, O7) show less adequate compatibility, either due to lower physical strength, suboptimal anthropometric fit, or higher postural risk scores.

These findings suggest that allocating operators such as O10 to the targeted workstation would minimize musculoskeletal strain and enhance operational efficiency, whereas others may require either task reassignment or workstation adjustment.

Overall, the results confirm that the proposed TOPSIS-based framework provides a reliable and transparent method for ergonomic operator–workstation assignment. It enables practitioners to quantify compatibility objectively and to make informed allocation decisions that balance both human well-being and production performance.

Sensitivity analysis

A sensitivity analysis was conducted to evaluate the impact of the criteria weight variation on the results. A total of fifteen scenarios (Sc1, Sc2, ..., Sc15) were considered. For scenarios (Sc1, sc2,...,sc12), each criteria weight was increased or decreased by ± 50%. As the results were robust, more scenarios (Sc11, ..., sc15) were added to test some extreme cases were each time a criterion is set to 0 and the remaining weight is distributed equally among the other criteria. For example, in scenario Sc11, the weights of criteria were [0, 0.25, 0.25, 0.25, 0.25].

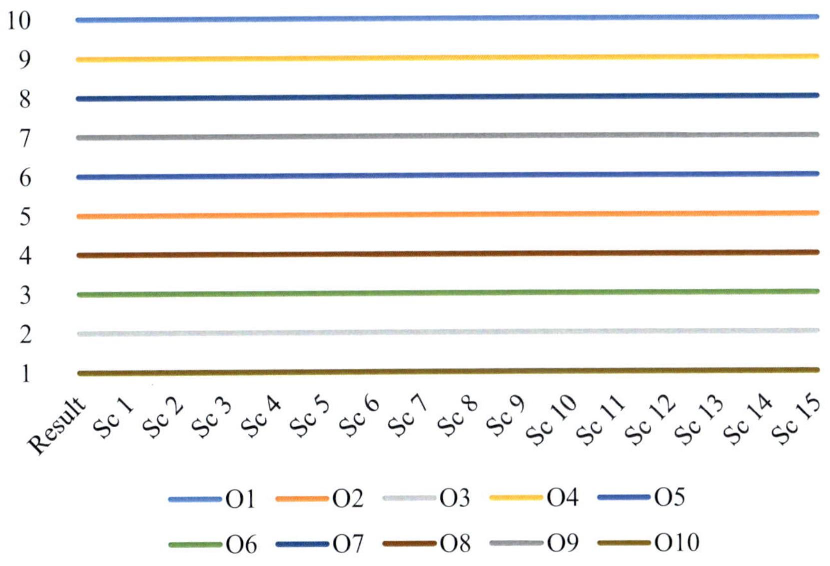

Fig. 1. Sensitivity analysis

Figure 1 shows that in all the scenarios the results are similar, which shows that the results are very robust.

Within the framework of Industry 5.0, human–machine integration emphasizes not only productivity but also sustainability and individual well-being [1, 2]. Traditional assignment methods often overlook human diversity, leading to mismatches that degrade both performance and health outcomes.

By contrast, multicriteria decision-making (MCDM) methods such as TOPSIS introduce transparency and scientific rigor into human-centered production decisions. Unlike single-criterion assessments (e.g., posture scores), MCDM integrates physical, cognitive, and organizational dimensions to generate a holistic representation of the operator–task fit.

This work contributes to the ongoing evolution toward personalized and adaptive manufacturing, where each worker is assigned not merely based on job availability but according to objective measures of capability, endurance, and ergonomic compatibility. As such, it supports the Industry 5.0 principle of empowering humans through technology.

Moreover, this model can serve as a foundation for future smart decision-support tools, where real-time data (from sensors or digital twins) continuously optimize task assignments according to operator fatigue, experience, and comfort.

4 Conclusion and Future Work

This study proposed a multicriteria decision-making (MCDM) framework to support ergonomic operator–workstation assignment within the context of Industry 5.0. By combining anthropometric, physiological, and experiential factors, the approach enables an objective and transparent selection of the most suitable operator for each workstation. The application of the TOPSIS method demonstrated its ability to handle complex, multidimensional decision problems in a transparent and reproducible manner.

The results of the case study confirmed that the method can help reduce ergonomic risks, enhance operator comfort, and improve both safety and productivity in assembly operations.

Unlike traditional ergonomic evaluations that rely solely on postural analysis, this approach provides a predictive and proactive mechanism, useful even before the physical deployment of workers on the line. It thus contributes to workforce planning and preventive ergonomics, supporting a smoother integration of human factors in manufacturing decision-making.

From a broader perspective, this work aligns with the human-centric philosophy of Industry 5.0, where technology serves as an enabler of well-being and inclusiveness rather than mere automation.

Future research will aim to enrich the model by integrating real-time physiological data (e.g., fatigue monitoring, motion tracking) and cognitive factors such as attention or task load. Combining TOPSIS with other MCDM methods such as AHP or VIKOR may also enhance the robustness of the weighting and improve decision accuracy. Ultimately, the integration of this model into digital twins or intelligent decision-support platforms could pave the way toward adaptive and self-learning ergonomic management systems.

From a system-level perspective, the proposed TOPSIS-based framework is intended to be compatible with computerized decision support systems. In particular, it may be integrated in future work into an existing CDSS for engineering change decision-making previously developed by Lahyani et al. [20], in order to support more holistic and human-centric decisions. Such integration could enable the combination of ergonomic assignment models with broader production and change-management decisions.

References

1. Breque, M., De Nul, L., Petridis, A.: Industry 5.0: Towards a Sustainable, Human-Centric and Resilient European Industry. Publications Office of the European Union (2021). https://doi.org/10.2777/308407
2. Nahavandi, S.: Industry 5.0—a human-centric solution. Sustainability. **11**(16), 4371 (2019). https://doi.org/10.3390/su11164371
3. Bernardes, J.M., Ruiz-Frutos, C., Moro, A.R.P., Dias, A.: A low-cost and efficient participatory ergonomic intervention to reduce the burden of work-related musculoskeletal disorders in an industrially developing country: an experience report. Int. J. Occup. Saf. Ergon. **27**(2), 452–459 (2021)
4. Realyvásquez-Vargas, A., Arredondo-Soto, K.C., Blanco-Fernandez, J., Sandoval-Quintanilla, J.D., Jiménez-Macías, E., García-Alcaraz, J.L.: Work standardization and anthropometric workstation design as an integrated approach to sustainable workplaces in the manufacturing industry. Sustainability. **12**(9), 3728 (2020). https://doi.org/10.3390/su12093728
5. Dul, J., Neumann, W.P.: Ergonomics contributions to company strategies. Appl. Ergon. **40**(4), 745–752 (2009)
6. Liu, S., Zhang, J., Yi, S., Gao, R., Mourtzis, D., Wang, L.: Human-centric systems in smart manufacturing. In: Manufacturing from Industry 4.0 to Industry 5.0, pp. 181–205. Elsevier (2024). https://doi.org/10.1016/B978-0-443-13924-6.00006-5
7. Fan, J., Zheng, P., Lee, C.K.: A vision-based human digital twin modeling approach for adaptive human–robot collaboration. J. Manuf. Sci. Eng. **145**(12), 121002 (2023). https://doi.org/10.1115/1.4062430
8. Kaya, İ., Kahraman, C.: A comparison of fuzzy multicriteria decision making methods for intelligent building assessment. J. Civ. Eng. Manage. **20**(1), 59–69 (2014). https://doi.org/10.3846/13923730.2013.801906
9. Yazdani, M., Gonzalez, E.D., Chatterjee, P.: A multi-criteria decision-making framework for agriculture supply chain risk management under a circular economy context. Manage. Decis. **59**(8), 1801–1826 (2021)
10. Rathore, B.: Analysis of enablers to implement ergonomic interventions in Indian manufacturing industry: a multi-criteria decision-making approach. Glob. Bus. Rev. **26**(4), 900–922 (2025). https://doi.org/10.1177/09721509211044293
11. Upadhyay, H.K., Juneja, S., Muhammad, G., Nauman, A., Awad, N.A.: Analysis of IoT-related ergonomics-based healthcare issues using analytic hierarchy process methodology. Sensors. **22**(21), 8232 (2022). https://doi.org/10.3390/s22218232
12. Stanujkic, D., Karabasevic, D., Zavadskas, E.K.: Selection of industrial operators using MOORA and COPRAS methods. Technol. Econ. Dev. Econ. **19**(3), 583–611 (2013)
13. Singh, R. et al.: A historical review and analysis on MOORA and its fuzzy extensions for different applications. Heliyon. **15**(10), 3 (2024)
14. Hwang, C.L., Yoon, K.: Multiple Attribute Decision Making: Methods and Applications. Springer, New York (1981). https://doi.org/10.1007/978-3-642-48318-9
15. Kose, Y., Civan, H.N., Ayyildiz, E., Cevikcan, E.: An AHP–TOPSIS integrated model for ergonomic assessment of setup processes under SMED. Mathematics. **10**(15), 13804 (2022)
16. Taherdoost, H.: A comprehensive review on TOPSIS and its applications. Decis. Anal. J. (2024)

17. Khandan, M., & Koohpaei, A. (2017). Assessment and Analysis of Ergonomics Conditions Based on TOPSIS in a Manufacturing Company.
18. Khandan, M., Koohpaei, A.R., Nili, M., Farjami, Y.: Occupational musculoskeletal disorders management using fuzzy TOPSIS assessment of repetitive tasks (ART). Work. **56**(2), 267–276 (2017)
19. Kose, Y., Civan, H.N., Ayyildiz, E., Cevikcan, E.: An interval valued Pythagorean fuzzy AHP–TOPSIS integrated model for ergonomic assessment of setup process under SMED. Sustainability. **14**(21), 13804 (2022)
20. Lahyani, H., Al Khatib, A., Sedqui, A.: Computerized decision support system (CDSS) for engineering change management in automotive wiring harness projects. Prod. Eng. Arch. **31**(3), 356–375 (2025). https://doi.org/10.30657/pea.2025.31.33
21. Karim, O., Lahyani, H., Khatib, A.A., Sedqui, A.: Implementation of a digital twin in an industrial production line: real-time visualization and monitoring. Lect. Notes Inform. Syst. Organ. **7**, 197–208 (2025). https://doi.org/10.1007/978-3-032-06725-8_17

Towards Intelligent Public Logistics Governance in Morocco: Integrating Internal Audit and Artificial Intelligence into an Applied Conceptual Model

Khaoula El Assassi(✉) and Redouan Daafi

ENCG Casablanca, Hassan II University of Casablanca, Casablanca, Morocco
Khaoula.elassassi2@gmail.com

Abstract. This paper proposes an integrated conceptual model for intelligent public logistics governance in Morocco, combining internal audit, artificial intelligence, and digital transformation within a regional digital hub. Drawing on Morocco's constitutional, legal, and strategic frameworks—alongside international standards such as COSO, INTOSAI, OECD, UNDP, and the World Bank—the model aims to strengthen traceability, transparency, and performance in managing public flows. By positioning internal audit as a strategic management tool and AI as a predictive mechanism, the study redefines public logistics as a driver of governance and territorial performance. Its practical relevance is illustrated through Morocco's preparation for the 2030 FIFA World Cup, which exemplifies large-scale coordination, data-driven decision-making, and sustainable public management.

Keywords: Public logistics · Internal audit · Artificial intelligence · Governance · Morocco · Digital hub

1 Introduction

Since the adoption of the 2011 Constitution, Morocco has engaged in a profound transformation of its institutional and governance architecture. This reform process is structured around three interrelated pillars: advanced regionalization, the modernization of public management, and the strengthening of transparency and accountability mechanisms. These orientations have been legally consolidated through Organic Law No. 111–14 on regions and Organic Law No. 130–13 related to the Finance Act (Loi Organique relative à la Loi de Finances, LOLF), which collectively promote a governance model grounded in performance management, risk control, and results-based accountability. Within this framework, the effectiveness of territorial public policies increasingly depends on the capacity of public institutions to plan, coordinate, and control complex public flows—financial, informational, and decision-making—that constitute the invisible yet critical infrastructure of public action.

Y. Benadada et al. (Eds.): GOL 2026, LNNS 2034, pp. 333–343, 2027.
https://doi.org/10.1007/978-3-032-29677-1_30

In this evolving context, the concept of public logistics acquires renewed analytical relevance. Traditionally, logistics has been defined as the planning, organization, and coordination of physical, informational, and financial flows to ensure that resources are delivered at the right place and time [2]. When transposed to the public sector, however, logistics extends beyond physical distribution to encompass the coordinated management of resources, information, and decisions in pursuit of coherence, continuity, and efficiency in public action [9, 14]. Public logistics governance thus refers to the set of institutional, organizational, and technological mechanisms designed to optimize, trace, and sustain these flows in accordance with the principles of good governance and sustainable territorial development.

This analytical perspective is closely aligned with Morocco's major strategic orientations, notably the National Logistics Competitiveness Strategy [11], the National Strategy for Sustainable Development [10], and the Morocco Digital 2030 program. Together, these initiatives highlight the growing role of digital transformation and artificial intelligence (AI) as enablers of public-sector performance, coordination, and data-driven decision-making. They reflect a strategic ambition to move beyond fragmented administrative practices toward more integrated and intelligent governance systems.

These ambitions take on a particularly concrete dimension in the context of Morocco's preparations for co-hosting the 2030 FIFA World Cup. This large-scale international event mobilizes substantial financial, infrastructural, and organizational resources, generating complex and interdependent public flows across multiple levels of governance. As such, it constitutes a critical testing ground for the implementation of integrated public logistics governance, where coordination, efficiency, transparency, and accountability become decisive success factors rather than abstract principles.

Within this setting, two complementary governance levers emerge as particularly central. On the one hand, internal audit plays a structuring role by ensuring the reliability, compliance, and control of public processes, while contributing to risk management and organizational learning. On the other hand, artificial intelligence functions as a predictive and analytical engine capable of enhancing anticipation, traceability, and decision-making through the exploitation of large volumes of data. The integration of these two mechanisms within a regional digital governance architecture opens new perspectives for the development of intelligent public logistics governance, grounded in performance monitoring, data transparency, and systemic coherence.

Against this backdrop, this article does not seek to empirically measure the effects of internal audit or artificial intelligence. Rather, it adopts a conceptual and analytical perspective aimed at structuring and clarifying their potential contribution to public logistics governance.

Accordingly, the central research question is formulated as follows:

How can an integrated conceptual model of intelligent public logistics governance be designed, incorporating internal audit and artificial intelligence, in order to strengthen the traceability, transparency, and performance of public flows in Morocco?

To address this question, the article relies on a conceptual and comparative approach, drawing on established national and international governance frameworks, including those developed by COSO, INTOSAI, the OECD, UNDP, and the World Bank.

The paper is organized as follows. It first presents the Moroccan regulatory and institutional framework. It then reviews the relevant literature and conceptual foundations. The methodological approach is subsequently outlined, followed by the presentation of the proposed integrated model. The paper concludes with a discussion of the main implications and directions for future research.

2 Moroccan Regulatory and Institutional Framework

Public logistics governance in Morocco is grounded in a robust regulatory and institutional framework that reflects the core principles of good governance, transparency, and performance-based management. This framework has progressively evolved to support the coordination and control of public flows within a decentralized and results-oriented governance model.

The 2011 Constitution constitutes a foundational milestone in this evolution. It establishes responsibility, accountability, and transparency as central pillars of public action, while enshrining advanced regionalization as a strategic choice aimed at promoting decentralized, participatory, and territorially balanced governance. In this configuration, regions are positioned at the core of planning, investment, and economic development, thereby assuming a growing responsibility in the management of public resources and public action.

This constitutional orientation is operationalized through Organic Law No. 111–14 on regions, which clarifies the division of powers between the State and territorial collectivities and institutionalizes a multilevel governance model based on the principles of subsidiarity and coordination. The law promotes integrated territorial planning, notably through instruments such as the Regional Spatial Planning Scheme (Schéma Régional d'Aménagement du Territoire, SRAT) and the Regional Development Program (Programme de Développement Régional, PDR). Within this framework, regions are required to manage their financial, informational, and material flows in coherence with national priorities, a requirement that is fully consistent with the logic of public logistics governance.

At the fiscal and managerial level, Organic Law No. 130–13 related to the Finance Act (LOLF) introduced performance-based budgeting, explicitly linking public expenditure to measurable objectives and results indicators. This shift has been reinforced by Decree No. 2-17-265 on internal control and risk management, which institutionalized the internal audit function within public administrations and public entities. Taken together, these instruments have strengthened the traceability and control of financial flows while fostering a culture of accountability, risk management, and results-driven public management.

From a strategic perspective, Morocco's public logistics governance is further supported by national policy frameworks such as the National Logistics Competitiveness Strategy [11] and the National Strategy for Sustainable Development [10]. These strategies aim to embed efficiency, sustainability, and coherence into both public and private management practices. In parallel, the Morocco Digital 2030 program promotes the digital transformation of public administration through the development of integrated information systems, the establishment of regional digital hubs, and the gradual incorporation of artificial intelligence into governance and decision-support tools.

Finally, Morocco's preparation for co-hosting the 2030 FIFA World Cup constitutes a large-scale and highly visible test for this public logistics governance framework. The project requires rigorous coordination across central, regional, and local levels of government, continuous monitoring of public investments, and enhanced interoperability between public information systems. As such, it offers a concrete opportunity to operationalize an integrated public logistics governance model anchored in transparency, performance, and sustainability—values that lie at the heart of Morocco's ongoing administrative and territorial reforms.

3 Literature Review and Conceptual Framework

The analysis of public logistics governance requires an interdisciplinary perspective situated at the intersection of governance theory, public management, internal auditing, and digital transformation. Recent literature increasingly emphasizes that the effectiveness of public institutions depends on their capacity to articulate these dimensions within a coherent, transparent, and performance-oriented governance system [4, 12]. In this sense, public logistics governance cannot be approached in isolation but must be understood as the outcome of interacting institutional, managerial, and technological mechanisms.

Internal audit as a pillar of good governance
Within this framework, internal audit emerges as a central pillar of good governance. According to the COSO framework [3], internal audit aims to ensure effective risk control, the reliability of information, and the continuous improvement of organizational processes. In the public sector, INTOSAI [4] defines internal audit as both an assurance and advisory mechanism designed to strengthen compliance, efficiency, and performance in public action.

In the Moroccan context, the institutionalization of internal audit through Decree No. 2-17-265 reflects a clear commitment to promoting transparency, accountability, and traceability within public administrations. Beyond its traditional ex post verification role, internal audit has progressively evolved into a strategic management instrument. It now contributes to an integrated vision of risk, supports evidence-based public decision-making, and enhances the overall coherence of public management systems.

Artificial intelligence as a lever of efficiency and foresight In parallel, artificial intelligence (AI) has emerged as a major driver of anticipation, rationalization, and efficiency in public governance. International organizations such as the OECD [12], the World Bank [15], and the United Nations Development Programme UNDP [13] highlight AI's potential to enhance transparency, improve predictive analytics, and support the personalization and optimization of public services.

When applied to governance processes, AI enables the exploitation of data generated by public information systems in order to model trends, forecast risks, and correct dysfunctions in near real time. Its integration into public policies signals a transition toward a data-driven and proactive governance model, in which predictive analysis reinforces institutional performance, supports strategic coordination, and contributes to the sustainability of public decisions.

Public logistics as the operational framework

Public logistics constitutes the operational framework through which internal audit and artificial intelligence can be effectively articulated. Drawing on the foundations of business logistics [2, 9], public logistics refers to the organization and coordination of public flows—financial, decision-making, material, and informational—necessary for the implementation of territorial public policies [14].

From a governance perspective, public logistics governance encompasses the institutional, organizational, and digital mechanisms designed to optimize these flows in accordance with principles of efficiency, transparency, and sustainability. It provides the structural link that connects control mechanisms and technological tools within a unified system of public action.

Toward an integrated conceptual model

Synthesizing these strands of the literature leads to the formulation of an integrated conceptual model structured around three interdependent components:

(i) Internal audit, which ensures compliance, risk management, and traceability
(ii) Artificial intelligence, which acts as a lever for analysis, anticipation, and prediction; and
(iii) Public logistics governance, which serves as the coordination and integration framework for public flows.

These components interact within a regional digital hub, conceived as a platform for the collection, processing, and dissemination of public data. Such an architecture supports the emergence of an intelligent public governance model capable of enhancing performance, strengthening transparency, and improving the coherence of territorial public action (Fig. 1).

This model is illustrated below:

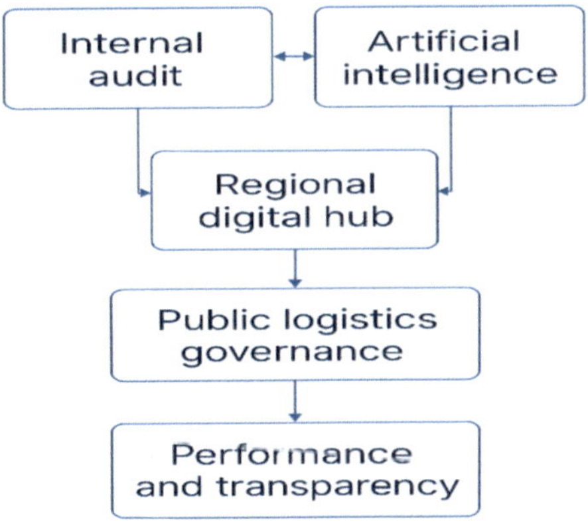

Fig. 1. Integrated conceptual model of intelligent public logistics governance. *Source* Author's elaboration based on COSO [3], INTOSAI [4], OECD [12], UNDP [13], and World Bank [15].

4 Methodology

This research adopts a conceptual and comparative approach aimed at constructing an integrated model of intelligent public logistics governance grounded in recognized theoretical and regulatory frameworks. In line with the conceptual nature of the study, the objective is not to generate or analyze empirical data, but rather to structure and synthesize existing knowledge in order to clarify the mechanisms through which internal audit and artificial intelligence can jointly support public logistics governance.

The methodological design is based on a comprehensive documentary analysis, combining national legislative and regulatory texts, institutional guidelines, and international conceptual frameworks related to governance, internal auditing, digital transformation, and artificial intelligence. This approach makes it possible to capture both normative orientations and analytical perspectives that inform contemporary public governance practices.

The study follows a deductive and systemic perspective. It begins with the identification of widely accepted principles of good governance, public performance, and accountability, from which an operational analytical framework is derived and subsequently adapted to the Moroccan institutional and territorial context. The main objective of this process is to identify the key conditions that enable the emergence of an integrated public logistics governance model based on transparency, coordination, and the traceability of public flows.

At the international level, the comparative analysis draws on several major reference frameworks. These include the COSO framework [3], which defines the essential components of risk management and internal control; the INTOSAI standards [4], which establish principles for internal control, audit, and performance in the public sector; the OECD guidelines [12] on the ethical and responsible use of artificial intelligence in public governance; the UNDP framework [13], which promotes transparency and accountability through the digitalization of public processes; and the World Bank's GovTech Index [15], which assesses governments' levels of digital maturity.

At the national level, the analysis relies on a coherent regulatory and strategic corpus, including the 2011 Constitution, Organic Law No. 111–14 on regions, Organic Law No. 130–13 related to the Finance Act (LOLF), and Decree No. 2-17-265 on internal control and risk management. This corpus is complemented by national strategic documents such as the National Logistics Competitiveness Strategy [11], the National Strategy for Sustainable Development [10], and the Morocco Digital 2030 program. Together, these texts provide an integrated framework linking territorial performance, digital transformation, and sustainability within public governance.

The conceptual modeling process relies on a systemic analysis of interactions among three main levers: internal audit, artificial intelligence, and public logistics governance. These components are analyzed in terms of their respective roles and complementarities, and are integrated within a regional digital hub, conceived as a central node for the collection, analysis, and dissemination of public data and public flows. The proposed model thus illustrates how audit mechanisms and digital technologies can jointly contribute to a more efficient, transparent, and sustainable system of territorial governance.

Source Author's elaboration based on COSO [3], INTOSAI [4], OECD [12], UNDP [13], World Bank [15], and official Moroccan texts.

Table 1. Sources and reference frameworks mobilized.

Domain	Main reference(s)	Objective/Contribution
Internal audit and control	[3, 4]	Risk management and transparency of public processes
Artificial intelligence and governance	[12, 13, 15]	Data-driven, ethical, and predictive governance
Moroccan national framework	Constitution [5], Law 111–14, LOLF 130–13, Decree 2-17-265	Institutional and budgetary framework for public governance
Strategies and programs	[1, 10, 11]	Sustainable logistics and digital transformation
Contextual application	FIFA World Cup 2030	Integrative project and laboratory for logistics governance

5 Integrated Conceptual Model

The proposed conceptual model is based on a systemic and integrated approach to public logistics governance. It brings together three interdependent pillars—internal audit, artificial intelligence, and public logistics governance—which are interconnected through a regional digital hub. This hub serves as a central platform for interconnection and analysis, where financial, decision-making, and informational flows generated by various public actors converge.

In the Moroccan context, this approach gains particular relevance in light of the provisions of the Organic Finance Law (LOLF), ongoing administrative digitalization strategies, and the preparation of major national projects such as the 2030 FIFA World Cup, with total investments estimated at over 160 billion MAD for the period 2024–2030 (Ministry of Economy and Finance 2024). These dynamics increase the volume and complexity of public flows, thereby reinforcing the need for integrated and transparent governance mechanisms.

Internal audit constitutes the cornerstone of compliance and traceability within the proposed model. It ensures the reliability of information, adherence to procedures, and coherence between strategic objectives and achieved results. In Morocco, the implementation of Decree No. 2-17-265 has led to the establishment of internal audit units within most ministries and public institutions, with approximately 60% operating under formal internal control plans (IGF 2023). These mechanisms allow continuous monitoring of public expenditures, notably within a national investment budget that reached 95.5 billion MAD in 2024. Within this model, internal audit extends beyond post-facto verification to support the production of structured data feeding the regional digital hub and enabling ongoing monitoring and evaluation of public policies.

Artificial intelligence, embedded at the core of the digital hub, functions as a mechanism of learning and anticipation. Early national initiatives—such as the *Hakam* platform in the justice sector and the hydrological forecasting system of the Ministry of Equipment—illustrate the potential of AI to model, analyze, and anticipate public needs. Applied to public logistics governance, AI enables the analysis of data derived from territorial information systems (GID, GIS, and databases produced by the High Commission for Planning), helping identify correlations between budgetary, temporal, and geographic variables and anticipate deviations or project delays. In this sense, AI contributes to transforming logistics governance into a more proactive and adaptive system.

Public logistics governance provides the overarching framework of the model. It aims to ensure the coordinated management of public flows, the sustainability of decisions, and the transparency of institutional interactions. Morocco already possesses several strategic instruments aligned with this vision, including the National Spatial Planning Scheme (SNAT), Regional Development Plans (PDR), and the National Logistics Competitiveness Strategy (SNCL). Logistics investments currently reach approximately 55 billion MAD per year, representing nearly 4% of national GDP (AMDL 2023). This logistical and territorial dimension reflects Morocco's gradual evolution toward multi-level governance, in which regions manage their flows coherently with national objectives of performance and sustainability.

As summarized in Table 1 (Sources and reference frameworks mobilized), the interaction among these three components creates an integrated public value chain in which internal audit supplies reliable and verified data, artificial intelligence analyzes and transforms these data into decision-support indicators, and public logistics governance uses them to guide action and evaluate performance.

To clarify the functional articulation of these three pillars and their respective theoretical foundations, Table 2 summarizes the main functional layers of the proposed model and their expected governance impacts.

Source Author's elaboration based on COSO [3], INTOSAI [4], OECD [12], UNDP [13], World Bank [15], and official Moroccan regulatory texts.

The regional digital hub thus acts as the nerve center of public performance, centralizing, processing, and disseminating information while ensuring data security, traceability, and transparency. This mechanism is consistent with the objectives of the *Morocco Digital 2030* program, whose administrative modernization budget amounts to approximately 6 billion MAD (Directorate-General for Digital Transformation 2024).

If operationalized, this conceptual model would support Morocco's transition toward an intelligent, predictive, and participatory governance system, capable of accompanying major national projects—such as the 2030 FIFA World Cup—while reinforcing the principles of performance and accountability that underpin ongoing state reform.

6 Conclusion and Future Perspectives

This research develops an integrated conceptual model for intelligent public logistics governance in Morocco, structured around the strategic synergy between internal audit, artificial intelligence (AI), and logistics coordination within a regional digital hub. The model reframes the management of public flows—financial, informational, and decision-based—through a logic of traceability, transparency, and performance, consistent with the strategic orientations of the New Development Model (NDM) 2035 and the Morocco Digital 2030 roadmap.

Table 2. Functional layers and governance impacts of the proposed intelligent public logistics governance model

Governance layer	Illustrative enablers	Governance objective	Expected contribution	Primary reference(s)
Data integration	Interoperable information systems, digital platforms	Transparency	Consolidated and real-time visibility of financial, informational, and logistical public flows	[12, 13]
Analytical and predictive capacity	Data analytics, predictive models, AI-assisted analysis	Anticipation	Early identification of risks, delays, and deviations in public projects and logistics operations	[12, 15]
Decision support	Performance dashboards, decision-support tools	Effectiveness	Evidence-based decision-making and optimized allocation of public resources	[3, 15]
Control and accountability	Internal audit mechanisms, traceability tools	Accountability	Strengthened compliance, secure monitoring of public expenditures, and reinforced institutional responsibility	[3, 4]

The scientific contribution of this study lies in its conceptual operationalization of "Governance 4.0" by explicitly linking digital and analytical capabilities to concrete governance functions. As summarized in the functional impact matrix (Table 2), integrating analytical layers (e.g., AI-assisted analytics) with traceability and control mechanisms strengthens the governance cycle from monitoring to anticipation. In this perspective, internal audit is positioned not only as a compliance mechanism, but also as a contributor to structured, reliable information flows, while AI supports data-driven analysis and forward-looking decision support.

The practical relevance of the model is particularly illustrated by the 2030 FIFA World Cup, a large-scale national project requiring multi-level coordination, continuous monitoring of public resources, and timely information sharing. Within this context, the regional digital hub can serve as an integration pivot for infrastructure, budgeting, and public–private partnership data. In parallel, audit mechanisms reinforce regulatory compliance, while AI-enabled tools can support anticipatory and adaptive management in a high-stakes environment.

Looking ahead, the effective operationalization of this conceptual model depends on several key conditions, including the development of an interconnected digital ecosystem linking national and regional administrations, the establishment of ethical and regulatory frameworks governing AI use in public management, and sustained investment in human expertise in auditing, data analytics, and risk governance. In the longer term, this vision aligns with NDM 2035 objectives advocating for a results-oriented, transparent, and efficient public administration.

Finally, the proposed model opens promising avenues for future applied research, including pilot implementation in selected regions—such as Béni Mellal-Khénifra and Tangier-Tétouan-Al Hoceima—and the development of a Public Logistics Governance Index (PLGI) to assess regional maturity. Comparative benchmarking with reform-driven countries (e.g., Rwanda or the United Arab Emirates) could further refine the model and contribute to positioning Morocco as a regional reference in intelligent public governance, combining strategic auditing and AI-enabled decision support within a unified vision of public value creation and accountability.

References

1. Agence de Développement du Digital (ADD): Morocco Digital 2030: National Roadmap for Digital Transformation. ADD, Rabat, Morocco (2023)
2. Ballou, R.H.: The evolution and future of logistics and supply chain management. Eur. Bus. Rev. **19**(4), 332–348 (2007). https://doi.org/10.1108/09555340710760152
3. Committee of Sponsoring Organizations of the Treadway Commission (COSO): Enterprise Risk Management: Integrating with Strategy and Performance. COSO (2017)
4. Guidelines for internal control standards for the public sector (INTOSAI GOV 9100). In: International Organization of Supreme Audit Institutions (INTOSAI). INTOSAI Publications (2020)
5. Kingdom of Morocco: Constitution of the Kingdom of Morocco. Official Gazette No. (2011) 5964 bis
6. Kingdom of Morocco: Organic law no. 111–14 on regions. Official Gazette No. 6358 (2015a)
7. Kingdom of Morocco: Organic Law No. 130–13 on the Finance Law (LOLF). Official Gazette. No. 6344 (2015b)
8. Kingdom of Morocco: Decree No. 2-17-265 on internal control and audit in public administrations. Official Gazette No. 6590 (2017)
9. Lambert, D.M., Stock, J.R.: Strategic Logistics Management, 5th edn. McGraw-Hill Education, New York, NY (2019)
10. Ministry of Energy Transition and Sustainable Development: National Sustainable Development Strategy (SNDD 2030). Rabat, Morocco (2017)
11. Ministry of Equipment, Transport and Logistics: National Logistics Competitiveness Strategy (SNCL). Rabat, Morocco (2010)
12. Organisation for Economic Co-operation and Development (OECD): Recommendation on the Governance of Artificial Intelligence. OECD Publishing, Paris, France (2022)
13. United Nations Development Programme (UNDP): Artificial Intelligence and Public Governance: Pathways for Transparency and Accountability. UNDP, New York, NY (2023)

14. United Nations Economic and Social Commission for Asia and the Pacific (UNESCAP): Public Sector Logistics and Resilient Supply Chains in Developing Economies. United Nations, Bangkok, Thailand (2021)
15. World Bank: GovTech Maturity Index: the State of Public Sector Digital Transformation. World Bank Group, Washington, DC (2022)

Towards Data-Driven Economic Intelligence to Optimize Logistics Competitiveness in the Age of Artificial Intelligence

Hiba Belgaid[1,2](✉), Fayssal Moukafi[1,2], and Amine Dafir[1,2]

[1] Hassan II University of Casablanca, Casablanca, Morocco
hiba.belgaid-etu@etu.univh2c.ma

[2] Laboratory for Research on the New Economy and Development (LARNED), Casablanca, Morocco

Abstract. This paper examines the strategic convergence between Economic Intelligence (EI) and Artificial Intelligence (AI) to enhance logistics competitiveness in the era of data-driven decision-making. It argues that the increasing digitalization of supply chains has transformed data into a critical strategic asset, reshaping logistics systems into intelligent, resilient, and predictive structures capable of anticipating disruptions and optimizing operational performance. Based on a conceptual and theoretical approach, the study introduces the Smart Logistic Intelligence Framework (SLIF), which integrates strategic knowledge (EI) and operational intelligence (AI) to optimize information flows and improve decision-making processes. While theoretical in nature, this study lays the foundation for future empirical research aimed at validating the framework in real-world logistics contexts. The paper contends that Economic Intelligence has entered a new era, where its relevance depends on a dynamic alliance with emerging technologies. The integration of AI, data analytics, and predictive modeling thus becomes a strategic prerequisite for the long-term competitiveness and agility of modern organizations.

Keywords: Economic intelligence · Artificial intelligence · Data governance · Supply chain resilience · Smart logistics

1 Introduction

The current international landscape is undergoing major economic, geopolitical, technological, and environmental shifts that have made global value chains increasingly fragile. At the same time, the growing influence of artificial intelligence and digital transformation is accelerating the transition from a resource-driven economy to one centered on knowledge and data. To this end, governments and companies that succeed in effectively collecting, processing, and leveraging this data strengthen their decision-making, anticipation, and influence on the international stage. Data is therefore becoming a strategic resource and now forms the basis of the information that is at the heart of economic

Y. Benadada et al. (Eds.): GOL 2026, LNNS 2034, pp. 344–356, 2027.
https://doi.org/10.1007/978-3-032-29677-1_31

intelligence. According to the Martre report economic intelligence is defined as all coordinated actions involving the research, processing, and distribution of information useful to economic actors for the purpose of its exploitation [1]. Historically, information has long been an essential element in political and economic decisions, but with the digital revolution and the rise of artificial intelligence, data has emerged as the new unit of measurement of power. Where information was once the result of a process of interpretation, data now represents the raw material used to construct that information, and its collection, processing, and exploitation determines the quality of public decisions and the competitiveness of economies. However, with the advent of artificial intelligence, the process of economic intelligence has become more complex, data has become more numerous, and its manual processing has become complicated, and this overabundance generates a massive, continuous, and multidimensional flow of information. Data has therefore become the key resource for any approach, feeding algorithms, supporting predictive analysis, and fueling the construction of decision-making knowledge. As a result, the ability to transform data into relevant information, and then into strategic intelligence, is now a major determinant of competitiveness. In the field of logistics, this dynamic takes on particular importance, as modern, interconnected, and globalized supply chains rely on the control of information flows that is as crucial as that of physical flows. The integration of economic intelligence thus paves the way for intelligent, data-driven logistics capable of improving operational performance, optimizing planning, and strengthening the competitiveness of both regions and companies. This raises a major question: how can data, as a basic element of information, contribute to strengthening the competitiveness of logistics systems in the era of artificial intelligence? To address this issue, the article will be structured in three main sections. The first section will offer a theoretical analysis through a review of the literature on economic intelligence, emphasizing its evolution and its role in creating informational value. The second section will focus on the nature of the collaboration between economic intelligence (EI) and artificial intelligence (AI). It will examine how these two disciplines interact to strengthen logistics competitiveness, particularly through the collection, processing, and strategic use of data. Finally, the third section will present a "smart logistics" model illustrating the circulation and exploitation of data from its collection to strategic decision-making. This section will be followed by a conclusion summarizing the main findings of the study and proposing prospects for future research and application in logistics hubs, ports, and emerging economies.

2 Literature Review and Theoretical Framework

2.1 Economic Intelligence: Conceptual Frameworks and Strategic Role

Economic intelligence (EI) has its origins in a predominantly French tradition. The concept was popularized in France in the 1990s, driven by the Martre report which defined it as the coordinated actions of research, processing, and distribution, with a view to exploiting information useful to economic actors [1]. This French interpretation emphasizes the state and collective dimension, integrating companies, public authorities, and regional actors into a coordinated system of monitoring, security, and influence. Moreover, in the Anglo-Saxon world, this discipline developed much earlier, finding its initial roots in the notion of organizational intelligence. This concept was notably introduced by Harold Wilensky, who defined intelligence as a process of gathering, processing, interpreting, and communicating information essential to decision-making [2]. It was only from the 1980s onward, according to Paturel and Huet, that practitioners such as Sammon et al. (1984), along with consultants like Fuld (1985) and Kelly (1987), appropriated and popularized the concept of economic intelligence under various related terms: Strategic Intelligence, Business Intelligence, and Competitive Intelligence [3].

Generally, the Anglo-Saxon notions of business intelligence and competitive intelligence have focused on the collection, processing, and analysis of strategic information aimed at enhancing organizational performance and achieving competitive advantage. To this end, information is the strategic resource for economic intelligence, which will enable any organization to gain a strategic advantage. Thus, according to Frédérique Péguiron, information refers to a process consisting of a series of actions through which one increases one's stock of data to develop knowledge [4, 5]. Therefore, the starting point for any economic intelligence system is data, which represents the raw state of information before it is analyzed and processed for strategic purposes. Economic intelligence is thus a process in which data is first identified and then transformed into useful knowledge using advanced analytical tools. This mastery of data enables public and private actors to anticipate risks, detect opportunities, and orient their strategies according to constantly evolving economic and geopolitical dynamics. Economic intelligence is commonly structured around three research to consist of 3 interrelated dimensions. The first dimension, strategic monitoring is concerned with the systematic gathering and analyzing of information for the purpose of prediction and decision-making. The second dimension, economic security, deals with safeguarding the strategic assets, information, and knowledge from threats and vulnerabilities. The third dimension, influence, indicates the power of the economic actors to create the environment by affecting the norms, standards, perceptions, and strategic choices [1]. Altogether, these three dimensions constitute the conceptual backbone of economic intelligence, however, their implementation in complex and data-rich contexts increasingly relies on advanced analytical technologies, such as artificial intelligence, to support decision-making. In literature, the impacts of artificial intelligence in economic intelligence process are increasingly examined through the lens of decision-making. Existing studies show that economic intelligence provides the strategic framework for anticipation, monitoring, and interpretation, while artificial intelligence strengthens these processes by enabling advanced analytical capabilities and real-time data processing. Table 1 synthesizes the main contributions that explicitly link economic intelligence and artificial intelligence in support of decision-making, highlighting key analytical dimensions and major insights from existing literature.

The table above presents a selection of studies examining competitive intelligence and the contribution of artificial intelligence (AI) to strategic decision-making. Competitive intelligence transforms raw data into actionable insights, thereby supporting strategy formulation and decision processes [13]. As Rothberg and Erickson [9] emphasize, an organization's strategic value lies less in data accumulation than in its intelligent activation [9]. However, the effectiveness of competitive intelligence in today's technological

Table 1. Literature synthesis: links between economic intelligence, artificial intelligence (synthesized from the literature)

Analytical dimension	Key insights from the literature	References
Strategic monitoring	Economic Intelligence (EI) enables structured environmental scanning and early detection of weak signals, supporting anticipation and strategic decision-making	[2, 6–8]
Artificial intelligence in EI processes	Artificial Intelligence (AI) enhances EI by enabling advanced analytics, real-time data processing, and predictive capabilities	[9–12]
Decision support	The combination of EI and AI strengthens decision-support processes by reducing uncertainty, shortening decision cycles, and improving coordination between strategic objectives and operational actions	[13–16]

landscape depends on the synergy between technological capabilities and human expertise. AI enhances the speed, scope, and depth of competitive analyses, while human judgment remains essential to interpret results, contextualize information, and translate analyses into relevant strategic decisions [11, 12].

2.2 Economic Intelligence and Artificial Intelligence: Conceptual Insights

Recent studies point out that the effectiveness of economic intelligence increasingly depends on advanced analytical capabilities and data-driven approaches. Indeed, numerous contributions highlight that AI-enabled competitive intelligence improves decision-making quality by facilitating the systematic analysis of extensive and diverse datasets [17]. Furthermore, in the era of artificial intelligence, economic intelligence is undergoing a profound transformation, with monitoring, analysis, and strategic anticipation systems being revamped to incorporate technologies capable of processing vast volumes of data in real time. This development offers new perspectives in terms of decision-making. Nevertheless, the growing abundance of data presents a twofold challenge: it represents both an opportunity to increase the informational performance of organizations and a challenge in terms of selection, reliability, and security. In this new digital ecosystem, data has become a key strategic lever and a structuring factor in economic competitiveness. In the conceptual framework presented by Farhat Mizen, the strategic economic intelligence model (Fig. 1) is developed in accordance with T.S. Eliot's epistemological sequence (1934), which illustrates the progressive transformation of data into information, information into knowledge, knowledge into understanding, and understanding into wisdom [18]. According to this model, the process of transforming data into strategic decisions follows a structured and cumulative progression. Inspired by Eliot's famous reflection, "Where is the wisdom we have lost in knowledge? Where is the knowledge we have lost in information?" This model highlights the value creation that accompanies the evolution of raw data into actionable knowledge. The first stage, observation, consists of defining the search criteria and collecting the relevant data, followed by the monitoring phase, during which the information is collected, processed, validated, and disseminated to ensure its reliability. The economic intelligence phase aims to analyze and contextualize this information to extract strategic knowledge, and finally, decision-making translates this knowledge into concrete actions that guide, share, adapt, and improve organizational strategies [19]. This model emphasizes that the true power of economic intelligence lies not in the amount of data available, but in the organization's ability to interpret, synthesize, and integrate it into the decision-making process, thus transforming information into strategic value.

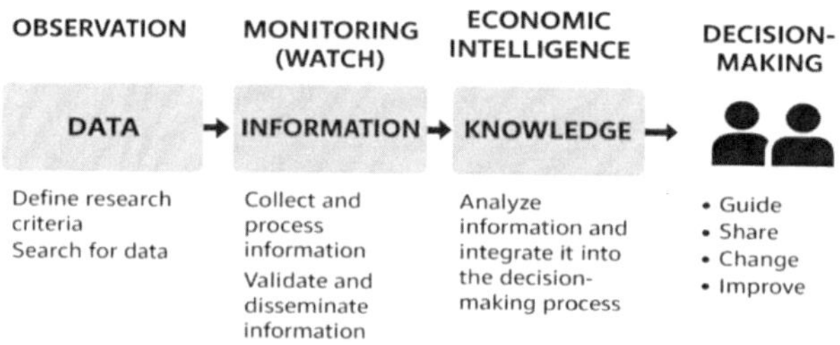

Fig. 1. T.S. Eliot's strategic economic intelligence model [18].

Raw data (trade flows, logistics indicators, financial transactions, technical reports, etc.) are now the raw material of economic intelligence. Their processing and analysis using statistical methods, artificial intelligence tools, or data analytics platforms transforms this data into useful information for decision-making. Economic intelligence is therefore based on leveraging data to transform it into information. However, with the emergence of AI, this dynamic has intensified. This approach now enables the processing of massive volumes of data, the detection of patterns that would otherwise escape human analysis, and the generation of actionable insights to support decisionmaking. However, data management is now an essential prerequisite for any strategic approach, as the growing complexity of economic environments and the proliferation of information sources exceed the capabilities of traditional human analysis. It is in this context that artificial intelligence is emerging as indispensable support for economic intelligence activities, enabling the automated processing and predictive interpretation of large amounts of data. This evolution marks the transition to augmented intelligence. Artificial intelligence plays a structuring role in supporting, analyzing, and transforming the disciplines of economic and decision-making intelligence [9]. AI acts as a catalyst, capable of improving the informational performance of organizations by optimizing business intelligence, big data analytics, competitive intelligence, and knowledge management. Through these interactions, it facilitates the transition to integrated decision-making intelligence, where data from multiple sources is collected, processed, and interpreted in real time to support strategic decision-making. This evolution reflects the shift from economic intelligence based on traditional monitoring to augmented intelligence, where machine learning algorithms amplify human capabilities for analysis and anticipation. Thus, Artificial Intelligence does not constitute a separate intelligence, but a technological enabler that strengthens the core functions of economic intelligence by accelerating data processing, enhancing anticipation, and supporting strategic interpretation.

3 Economic Intelligence for Logistics Competitiveness in the Age of Artificial Intelligence

3.1 Logistics in a Context of Uncertainty and Complexity

In a context marked by numerous disruptive factors, organizations are operating in areas of uncertainty. These uncertainties, whatever their nature, challenge traditional models of competitiveness. Economic intelligence thus presents itself as a structuring approach in this context, enabling the collection, processing, and analysis of strategic information to guide decision-making and strengthen organizational resilience. Today, this approach is part of an environment marked by the emergence of artificial intelligence, which is profoundly transforming the modes of production, circulation, and interpretation of information. AI thus complements and amplifies the analytical capabilities of economic intelligence by offering predictive, automated, and algorithmic tools capable of extracting meaning from massive volumes of data (Big Data). Its field of application now extends to highly operational domains such as logistics, where the mastery, analysis, and valorization of data have become essential drivers of competitiveness and sustainable performance. Logistics, serving as a central link between information, material flows, and organizational performance, is especially influenced by these ongoing changes. It significantly guides both strategic choices and operational actions. Considered as a holistic process, logistics calls for strong abilities in anticipation, coordination, and effective planning. In this regard, adopting a complementary economic intelligence approach can effectively support decision-making at all levels of the organization. Artificial intelligence (AI) is now emerging as a cross-functional solution with disruptive capabilities, capable of optimizing flows, improving forecasts, reducing lead times, and strengthening the resilience of supply chains. By leveraging machine learning, warehouse management systems (WMS), business intelligence (BI) platforms, and the Internet of Things (IoT), AI is paving the way for more predictive, connected, and responsive logistics. To this end, AI enables organizations to improve their performance by offering tools for better data collection and analysis, more accurate trend prediction, and increased process automation. By using AI to enhance their dynamic capabilities, companies and organizations can not only respond more effectively to various current challenges, but also anticipate and adapt to future changes, as perfectly explained by the theory of dynamic capability, which emphasizes the need for an organization to adapt in a changing environment [20, 21]. This provides a theoretical framework for understanding how the integration of AI into logistics can contribute to the creation of an agile, resilient, and competitive supply chain. Now, with the use of AI, problems are solved with greater precision thanks to increased speed and a larger volume of information [22]. With widespread access to information, the world has entered a new technological era marked by advances such as blockchain, the Internet of Things, and artificial intelligence. The introduction of these technologies will improve access to information, enable sophisticated analysis, support decision-making, and facilitate process automation [23]. To this end, the integration of artificial intelligence into the logistics sector represents one of the most significant advances in the digital transformation of supply chains. It is leading to a reconfiguration of traditional logistics models using intelligent systems capable of processing large volumes of data, making automated choices, and continuously optimizing operational flows. Several studies have highlighted the scale of this technological evolution and its impact on logistics performance. According to Chen et al., different AI models improve operational efficiency and enable the integration of sustainability criteria into the optimization of logistics flows [24] (Fig. 2).

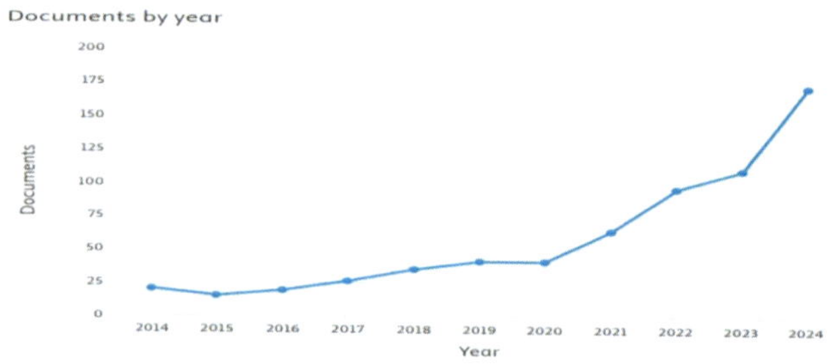

Fig. 2. Evolution of scopus publications on artificial intelligence and logistics between 2014 and 2024

Similarly, Culot et al. proposed a systematic review of AI in supply chains, demonstrating that AI improves visibility, decision-making, and operational performance in supply chains [25]. For their part, Kazancoglu et al. emphasized the importance of machine learning and predictive analytics techniques in operational planning, particularly for anticipating supply needs and automating repetitive tasks [26]. Furthermore, the steady annual increase in the number of scientific articles published in the cross-disciplinary fields of AI and logistics reflects a growing interest in research [27]. As shown in Fig. 2, in recent years, the number of articles including the terms "artificial intelligence" and "logistics" has increased significantly in databases such as Elsevier Scopus.

3.2 From Operational Intelligence to Strategic Logistics Competitiveness

Drawing from the literature, economic intelligence constitutes a strategic decision-support system aimed at reducing uncertainty through the structured collection, analysis, and exploitation of information in complex and competitive environments. In contexts characterized by high volatility and informational intensity, economic intelligence enables organizations to anticipate risks, identify opportunities, and guide strategic choices. These characteristics strongly resonate with contemporary logistics systems, which operate within globalized, interconnected, and uncertainty-prone supply chains. The competitiveness of logistics is becoming more dependent on the capability of handling not only physical but also informational flows, thus making economic intelligence an essential strategic instrument for logistics decision-making and coordination. Human capabilities, developed through Big Data Analytics (BDA) capabilities, also enhance organizations' ability to respond to disruptions and disturbances within supply chains [28]. These human-centered capabilities play a critical role in interpreting analytical outputs, assessing contextual risks, and translating data-driven insights into effective operational and strategic responses. As a result, recent literature highlights that AI is now establishing itself as a tool for logistical competitiveness. Analysis of existing work shows that the main functions of economic intelligence, such as data collection and analysis, risk prediction, and supply chain security, now have direct applications in intelligent logistics systems. In this regard, several studies emphasize that digital transformation should primarily be understood as a strategic issue rather than a purely technological one as cited by Bahrami et al. based on Rogers (2016) [29]. It is therefore essential to design an integrated model of smart logistics, based on the complementarity between economic intelligence and AI, to enhance the performance of supply chains. To this end, the effective contribution of AI in logistics is not limited to process automation, but also includes strategic aspects related to the collection, analysis, and use of information, which is a characteristic of economic intelligence.

4 Smart Logistic Intelligence Framework (SLIF)

4.1 Conceptualization of the Smart Logistic Intelligence Framework (SLIF)

Modern logistics is evolving from a reactive field to a predictive one, thanks to tools like artificial intelligence (AI) and the Internet of Things (IoT). To frame this shift, we propose the Smart Logistic Intelligence Framework (SLIF). This model conceptualizes logistics decision-making through two interdependent layers: *Strategic Logistic Knowledge (SLK)*, which supports long-term strategic planning and anticipation, and *Operational Logistic Intelligence (OLI)*, which enables short-term and real-time operational execution. The doer, Operational Logistic Intelligence (OLI), is powered by AI and acts on real-time data, handling the immediate, on-the-ground challenges. The real strength of this model is their synergy. The AI-driven doer generates predictions from live data, and the strategic planner interprets them to guide major decisions. This coordination allows us to anticipate disruptions and manage risks before they happen. Tools like blockchain provide trust and traceability, while digital twins allow us to simulate scenarios safely. Ultimately, SLIF isn't just a technological add-on, it's a blueprint for a living system where data, knowledge, and strategy work together to create sustainable development.

Within logistics systems, the relationship between economic intelligence and AI operates through a dynamic interaction mechanism. Artificial Intelligence enhances operational intelligence by processing large volumes of real-time logistics data and generating predictive insights. Economic intelligence then transforms these AI-generated outputs into strategic knowledge by contextualizing them within economic, geopolitical, and organizational environments. In return, Economic Intelligence guides the deployment and orientation of AI tools by defining strategic priorities, governance rules, and intelligence objectives. This reciprocal interaction creates a feedback loop that strengthens anticipation, resilience, and logistics competitiveness.

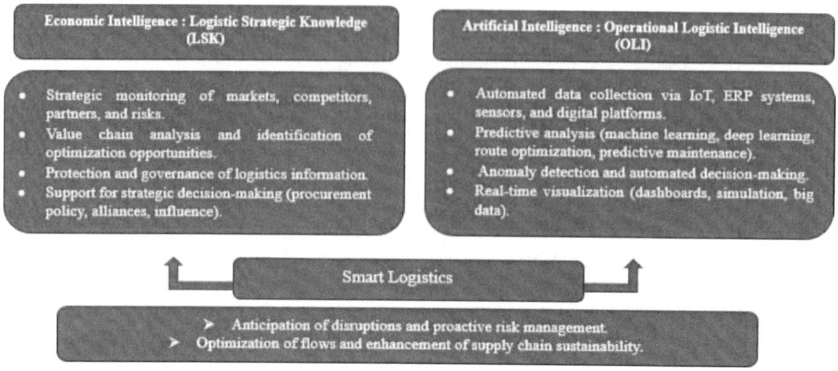

Fig. 3. Smart logistic intelligence framework: integrating economic and artificial intelligence.

4.2 Discussion

The main purpose of this model is not only to automate processes, but also to equip logistics with a reflexive and adaptive capacity based on understanding weak signals, data governance, and leveraging strategic knowledge. Within the Smart Logistic Intelligence Framework, data is the original and structuring element of the decision-making process, The rigorous collection, intrinsic quality, and contextual relevance of data are essential prerequisites for any strategic analysis. Before AI tools are used for processing, aggregation, and insight extraction, data allows us to capture operational reality, identify emerging trends, and reveal logistical risks and opportunities.

It thus forms the foundation on which AI transforms raw information into actionable knowledge and on which IE supports strategic decisions aimed at the efficiency, agility, and competitiveness of the logistics system. The Smart Logistic Intelligence Framework (SLIF) formalizes the integration between Economic Intelligence and Artificial Intelligence within logistics by structuring the decision-making process across three interdependent layers. At the foundational level, the data layer captures logistics-related data generated by physical and informational flows, constituting the raw material of intelligence. These data are processed at the operational level through Artificial Intelligence and Big Data Analytics, which form the Operational Logistic Intelligence (OLI) layer. This layer enables real-time analysis, prediction, and optimization of logistics operations, producing indicators and insights that enhance operational efficiency and responsiveness. At the strategic level, Strategic Logistic Knowledge (SLK), grounded in economic intelligence, transforms these analytical outputs into contextualized strategic knowledge by integrating economic, geopolitical, regulatory, and organizational dimensions. SLIF operates through a bidirectional mechanism: a bottom-up analytical flow that converts data into strategic intelligence, and a top-down strategic flow through which strategic orientations guide the configuration and prioritization of AI-driven operational processes. This interaction is further differentiated by its temporal structure, whereby OLI primarily addresses short-term and real-time operational challenges, while SLK focuses on medium- and long-term anticipation, resilience, and competitiveness.

Table 2 shows how SLIF works as a series of steps that help turn information into a strong logistics system. First, Economic Intelligence (EI) structures the upstream layer by

Table 2. The SLIF mechanism: from economic intelligence and AI capabilities to logistics competitiveness

SLIF component	Core mechanism	Contribution to logistics competitiveness
EI-driven data governance (Data layer)	Defines priorities, filters weak signals, ensures data quality, traceability and confidentiality	Improves reliability of analytics and reduces information asymmetries
AI/Big data analytics capabilities	Transforms data into operational insights	Enhance efficiency, speed, and operational responsiveness
Operational logistic intelligence (OLI)	Converts insights into timely operational decisions	Improves service reliability and agility under uncertainty
Strategic logistic knowledge (SLK)	Contextualizes OLI into strategic interpretation and governance (regulatory, geopolitical, market)	Strengthens anticipation, coherence, and long-term resilience
Recursive SLIF loop (bottom-up/top-down)	Learning cycle: data → OLI → SLK and strategic steering SLK → OLI → data	Sustains adaptive capability and resilience over time

defining what should be monitored, qualifying sources, and enforcing information governance. This foundation is crucial for the steps, which involve using Artificial Intelligence and Big Data to analyze the information and turn it into Operational Logistic Intelligence (OLI), through real-time monitoring, forecasting, optimization, and anomaly/risk detection, thereby producing actionable operational decisions. EI then re-enters at the strategic level to contextualize these operational outputs and consolidate them into Strategic Logistic Knowledge (SLK) that supports anticipation, arbitration, and governance under economic, regulatory, and geopolitical constraints. Crucially, SLIF is not a linear pipeline but a recursive learning loop: operational signals feed strategic interpretation, while strategic priorities continuously recalibrate analytics configurations, indicators, and data governance requirements. In this sense, logistics competitiveness is not merely a by-product of AI deployment, it emerges from an adaptive system in which EI secures meaning, reliability, and information security, while AI accelerates the translation of intelligence into operational performance, agility, and long-term resilience. SLIF highlights the strategic role of intelligence, data governance, and anticipation. It shifts the focus from short-term operational gains to long-term competitiveness and resilience. Nevertheless, the framework remains conceptual and presents several limitations. It does not provide empirical testing, nor does it specify sector-specific implementations. These limitations open promising avenues for future research, including empirical validation in ports, logistics hubs, and emerging economies, as well as comparative studies across different logistics ecosystems.

5 Conclusion

The world today is undergoing profound changes that are shaking the economic, geostrategic, and technological foundations. Technological disruption has particularly disrupted the functioning of organizations. In this dynamic, the economic environment has become increasingly complex and unstable, and strategic decisions are now more problematic due to the multiplicity of available data and its growing heterogeneity. The structuring, reliability, and relevance of this data pose major challenges for organizations, as its processing requires advanced analytical tools capable of extracting meaning and transforming raw information into usable knowledge. Logistics is now no exception to these transformational dynamics. Due to its dual operational and decision-making nature, it is at the heart of the changes brought about by the digital revolution. To this end, the study presented highlights the convergence between economic intelligence (EI) and AI as a lever for transforming and optimizing logistics systems in a global environment marked by widespread uncertainty. Analysis has shown that integrating AI into economic intelligence systems enables more detailed and predictive use of information, facilitating risk detection, operations planning, and supply chain governance, this synergy gives rise to logistics intelligence capable of anticipating and adapting through the strategic use of data flows. Logistics competitiveness in the age of artificial intelligence cannot be understood as the mere outcome of advanced analytical or technological capabilities. It depends on organizations' ability to structure, interpret, and govern AI-generated information within a strategic intelligence perspective. In this regard, the *Smart Logistic Intelligence Framework (SLIF)* provides a conceptual contribution by clarifying the mechanisms through which Artificial Intelligence, when embedded in Economic Intelligence processes, transforms operational insights into strategic logistics knowledge that supports performance and resilience. This perspective extends existing research by highlighting the central role of intelligence governance, anticipation, and temporal alignment between operational decisions and long-term strategic orientations. This theoretical framework highlights that logistics performance no longer relies solely on speed or cost reduction, but on the dynamic ability to transform data into a competitive advantage. However, although this article primarily proposes a theoretical model, it paves the way for future empirical research aimed at validating, testing, and adapting it to different logistics contexts (port, urban, industrial, or territorial). Accordingly, the strategic value of Artificial Intelligence in logistics lies not solely in data processing, but in the intelligence frameworks that enable its conversion into sustainable competitive advantage.

Acknowledgments. This research project was conducted with the support of the National Center for Scientific and Technical Research (CNRST) under the "Phd-Associate Scholarship-PASS."

References

1. Martre, H., Clerc, P., & Harbulot, C.: Intelligence économique et stratégie des entreprises: Travaux du groupe « Intelligence économique et stratégie des entreprises » présidé par Martre, H. La Documentation Française, Office of the General Planning Commission, Paris (1994)

2. Wilensky, H.L.: Organizational Intelligence: Knowledge and Policy in Government and Industry. Basic Books, New York (1967)
3. Paturel, R., Richomme-Huet, K.: Vers la mise en œuvre d'une démarche d'intelligence économique pour l'artisanat en Région PACA. Manage. Avenir **40**, 158–172 (2010). https://doi.org/10.3917/mav.040.0158
4. Péguiron, F., Thiéry, O.: Méta-modélisation et systèmes d'informations stratégiques. LORIA—Université de Lorraine, Nancy (2009)
5. Rahouti, B., Lahrach, R.: Nouveau regard sur le concept d'intelligence économique dans un monde VUCA: l'homme au cœur de la démarche. Revue Marocaine de Manage. et d'Innov. **2**(5), 1 (2021)
6. Aguilar, F.J.: Scanning the business environment. Macmillan, New York (1967)
7. Choo, C.W.: Environmental scanning as information seeking and organizational learning. Inform. Res. **7**(1), 7–1 (2001)
8. Calof, J.L., Wright, S.: Competitive intelligence: a practitioner, academic and inter-disciplinary perspective. Eur. J. Mark. **42**(7–8), 717–730 (2008)
9. Rothberg, H.N., Erickson, G.S.: Intelligence in action: strategically managing knowledge assets. Palgrave Macmillan, Cham (2019)
10. Mikalef, P., Krogstie, J., Pappas, I.O., Giannakos, M.: Investigating the effects of big data analytics capabilities on firm performance: the mediating role of dynamic capabilities. Inform. Manage. **57**(2), 103169 (2020). https://doi.org/10.1016/j.im.2019.103169
11. Cekuls, A.: AI-Driven competitive intelligence: Enhancing business strategy and decision making. J. Intell. Stud. Bus. **12**(3), 4–5 (2022). https://doi.org/10.37380/jisib.v12i3.961
12. Cekuls, A.: The complexity of competitive intelligence in the age of data ambiguity and artificial intelligence. J. Intell. Stud. Bus. **14**, 86 (2024). https://doi.org/10.37380/jisib.v14.si1.2417
13. El Fadili, S., Gmira, F.: Competitive intelligence: leaven of a new managerial device for decision support. TELKOMNIKA Indonesian J. Elect. Eng. **16**(1), 167–175 (2015). https://doi.org/10.11591/telkomnika.v16i1.8750
14. Ranjan, J., Foropon, C.: Big data analytics in building the competitive intelligence of organizations. Int. J. Inform. Manage. **56**, 102231 (2021). https://doi.org/10.1016/j.ijinfomgt.2020.102231
15. Alfi, M., Riyanta, S., Ras, A.R.: The role of competitive intelligence in strategic decision-making: a literature review. Asian J. Eng. Soc. Health **3**(10), 2307–2324 (2024)
16. van den Berg, L., Coetzee, B., Mearns, M.: Establishing competitive intelligence process elements in sport performance analysis and coaching: a comparative systematic literature review. Int. J. Inform. Manage. **52**, 102103 (2020). https://doi.org/10.1016/j.ijinfomgt.2020.102103
17. Elkaabi, A., Mamouny, A., Elmaallam, M.: The impact of artificial intelligence tools on the competitive intelligence process: a systematic literature review. In: 2025 International Conference on Circuit, Systems and Communication (ICCSC). Fez, Morocco, pp. 1–7 (2025)
18. Farhat, M., Henchiri, S.: L'ancrage territorial de l'intelligence économique: développe—ment et expérimentation d'un outil d'aide à la décision—Cas de la commune Houmet Essouk (Djerba, Tunisie). In: Proceedings of the International Conference on Intelligence Économique et Développement Territorial. Djerba (2022)
19. Monino, J.-L.: L'information au cœur de l'intelligence économique stratégique: Marché et organisations. Marché et Organ. **18**(2), 25–39 (2013)
20. Salim, G.: L'adoption de l'IA et du smart logistique à l'optimisation des coûts d'approviionnement: Adopting AI and smart logistics in optimizing procurement costs. Revue Logistique et Perform. **8**(4), 1 (2024)
21. Teece, D.J., Pisano, G., Shuen, A.: Dynamic capabilities and strategic managementment. Strateg. Manage. J. **18**(7), 509–533 (1997)

22. Kumar, A., Shankar, R., Thakur, L.S.: A big data driven sustainable manufacturing framework for condition-based maintenance prediction. J. Comput. Sci. **27**, 428–439 (2018). https://doi.org/10.1016/j.jocs.2017.06.006
23. Idrissi, Z.K.: Blockchain, IoT and AI in logistics and transportation. Digit. Bus. **5**(1), 100018 (2024). https://doi.org/10.1016/S2949-8962(24)00018-2
24. Chen, W., Men, Y., Fuster, N., Osorio, C., Juan, A.A.: Artificial intelligence in logistics optimization with sustainable criteria: a review. Sustainability **16**(21), 9145 (2024)
25. Culot, G., Podrecca, M., Nassimbeni, G.: Artificial intelligence in supply chain management: a systematic literature review of empirical studies and research directions. Comput. Ind. **162**, 104132 (2024)
26. Kazançoglu, Y., Mangla, S.K., Luthra, S., Agarwal, M.R.: Role of emerging technologies for empowering resilience and transparency in supply chains. Ann. Oper. Res. (2025). https://doi.org/10.1007/s10479-025-06741-2
27. Toorajipour, R., Sohrabpour, V., Nazarpour, A., Oghazi, P., Fischl, M.: Artificial intelligence in supply chain management: a systematic literature review. J. Bus. Res. **122**, 502–505 (2021)
28. Mandal, S.: An examination of the importance of big data analytics in supply chain agility development. Manage. Res. Rev. **41**(10), 1201–1219 (2018)
29. Bahrami, M., Shokouhyar, S., Seifian, A.: Big data analytics capability and supply chain performance: the mediating roles of supply chain resilience and innovation. Int. J. Product. Perform. Manage. **71**(7), 2123–2147 (2022)

Optimization and Innovative Decision support Tools for Managing Smart and Sustainable Supply Chain in Smart Cities

An Exact Solution for the Multi-compartment Multi-product Vehicle Routing Problem: A Case Study from a Tunisian Transport Company

Lina Becha(✉), Manel Kammoun, and Taicir Loukil

MODILS: Laboratory of Modeling and Optimization for Decisional Industrial and Logistic Systems, Faculty of Economics and Management Sciences, Sfax, Tunisia
bechalina84@gmail.com

Abstract. Optimizing logistics chain efficiency requires a focus on transport operations. Fuel distribution represents a complex logistical challenge, where the simultaneous management of multiple products within multi-compartment vehicles is crucial. Consequently, the Multi-Compartment Multi-Product Vehicle Routing Problem (MCMPVRP) has received considerable attention in recent years. In practice, delivery schedules are often generated manually by logistics planners. However, this task is tedious and time-consuming. In this article, a mathematical model is proposed to solve the MCMPVRP for a real-case application from the Tunisian International Road Transport (TTRI) company. The proposed approach enables optimal planning, making it possible to meet operational constraints, integrate routing and product allocation decisions, and minimize distribution costs. Numerical results confirm the effectiveness and performance of our proposed approach compared to the current manual planning.

Keywords: Vehicle routing problem · Multi-compartment vehicles · compartment vehicles · Multi-product distribution · product distribution · Oil and gas logistics · Mixed-integer linear programming · integer linear programming · Optimization

1 Introduction

This study is motivated by a pressing operational challenge faced by the Tunisian International Road Transport (TTRI) company, which specializes in fuel distribution. The company's delivery routing currently relies on the empirical expertise of its dispatchers. Although practical, this manual method is inherently suboptimal. It struggles to manage the complexity arising from a homogeneous fleet of multi-compartment vehicles, the strict dedication of compartments to specific products, and diverse multi-product customer demands. This often results in elevated operational costs, suboptimal fleet utilization, and increased planning time.

In recent decades, supply chain management has become a cornerstone of competitive advantage for modern organizations [12, 14]. Within this framework, physical distribution is a critical and costly component, where route optimization is paramount for

Y. Benadada et al. (Eds.): GOL 2026, LNNS 2034, pp. 359–367, 2027.
https://doi.org/10.1007/978-3-032-29677-1_32

reducing operational expenditures and enhancing resource utilization [7, 9]. Dantzig and Ramser [4] laid the groundwork for the Vehicle Routing Problem (VRP), is a fundamental combinatorial optimization problem concerned with identifying the most cost-effective routing plan for a fleet serving a group of customers. This seminal work has spawned extensive literature and numerous variants tailored to specific logistical constraints [5, 9], as documented in recent surveys [15].

Among these variants, the Multi-Compartment Vehicle Routing Problem (MCVRP) addresses the challenge of transporting multiple, often incompatible, products using a single vehicle equipped with dedicated compartments [2]. The distribution of petroleum products is a quintessential application, involving the simultaneous delivery of various fuels (e.g., gasoline, diesel) to service stations using compartmentalized tank trucks [2, 6]. This specific problem is formally categorized as the Multi-Compartment Multi-Product Vehicle Routing Problem (MCMPVRP). The literature proposes a range of solution approaches for VRPs, from heuristics to exact mathematical programming models [11]. Although the academic literature on the VRP and its variants is extensive, a distinct research gap persists. While recent studies have proposed exact algorithms for complex variants of the MCVRP [10], studies that tackle the MCMPVRP using these approaches on real-world operational data and provide a direct, quantitative comparison against established manual planning practices remain relatively scarce [1, 3, 4, 13]. It is this gap between theoretical models and practical validation that motivates the present research.

To address this gap, this paper focuses on modeling and solving the MCMPVRP within the fuel distribution sector, using a case study from TTRI. The main contributions of this work are three-fold:

- To formulate a new MILP model for the MCMPVRP that adequately represents the distinctive constraints encountered in fuel distribution operations.
- To validate the model's efficacy by comparing its optimal solutions for real-world instances against the company's current manual plans, quantifying the potential gains in cost reduction and efficiency.
- To provide a tangible and practical decision-support tool capable of generating significant operational efficiencies and cost savings for TTRI and similar logistics enterprises.

The analytical framework is presented in the following sequence. Section 2 presents the mathematical formulation of the problem, detailing the constraints and objective function. The numerical experiments based on the case study are presented and analyzed in Sect. 3. Finally, conclusions and perspectives for future research are drawn in Sect. 4.

2 Mathematical Model

Our approach draws inspiration from the cost structure and problem formulation applied to fuel logistics by Benantar et al. [2]. The ensuing section provides the mathematical modeling of the Multi-Compartment Multi-Product Vehicle Routing Problem (MCMPVRP). The model incorporates all the standard constraints of the MCVRP, which are critical for the practical feasibility of the solutions. These include:

- Vehicle capacity constraints for both the entire truck and its individual compartments.
- The requirement that each compartment can carry only a single, non-mixed product.
- The necessity to fully satisfy all customer demands.
- The depot-circularity constraint for all vehicle routes.

Sets:

- N The set of customers $N = \{1, 2, \ldots, n\}$
- V The set of customers including the central depot, $V = N \cup \{0\}$.
- A The set A represents all possible arcs connecting pairs of nodes in V
- T The Set of compartments in each vehicle, indexed by t.
- P The set of products, $P = \{1, 2, \ldots, p\}$.
- K The set of homogeneous multi-compartment trucks, $K = \{1, 2, \ldots, k\}$.

Parameters:

- C_k The capacity of truck K
- $.C_{tk}$ The capacity of tank t in truck K.
- d_{ij} The distance metric between vertices i and j, where $(i, j \in V)$.
- CV_k The variable cost of truck K.
- CF_k The fixed cost of truck K.
- q_{ip} The daily demand of customer i for product P.

Decision Variables:

$$x_{ijk} \begin{cases} 1 if truck k serves j immediately after i, \\ 0 otherwise \end{cases}$$

This decision variable is used to construct the routes by indicating if the arc from node V_i to node V_j is traversed by vehicle K.

$$y_{ipk} \begin{cases} 1 if product P for customer i is delivered by truck K, \\ 0 otherwise \end{cases}$$

This decision variable is used to assign the delivery of product P to customer V_i to vehicle K.

$$z_{ptk} \begin{cases} 1 if product P is allocated to tank t in truck K, \\ 0 otherwise \end{cases}$$

This decision variable is used to allocate each product to a specific compartment within a vehicle.

$$E_k \begin{cases} 1 if truck K is used, \\ 0 otherwise \end{cases}$$

This decision variable is used to indicate whether a vehicle is mobilized, thereby determining the total fixed cost.

$$u_i \{\text{Continuous variable } \geq 0$$

This auxiliary variable is used to establish a visit sequence for each customer node V_i to prevent the formation of sub-tours. This variable used exclusively by the Miller-Tucker-Zemlin (MTZ) method to eliminate illegal subtours.

Objective Function:

The model's primary objective is the minimization of Total Cost (TC), which comprises a fixed cost for each vehicle used and a variable cost proportional to the total distance covered by the fleet.

$$MinimizeTC = CF_k \sum_{k \in K} E_k + CV_k \sum_{i \in V} \sum_{j \in V} \sum_{k \in K} d_{ij} x_{ijk} \tag{1}$$

Constraints:

$$\sum_{j \in \mathrm{V}} \sum_{k \in \mathrm{K}} x_{ijk} = 1, \ \forall i \in N \tag{2}$$

$$\sum_{i \in \mathrm{V}} x_{ihk} - \sum_{\mathrm{j} \in \mathrm{V}} x_{hjk} = 0, \ \forall \, \mathrm{h} \in N, \mathrm{k} \in K \tag{3}$$

$$u_i - u_j + |N| x_{ijk} \le |N| - 1, \ \forall \, \mathrm{k} \in K, \forall i, j \in N, \mathrm{i} \ne \mathrm{j} \tag{4}$$

$$\sum_{j \in \mathrm{N}} x_{0jk} \le 1, \ \forall \, \mathrm{k} \in K \tag{5}$$

$$\sum_{i \in \mathrm{N}} x_{i0k} \le 1, \ \forall \, \mathrm{k} \in K \tag{6}$$

$$\sum_{k \in \mathrm{K}} y_{ipk} = 1, \ \forall i \in N, \mathrm{p} \in P \tag{7}$$

$$y_{ipk} \le \sum_{j \in \mathrm{v}} x_{jik}, \ \forall \mathrm{i} \in N, \ \mathrm{p} \in P, \mathrm{k} \in K \tag{8}$$

$$\sum_{i \in \mathrm{N}} \sum_{p \in \mathrm{P}} y_{ipk} . q_{ip} \le C_k, \ \forall \, \mathrm{k} \in K \tag{9}$$

$$\sum_{i \in \mathrm{N}} y_{ipk} . q_{ip} \le \sum_{t \in \mathrm{T}} C_{tk} . z_{ptk}, \ \forall \mathrm{p} \in P, \ \mathrm{k} \in K \tag{10}$$

$$\sum_{p \in \mathrm{P}} z_{ptk} \le 1, \ \forall \, \mathrm{t} \in T, \ \mathrm{k} \in K \tag{11}$$

$$z_{ptk} \le E_k, \ \forall \mathrm{p} \in P, \mathrm{t} \in T, \ \mathrm{k} \in K \tag{12}$$

$$\sum_{k=1} E_k \le |K|, \ \forall k \in K \tag{13}$$

$$x_{ijk} \in \{0, 1\}, \ \forall \, \mathrm{i} \in V, \ \mathrm{j} \in V, \ \mathrm{k} \ \in K \tag{14}$$

$$y_{ipk} \in \{0, 1\}, \ \forall i \in N, \ p \in P, \ k \in K \tag{15}$$

$$z_{ptk} \in \{0, 1\}, \ \forall p \in P, t \in T, \ k \in K \tag{16}$$

$$E_k \in \{0, 1\}, \ \forall k \in K \tag{17}$$

The objective function (1) aims to minimize the total cost of fuel distribution, composed of fixed and variable costs. This objective is subject to the following set of constraints: Constraint (2) guarantees that every customer is serviced by a single truck. Constraint (3) states that a truck visiting node h must also leave that same node. Constraint (4) prevents sub-tour formation using the classical approach. Constraints (5) and (6) ensure that each truck starts its route from the depot and returns to it. Constraint (7) ensures that each product requested by a customer is delivered by exactly one vehicle. Constraint (8) ensures consistency between delivery assignments and the routes of the trucks, preventing a truck from being assigned to deliver a product to a customer it does not actually visit. Constraint (9) ensures that no truck is loaded beyond its total capacity. Constraint (10) ensures that the compartment capacity is respected for each product. Constraint (11) states that a compartment can contain only a single product. Constraint (12) ensures that a product can only be loaded into a compartment if the corresponding vehicle is actually used. Constraint (13) restricts vehicle usage to the available fleet size. Finally, Constraints (14) through (17) enforce a binary domain for the decision variables.

3 Numerical Experiments and Results

The proposed approach was validated using real-world operational data from the Tunisian International Road Transport (TTRI) company. The mathematical model, formulated as a Mixed-Integer Linear Program (MILP), was implemented in Python. The solutions were obtained using the IBM ILOG CPLEX solver (version 22.1.1) accessed via its Python API. All computational experiments were executed on a Hp computer equipped with an Intel Core i7 processor and 16 GB of RAM.

3.1 Experimental Setup

The model was tested and validated using a set of 9 real-world instances derived from the operational records of Tunisian International Road Transport (TTRI). These instances, representing Saturday delivery operations for August and September, were selected due to their operational clarity; the company's single-shift policy on these days creates a self-contained planning horizon, providing a well-defined problem scope. These operational scenarios vary in complexity, which is primarily driven by two factors: the number of customers (N), ranging from 4 to 6, and the number of products (P), ranging from 3 to 4. The product mix, a core aspect of the problem's complexity, involves the transport of up to four distinct and incompatible petroleum products, namely P1 (Shell Vpower Mogas), P2 (Automotive Gasoil 1000 PPM), P3 (Shell Fuelsave Diesel 10ppm), and P4 (White Spirit). Each instance is identified by *SAT-N-P-D*, where SAT stands for Saturday, *N* for the customer count, *P* for the product count, and *D* for a specific data snapshot or daily scenario. A summary of these instances is presented in Table 1.

Table 1. Summary of key characteristics for the 9 test instances.

Instance	# Customers	# Number of products	#Type of product
SAT-N4-P3-D1	C1, C2, C3, C4	3	P1, P2, P3
SAT-N4-P4-D1	C1, C2, C3, C4	4	P1, P2, P3, P4
SAT-N4-P4-D2	C1, C2, C3, C4	4	P1, P2, P3, P4
SAT-N5-P4-D1	C1, C2, C3, C4, C5	4	P1, P2, P3, P4
SAT-N5-P4-D2	C1, C2, C3, C4, C5	4	P1, P2, P3, P4
SAT-N5-P4-D3	C1, C2, C3, C4, C5	4	P1, P2, P3, P4
SAT-N5-P4-D4	C1, C2, C3, C4, C5	4	P1, P2, P3, P4
SAT-N5-P4-D5	C1, C2, C3, C4, C5	4	P1, P2, P3, P4
SAT-N6-P4-D1	C1, C2, C3, C4, C5, C6	4	P1, P2, P3, P4

Beyond these variables, all instances share a common set of fixed operational parameters. The fleet consists of 4 homogeneous trucks, each with a 32,000 kg capacity distributed across 6 heterogeneous compartments. The objective function is strategically structured to capture the fundamental economic trade-off in logistics. First, a fixed cost of 400.7 TND is incurred for each mobilized vehicle. This term represents major, distance-independent expenditures like driver wages and vehicle depreciation, and its primary role is to drive fleet consolidation, which means the goal is to use the fewest possible trucks by filling each one as much as possible. By penalizing the use of each truck, it forces the model to maximize vehicle utilization and deploy the absolute minimum number of assets required. Second, a variable cost of 1.049 TND/km models the operational expenses tied directly to distance, such as fuel and maintenance. This component acts as the primary driver for routing efficiency, pushing the model to design the most direct and economical paths for the active vehicles. This duality is central to the problem: the model must find the optimal balance, accepting potentially longer routes (higher variable cost) if doing so allows for the elimination of a vehicle (a major fixed cost saving).

3.2 Results and Comparative Analysis

The model was able to solve all 9 instances to optimality. A direct comparative analysis was conducted between the solutions generated by our model and the actual plans used by TTRI's dispatchers. Figure 1 graphically illustrates this comparison, highlighting the differences in total costs between our optimized approach and the company's current operational solution for each instance. The detailed results of this comparison are also summarized in Table 2.

The column headings in Table 2 are as follows:

- K: Number of vehicles used
- DT: Total distance traveled (km)
- Cost (TND): Total cost in Tunisian Dinar
- CPU(s): Computation time in seconds
- GAP (%): Percentage improvement of our model compared to TTRI's solution

Table 2. Comparative results of the proposed model and TTRI's current operational solution.

Data	Our model				Company's solution (TTRI)			Improvement
Instances	K	DT	Cost (TND)	CPU(S)	K	DT	Cost (TND)	GAP %
SAT-N4-P3-D1	3	340	1558.8	0.11	3	340	1558.8	0.0
SAT-N4-P4-D1	4	308	1925.9	0.05	4	308	1925.9	0.0
SAT-N4-P4-D2	4	436	2060.2	0.27	4	436	2060.2	0.0
SAT-N5-P4-D1	4	334	1953.2	0.59	4	334	1953.2	0.0
SAT-N5-P4-D2	4	366	1986.7	0.06	4	460	2085.3	4.7
SAT-N5-P4-D3	4	324	1942.7	0.06	4	324	1942.7	0.0
SAT-N5-P4-D4	4	398	2020.3	0.06	4	500	2127.3	5.0
SAT-N5-P4-D5	4	324	1942.7	0.08	4	324	1942.7	0.0
SAT-N6-P4-D1	4	350	1969.9	0.44	4	350	1969.9	0.0

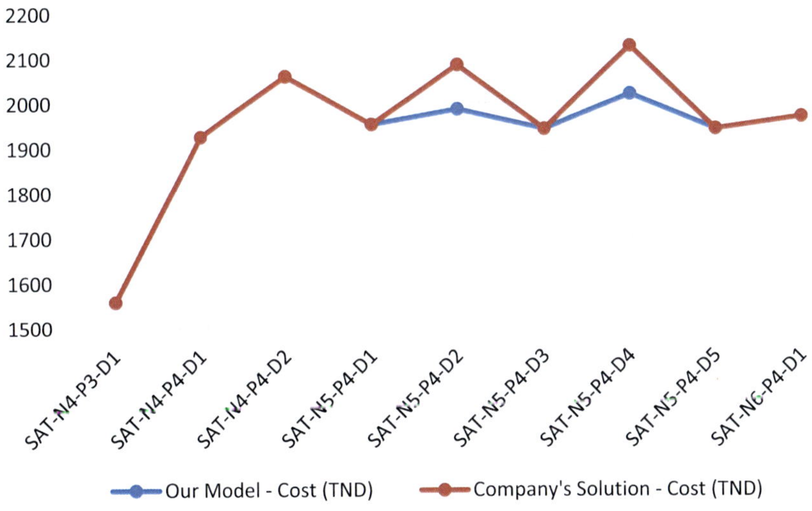

Fig. 1. Performance of the MCMPVRP model compared to TTRI's operational solution

The results, summarized in Table 2 and illustrated by Fig. 1, reveal a highly significant finding: for 7 out of the 9 instances (78%), our model produced solutions that were identical to the company's manual plans in terms of vehicles used, distance traveled, and total cost. This validates the high level of expertise of TTRI's logistics planners.

However, for two complex instances (*SAT-N5-P4-D2* and *SAT-N5-P4-D4*), the model identified significantly more efficient routes, achieving cost reductions of 4.7% and 5.0%, respectively. On average across all instances, the model provided a 1.1% reduction in total costs.

4 Conclusion

In this work, we addressed the Multi-Compartment Multi-Product Vehicle Routing Problem (MCMPVRP) through the development of an exact Mixed-Integer Linear Programming (MILP) model. Our approach was rigorously motivated and validated by a real-world case study utilizing operational data from TTRI, a Tunisian transport company specializing in fuel distribution.

The numerical experiments demonstrated that the proposed model can solve realistic small to medium-sized instances to optimality. The comparative analysis with the company's manual plans yielded two primary and significant outcomes: first, a robust validation of TTRI's operational expertise, as our model replicated their optimal solutions in 78% of the cases; and second, the identification of concrete optimization opportunities, where the model achieved cost reductions of up to 5% in complex scenarios by finding more efficient routes.

Beyond minimizing total cost, our model's bi-component objective function explicitly captures the fundamental logistics trade-off between fleet consolidation (minimizing fixed costs) and routing efficiency (minimizing variable costs). This provides decision-makers with a transparent framework for strategic analysis and informed planning.

For future research, several promising avenues can be explored. Given the NP-hard nature of the MCMPVRP, the development of efficient metaheuristics, is essential for solving larger-scale instances. Extending the model to incorporate practical constraints such as customer time windows (MCMPVRPTW), stochastic demands, or environmental sustainability measures would enhance its applicability. This aligns with recent trends in green logistics [5], such as considering traffic congestion and carbon policies in MCVRP models [8]. Finally, integrating this tactical planning model with strategic decisions like fleet sizing or depot location could offer a more holistic decision-support system for logistics enterprises.

References

1. Baptista, G., Vieira, M., Pinto, T.: An exact approach to the multi-compartment vehicle routing problem: the case of a fuel distribution company. Mathematics **12**(4), 527 (2024)
2. Benantar, A., Ouafi, R., Boukachour, J.: A combined vehicle loading and routing problem: a case study of fuel logistics. Int. J. Logistics Syst. Manage. **32**(3/4), 346–371 (2019)

3. Cornillier, F., Boctor, F.F., Renaud, J.: Heuristics for the multi-depot petrol station replenishment problem with time windows. Eur. J. Oper. Res. **220**(2), 361–369 (2012)
4. Dantzig, G.B., Ramser, J.H.: The truck dispatching problem. Manage. Sci. **6**(1), 80–91 (1959)
5. Demir, E., Bektaş, T., Laporte, G.: The green vehicle routing problem: a systematic literature review. J. Clean. Prod. **279**, 123691 (2020)
6. Derigs, U., Gottlieb, J., Kalkoff, J., Piesche, M., Rothlauf, F., Vogel, U.: Vehicle routing with compartments: applications, modeling and heuristics. OR Spectrum **33**(4), 885–914 (2010)
7. Dubey, R., Bryde, D.J., Foropon, C., Graham, G., Giannakis, M.: Bilateral buyer–supplier relational stability and firm sustainability performance: the role of digitalization and network complexity. Int. J. Oper. Prod. Manage. **43**(11), 1719–1748 (2023)
8. Fan, X., Yao, G., Yang, Y.: Multi-compartment vehicle routing problem considering traffic congestion under the mixed carbon policy. Appl. Sci. **13**(17), 10304 (2023)
9. Figliozzi, M., Jennings, D.: A review of vehicle routing problems and models for last-mile delivery with modes such as drones, autonomous vehicles, and bikes. Transp. Res. Part C: Emerg. Technol. **155**, 104323 (2023)
10. Heßler, K.: Exact algorithms for the multi-compartment vehicle routing problem with flexible compartment sizes. Eur. J. Oper. Res. **294**(1), 188–205 (2021)
11. Ibrahim, A., Abdulaziz, R.O., Ishaya, J.A., Sowole, S.O.: Vehicle Routing Problem with Exact Methods. IOSR J. Math (IOSR-JM). **15**(1), 5–15 (2019)
12. Ivanov, D., Dolgui, A.: Supply chain dynamics and resilience: frameworks and review. Int. J. Prod. Res. **60**(11), 3467–3488 (2022)
13. Koc, Ç., Jabali, O., Mendoza, J.E.: A hybrid adaptive large neighborhood search heuristic for the multi-compartment vehicle routing problem. Comput. Oper. Res. **103**, 198–207 (2019)
14. Lambert, D.M., Enz, M.G.: Issues in supply chain management: progress and potential. Ind. Mark. Manage. **62**, 1–16 (2017)
15. Mor, A., Speranza, M.G.: Vehicle routing problems over time: A survey. Ann. Oper. Res. **314**, 255–275 (2022)

Blockchain-Enabled Consent-Based Data Governance for Smart Territories

Mongetro Goint[1], Cyrille Bertelle[1(✉)], and Claude Duvallet[2]

[1] Université Le Havre Normandie, LITIS, UR 4108, Le Havre, Rouen F-76000, France
cyrille.bertelle@univ-lehavre.fr

[2] Université Le Havre Normandie, LITIS, UR 4108, Le Havre, Rouen F-76000, France
claude.duvallet@univ-lehavre.fr

Abstract. The digital transformation of territories has fostered the rise of smart environments, where heterogeneous public and private actors generate massive data volumes. While data-driven services are critical for optimizing urban infrastructures, mobility, logistics, and citizen services, the lack of transparent and secure data governance remains a major adoption barrier. In particular, establishing digital trust and privacy-preserving consent for data access are central challenges in smart territories. We propose a blockchain-based architecture enabling secure, verifiable, and decentralized management of data access consent. Our model introduces a dual-level consent mechanism—Automatic Consent (AC) for research or public use, and Non-Automatic Consent (NAC) for sensitive or commercial exploitation. We extend ADA-M (Automatable Discovery and Access Matrix) to formalize consent policies and access requests, immutably anchored on blockchain. Large-scale data is stored off-chain and protected by symmetric cryptography, while data sharing is based on encryption and access control with proxy re-encryption. The architecture provides a foundation for trusted, transparent, and privacy-preserving data exchange among stakeholders. This work presents an architectural blueprint with explicit security requirements and design principles; quantitative validation through prototype implementation is identified as essential future work before deployment.

Keywords: smart territories · digital trust · consent management · blockchain · privacy-preserving · data sharing · data economy

1 Introduction

The rise of smart territories marks a paradigm shift in how urban and regional systems leverage data and digital technologies to improve citizens' quality of life, optimize public services, and support sustainable development [3,5]. These territories integrate heterogeneous infrastructures—mobility, energy, logistics, governance, and environment—through intensive use of data and interconnected

Y. Benadada et al. (Eds.): GOL 2026, LNNS 2034, pp. 368–381, 2027.
https://doi.org/10.1007/978-3-032-29677-1_33

digital services. This transformation relies on the cooperation of diverse stakeholders: public administrations, private companies, service providers, and citizens. However, as data sharing and processing intensify, significant concerns emerge regarding privacy, transparency, and digital trust.

Establishing digital trust is critical, yet centralized platforms create vulnerabilities. GDPR mandates explicit consent and traceability, requiring decentralized alternatives. Blockchain provides immutable coordination explored in smart cities and healthcare, but existing approaches remain domain-specific or single-layer, failing to reflect smart territories' multi-stakeholder complexity.

In this work, we propose a blockchain-based consent-driven data governance model that establishes digital trust among smart territory stakeholders. This work is primarily architectural and proposes design principles rather than a fully implemented system. We identify critical technical requirements, security properties, and integration challenges, with quantitative evaluation deferred to future implementation work. Our model introduces a dual consent mechanism, distinguishing **Automatic Consent (AC)** for non-commercial or research purposes from **Non-Automatic Consent (NAC)** for commercial or sensitive uses. Consent declarations and access policies are formalized using an extended **Automatable Discovery and Access Matrix (ADA-M)**, providing a structured representation of rules for secure and transparent data exchange. All consents are immutably anchored on the blockchain and can be revoked at any time, while large-scale data is stored off-chain under cryptographic protection.

The rest of this paper is structured as follows: Section 2 reviews background related work; Sect. 3 presents our main contribution, the proposed architecture, the consent management mechanism and the encryption and access control process for data access and sharing in smart territories. Finally, Sect. 4 concludes the paper with future research directions.

2 Background and Related Work

2.1 Data Governance and Consent in Smart Territories

Smart territories require intensive data sharing across domains (mobility, energy, environment) involving public authorities, private operators, and citizens [8,24], yet existing platforms (FIWARE, open data) remain centralized and lack transparent, GDPR-compliant consent mechanisms [20].

Consent is a cornerstone of GDPR compliance [13], requiring explicitness, granularity, and revocability [15]. Blockchain-based consent solutions have emerged: Zyskind et al. [31] use off-chain storage with on-chain access control; Jaiman et al. [18] leverage ADA-M for health data with single-layer consent; Agarwal et al. [2] introduce role-based access. In smart territories, Makhdoom et al. [22] propose PrivySharing, while ADvoCATE [26] manages IoT consent via on-chain hashes.

Comparative positioning. Threshold PRE (NuCypher/Umbral [12]) offers robustness via distributed re-encryption but adds protocol complexity; distributed coordination overhead increases end-to-end latency vs. single-TEE deployment ($\sim$ 20 ms AFGH in SGX). ABE schemes [30] embed policies in ciphertext but impose $O(|policy|)$ decryption overhead. Our TEE-centralized PRE balances latency, revocability, and operational simplicity for municipalities' asynchronous use cases.

Our approach addresses three gaps: (1) **Dual-consent granularity** (AC/NAC) vs. single-layer systems; (2) **Owner-centric PRE** with per-requester key rotation; (3) **Multi-domain applicability** beyond healthcare or IoT. Recent systems provide quantitative baselines: PrivacyGuard [11] reports $<$ 50 ms end-to-end latency for consent verification on Ethereum testnets; PRE-based e-prescriptions [19] achieve $\sim$ 10 ms re-encryption with pyUmbral. Consentio [2] demonstrates 3000+ Transactions per second (TPS) on Hyperledger Fabric for consent logging. ABE-based approaches (MA-CP-ABE) offer stronger receiver privacy via cryptographic policy hiding but impose $O(|policy|)$ decryption overhead vs. PRE's constant-time $O(1)$ with explicit rk. Our architecture targets a middle ground: owner-centric revocation (PRE advantage) with TEE-based trust (simpler than threshold coordination), accepting higher latency than Umbral for formal security guarantees. Empirical head-to-head evaluation against these baselines is critical future work.

2.2 Blockchain and Smart Contracts

Blockchain provides decentralized, immutable coordination without trusted intermediaries [9,23], with smart contracts enabling programmable consent enforcement [7,27]. However, limited on-chain storage and oracle constraints require hybrid architectures for large-scale data sharing.

2.3 Automatable Discovery and Access Matrix (ADA-M)

ADA-M [29] is a machine-readable framework for describing data access conditions, initially designed for healthcare. It structures metadata into profiles, conditions, and usage constraints to enable automated request screening. In this work, we extend ADA-M beyond its original scope to support dual-consent policies and blockchain-based enforcement in multi-domain smart territories.

3 A Dual-Consent Blockchain Platform Model for Improved Data Governance in Smart Territories

We propose a blockchain-based data platform for secure consent management in smart territories, enabling automatic consent (AC) and non-automatic consent (NAC). The architecture leverages an extended ADA-M, Data Usage Policy Statements (DUPoS)/Data Usage Purpose Statement (DUPuS), blockchain anchoring, and proxy re-encryption (PRE) to ensure digital trust and GDPR compliance.

3.1 Overview and Contributions

Our platform introduces three key design contributions:

1. A dual-consent protocol distinguishing AC (public-interest, automatic) and NAC (commercial, owner-approved).
2. An extended ADA-M with DUPoS (owner policy) and DUPuS (requester purpose) for machine-readable, verifiable consent.
3. A tamper-proof blockchain system anchoring declarations and consents, eliminating centralized trusted intermediaries for data access control.

These contributions are presented as architectural principles. Practical deployment would require addressing implementation challenges including TEE side-channel mitigations, IPFS persistence strategies, and governance mechanisms for role authorities—aspects we identify as critical future work.

3.2 Extended ADA-M with DUPoS and DUPuS

We extend ADA-M [29] to support Data Usage Policy Statement (DUPoS) and Data Usage Purpose Statement (DUPuS). In DUPoS, owner-defined access policy (Table 1), while in DUPuS, requester-defined usage intent (Table 2). Smart contracts automatically match DUPoS and DUPuS to establish consent.

Table 1. DUPoS Structure

Field	Description
DUPoS Id	DUPoS unique Identifier
Data category	Type of data (e.g., mobility, health, ...)
DO wallet	Data owner wallet address
UDG id	Unique data category identifier
Possible accessor	Eligible requester roles
Consent type	`Automatic` or `NonAutomatic`
Data purpose	Reason why the data will be used
Remuneration Required	`true/false`

Matching Semantics. A DUPuS satisfies a DUPoS when: (1) Data Category matches or subsumes (via domain ontology, e.g., "traffic" $\subseteq$ "mobility"), (2) Requester Role $\in$ Possible Accessor set, (3) Data Purpose is compatible (exact match or subsumption via ISO/IEC 19944 vocabulary, e.g., "traffic analysis" $\subseteq$ "research"), (4) Remuneration constraint satisfied (if DUPoS requires payment, DUPuS must offer it). If multiple DUPoS match, the most restrictive policy applies (NAC $>$ AC). Smart contracts implement matching via on-chain ontology references (IPFS-hosted, hash-anchored).

Table 2. DUPuS structure

Field	Description
DUPuS id	DUPuS unique Identifier
Data category	Requested data type
DA wallet	Requester's address
User role	Requester's certified role
Consent type	`Automatic` or `NonAutomatic`
Data purpose	Data intended use
Remuneration offered	Payment willingness indicator

3.3 Consent Model

Consent is a verifiable agreement defined between actors to share and access data, and its defining characteristics are shown in the Table 3. We identify two categories in our work:

Table 3. Consent structure

Field	Description
DO wallet	Data owner wallet address
DA wallet	Data accessor wallet address
DUPoS id	Data usage policy statement reference
DUPuS id	Data Usage Purpose statement reference
UDG id	User Data group identifier
Consent deadline	Expiry date for the consent
Consent state	Active/Revoked

Automatic Consent (AC)

AC operates via *owner-delegated automation*: owners pre-authorize policies (DUPoS), and a client-side agent automatically generates re-encryption keys when valid requests occur. Requirements: (1) DUPuS/DUPoS match verified by smart contracts, (2) Requester role certified by competent authority. The owner's agent monitors blockchain events and computes $rk_{owner \rightarrow req}$ without manual approval, ideal for research/public-interest use.

Non-automatic Consent (NAC)

NAC requires owner approval after DUPoS/DUPuS matching. Used for commercial applications or sensitive data and may involve fees or special conditions.

3.4 Three-layer architectural model

The proposed platform model adopts a three-layer architecture (Fig. 1):

1. **User Interaction Layer:** This layer provides citizens and organizations with access to decentralized applications (dApps) such as consent dashboards and DUPoS/DUPuS submission interfaces. It ensures transparent participation and interaction with the underlying blockchain infrastructure.
2. **Encryption and Storage Layer:** Sensitive datasets are encrypted using AES-256 and stored off-chain in a decentralized service, like the InterPlanetary File System (IPFS). AES algorithm was chosen for its superior performance and security compared to other symmetric methods such as DES or 3DES[1]. It offers fast encryption, resistance to cryptanalytic attacks, and allows for the efficient processing of large volumes of data with low resource consumption—essential advantages for secure and scalable data management in smart territories. IPFS provides decentralized, content-addressed storage, ensuring data integrity, fault tolerance, and immutability without relying on centralized servers.
   IPFS persistence: In a consortium deployment, designated nodes (municipal infrastructure providers) act as pinning services, ensuring dataset availability with redundancy. Pinning policies, garbage collection rules, and incentive mechanisms (e.g., payment for storage) require governance agreements. Private IPFS swarms may be used for sensitive data to restrict access to consortium members. Detailed availability and retrieval performance analysis is future work.
3. **Blockchain Anchoring Layer:** This layer maintains immutable decentralized records of consent events, hash references, and trigger Proxy Re-Encryption (PRE) operations.

For data access protection and sharing, we use a proxy encryption (PRE) system, as described in Sect. 3.5. This mechanism ensures the confidentiality of user data.

3.5 Encryption and Access Control with Proxy Re-Encryption (PRE)

To enable secure and verifiable data sharing, the system employs Proxy Re-Encryption (PRE) [4,6]. PRE enables a proxy running in a Trusted Execution Environment (TEE)—a hardware-isolated, tamper-resistant enclave—to transform a ciphertext encrypted under the data owner's public key into one decryptable with the requester's private key, without learning the plaintext or any private keys [6].

Consensus Finality and Oracle Safety

Finality-Aware PRE Triggering. PRE operations occur only after blockchain finality to prevent oracle attacks. *Tiered strategy:* (1) **Ethereum**

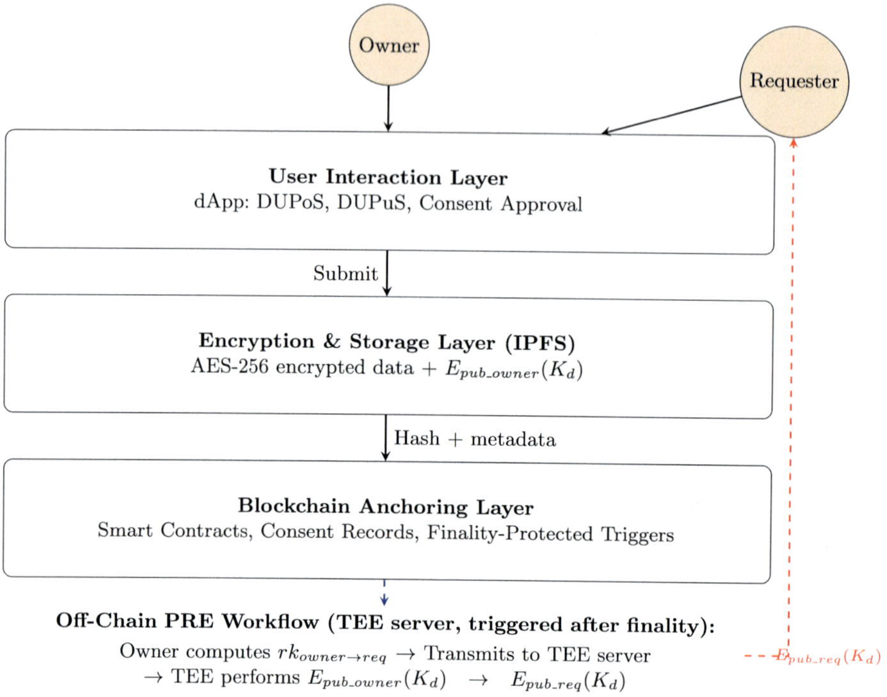

Fig. 1. Three-layer architecture for consent-driven data governance in smart territories. The PRE workflow is triggered off-chain only after blockchain consensus finality (see text).

PoS: 32 blocks ($\sim$ 6.4 min economic finality) for NAC (commercial use, latency-tolerant); 12 blocks ($\sim$ 2.5 min, >99.9% probabilistic) acceptable for low-risk AC after risk assessment. (2) **Hyperledger Fabric:** Immediate finality (1–2 s via Raft) for latency-sensitive AC (real-time services). *Recommended deployment:* Hybrid approach—Fabric for AC ($<$ 30 s consent-to-access), Ethereum mainnet for NAC audit ($<$ 10 min acceptable). Smart territory use cases are predominantly asynchronous (research analytics, planning), not real-time control, making minute-scale latencies acceptable.

Latency considerations: Ethereum's 12-min finality is impractical for real-time applications. We recommend hybrid deployment: permissioned networks (Fabric, 1–2 s) or Layer-2 solutions for low-latency cases; Ethereum mainnet for high-value NAC where security justifies latency. Quantitative trade-off analysis is future work.

Governance roles and responsibilities: (1) *Consortium validators:* Municipal infrastructure providers operate blockchain nodes and TEE servers under service-level agreements. (2) *Role authorities:* Certified institutions (medical boards, research ethics committees) issue W3C Verifiable Credentials anchored on-chain; revocation via on-chain registry updates. (3) *Ontology curators:*

Domain experts maintain ISO/IEC 19944-aligned taxonomies on IPFS with versioned hash pointers updated via multi-signature smart contracts. (4) *Audit entities:* Independent third parties verify compliance through on-chain event logs and periodic TEE attestation audits.

PRE Node and TEE Integration

The PRE node is deployed on a server (e.g., within an Intel SGX enclave), providing a hardware-isolated, tamper-proof environment with remote attestation [10]. Although represented in the blockchain layer for clarity, the PRE operates **off-chain**. Its integration is indicated by being **triggered by on-chain smart contracts** upon consent validation and after finality is achieved. We emphasize that the PRE node is not a trusted party in the traditional sense: it is trusted only for correct execution of the re-encryption algorithm within a verified enclave, and never entrusted with plaintext data, symmetric keys, or private keys. Compromise of the PRE node does not enable data decryption without valid requester private keys.

SGX Side-Channel Mitigation: We acknowledge known vulnerabilities (Spectre, Foreshadow) and implement: (a) constant-time crypto libraries (Intel IPP, mbedTLS SGX), (b) minimal enclave footprint (only re-encryption, no plaintext/long-term keys), (c) TCB level ≤ 2 enforcement via attestation. **Residual single-point-of-failure risk:** The TEE proxy remains an availability bottleneck. Mitigation strategies include: (1) active-active SGX replicas with state synchronization via secure channels, (2) fallback to manual owner approval if TEE unavailable, (3) distributed threshold PRE (e.g., Shamir secret sharing of rk across n enclaves, requiring $t < n$ for re-encryption). Production deployment must implement redundancy; single-node design is acceptable for proof-of-concept only. Residual risks (EM side-channels, unpatched exploits) are mitigated by rk ephemerality (used once, immediately deleted). Alternative TEEs (ARM TrustZone, AMD SEV-SNP) can substitute SGX as deployment policies dictate.

Key Elements and Notation

- K_d: AES-256 symmetric key encrypting the dataset.
- $E_{pub\_owner}(K_d)$: K_d encrypted with the data owner's public key.
- sk_{owner}: private key of the data owner, used to generate signatures and compute re-encryption keys.
- pk_{owner}: public key of the data owner (notation: pub_owner in ciphertexts), used to encrypt K_d as $E_{pub\_owner}(K_d)$.
- pk_{req}: public key of the requester, used in re-encryption key generation.
- sk_{req}: private key of the requester, used to decrypt re-encrypted data.
- $E_{pub\_req}(K_d)$: K_d encrypted with the requester's public key.
- $rk_{owner \rightarrow req}$: re-encryption key computed by the owner, allowing the PRE node to transform $E_{pub\_owner}(K_d)$ into $E_{pub\_req}(K_d)$ without accessing K_d or any private keys.

Process Overview

The platform adopts a unidirectional, single-hop PRE scheme (AFGH [4]), ensuring that re-encryption occurs only from the owner to authorized requesters.

Unlike Blaze et al. [6], where bidirectional re-encryption is possible, this design prevents chained or reverse transformations. The detailed PRE flow is illustrated in Fig. 2.

1. **Data Encryption:** Owner encrypts data with K_d, then encrypts K_d under their public key to produce $E_{pub\_owner}(K_d)$.
2. **Off-Chain Storage:** Encrypted data and $E_{pub\_owner}(K_d)$ are stored on IPFS. Only the hash (CID) and metadata are anchored on-chain.
3. **Consent Validation and Re-Encryption:**
   - Requester submits DUPuS describing intended data use.
   - Smart contracts verify DUPuS/DUPoS match.
   - *Automatic Consent (AC)*: Upon validation and consensus finality (e.g., 32 blocks on Ethereum PoS), the data owner computes the re-encryption key locally: $rk_{owner \to req} = \text{ReKeyGen}(sk_{owner}, pk_{req})$.
     Before transmission, the Owner's client software performs remote attestation of the PRE Node's TEE (Intel SGX enclave) using EPID or DCAP protocols. This cryptographically verifies that: (i) the enclave runs the authorized PRE code (validated via MRENCLAVE measurement against a consortium-maintained whitelist), and (ii) the hardware environment is genuine and uncompromised. Only after successful attestation verification does the owner transmit rk through the secure channel.
     The TEE receives rk, performs $E_{pub\_req}(K_d) = \text{ReEncrypt}(E_{pub\_owner}(K_d), rk)$, and securely deletes rk from enclave memory immediately after use. This workflow ensures that sk_{owner} never leaves the owner's environment, and rk exists only transiently within the hardware-protected TEE.
   - Requester retrieves $E_{pub\_req}(K_d)$ from the TEE and decrypts data with their private key.
   - *Non-Automatic Consent (NAC)*: Owner must manually approve $rk_{owner \to req}$ generation for commercial, sensitive or remuneration-based cases.
4. **Blockchain Anchoring:** Hashes of data, key references, and consent events are immutably recorded on-chain. No plaintext or private keys are stored on-chain.

AFGH Selection Rationale. We adopt AFGH [4] for: (1) CCA-security proofs under DBDH, (2) Unidirectional single-hop enabling owner-centric revocation, (3) Mature JPBC implementation ($\sim$ 20 ms ReEncrypt). Umbral ($\sim$ 5 ms) requires threshold coordination; ABE imposes $O(|policy|)$ overhead. AFGH suits < 500 consent/hour deployments prioritizing auditability; empirical comparison with alternatives is future work.

Security and Compliance

- **End-to-End Confidentiality:** The PRE node accesses only rk (computed by owner), not sk_{owner}, sk_{req}, or K_d. Blockchain stores only hashes and metadata. Private keys remain with owners/requesters; public keys are used as intended.

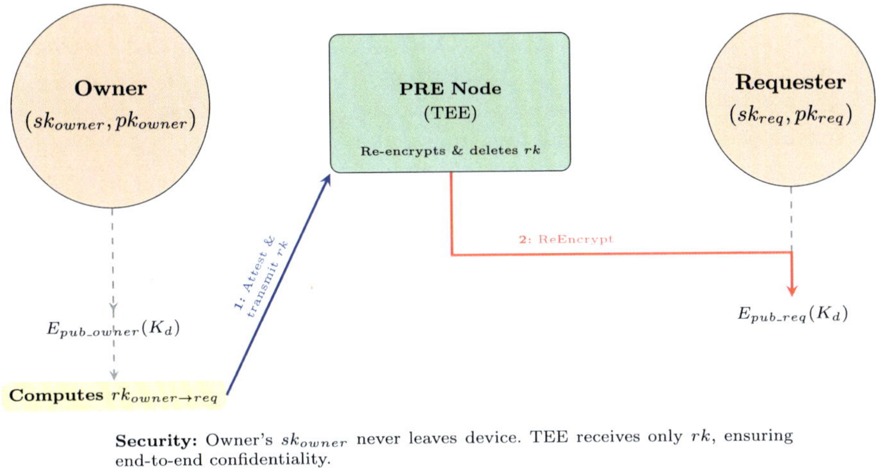

Fig. 2. PRE flow: secure delegation from Owner to Requester via TEE-based PRE node.

- **Revocability:** Access is revocable via key rotation (see below), ensuring forward secrecy.
- **Integrity and Auditability:** All consent events and key transformations are immutably logged on-chain.
- **GDPR Compliance (pending DPIA validation)**: The system *aims to implement* a GDPR joint controller model: data owners act as co-controllers (determine DUPoS, grant/revoke consent), consortium manages infrastructure (security, audits), requesters operate as processors (process per validated DUPuS). Legal bases: AC via Art. 6(1)(a) or 6(1)(e), NAC via Art. 6(1)(a)+Art. 9 for sensitive data. However, on-chain metadata may constitute pseudonymous personal data under GDPR Art. 4(1), requiring Data Protection Impact Assessment (DPIA) to validate compliance claims. Full GDPR conformity depends on governance agreements and jurisdictional legal review. No personal data stored on-chain; erasure enforced via off-chain deletion + key rotation.

Revocation and Forward Secrecy
To prevent continued access after consent withdrawal, the system implements key versioning. Each dataset is encrypted with $K_{d,v}$ (initially $v = 1$). Upon revocation:

1. Owner submits revocation transaction (on-chain: `consent.state = Revoked`).
2. Generate new key $K_{d,v+1}$, re-encrypt data: $C_{d,v+1} = \text{AES.Enc}(K_{d,v+1}, D)$.
3. Store on IPFS with new CID; update on-chain metadata.
4. For active (non-revoked) consents: owner computes new rk' using $K_{d,v+1}$.
5. Revoked requester retains $K_{d,v}$ (past access, non-repudiation) but cannot decrypt $K_{d,v+1}$ (forward secrecy).

Cost: $O(m)$ re-wrapping for m active consents. **Scalability:** For $m = 100$: $\sim$ 2.0 s (100 $\times$ 20 ms crypto + storage). **Optimizations:** (a) Batched rotations—daily window amortizes cost: 10 revocations/day $\rightarrow$ 1 rotation affecting $m-10$ consents instead of 10 separate m-wrappings; (b) Lazy re-encryption—mark revoked immediately, rotate at next maintenance window; (c) Hybrid KEM—encrypt K_d with session key K_s, wrap only K_s per requester: $\sim$ 200 ms for $m = 100$ vs. 2.0 s direct. These enable scalability to $m \leq 1000$ per dataset.

3.6 Security Analysis

We consider honest-but-curious blockchain validators, malicious requesters attempting unauthorized access, compromised IPFS nodes, and external attackers. We assume TEE (Intel SGX) correctness, cryptographic primitives (AES-256, PRE, ECDSA) security, and that owners/requesters protect private keys.

Security Properties
The system ensures: **Confidentiality** (AES-256 semantic security; PRE requires rk and sk_{req}), **Integrity** (tamper-proof on-chain records), **Authentication** (W3C Verifiable Credentials), **Forward Secrecy** (key rotation prevents post-revocation access), and **Auditability** (immutable event logs).

Threat Model and Trust Assumptions
We assume: (1) Honest-but-curious validators (BFT/PoS), (2) Malicious requesters/attackers, (3) Compromised IPFS nodes (mitigated by CID checks), (4) TEE correctness with SGX side-channel mitigations (constant-time crypto [17], minimal enclave, TCB $\leq$ 2 [16]). We do not defend against: state-level physical attacks, compromised endpoints, coercion. PRE node trusted only for algorithm correctness—compromise requires sk_{req} for decryption.

Attack Resistance
Collusion (requester + PRE node): PRE node holds only rk, not sk_{owner} or K_d—no advantage gained. Replay attacks: timestamps and deadlines prevent. Sybil attacks: roles certified by trusted authorities (medical boards, IRBs) anchored on-chain. IPFS tampering: content-addressed hashes detect modifications.

Metadata Linkage and Privacy Limitations
On-chain metadata (timestamps, CIDs, consent patterns) enables linkage attacks: unique CID + timestamp pairs allow de-anonymization [25]; repeated AC requests create behavioral profiles. Mitigations (ZK-SNARKs, temporal bucketing, CID encryption) trade privacy for latency (50–200 ms [14]) or auditability. Full anonymity requires Layer-2 private chains incompatible with public governance. Our architecture prioritizes transparency; GDPR compliance via DPIA-validated pseudonymization, accepting that strong anonymity is not guaranteed.

4 Conclusion and Future Work

This paper introduced a blockchain-based architecture for consent-driven data governance in smart territories, extending ADA-M with DUPoS/DUPuS for machine-readable consent policies. The dual-consent model (AC for public-interest, NAC for commercial use) offers granular, GDPR-aligned control with PRE-based cryptographic delegation.

PRE with TEE-based transformation and key rotation enables cryptographic delegation without exposing keys. The three-layer architecture (user interaction, encrypted IPFS storage, blockchain anchoring) achieves scalability, confidentiality, and decentralized trust, remaining agnostic to data types and blockchain platforms.

Limitations and Future Work. This work presents an architectural framework requiring prototype validation. Critical open questions: (1) *Performance*: measure end-to-end latency, gas costs, revocation overhead vs. baselines (direct wrapping, ABE); (2) *Security*: formal protocol verification, DPIA for metadata linkage, SGX side-channel testing; (3) *Governance*: operationalize consortium model (role registries, upgrade voting, IPFS SLAs); (4) *Usability*: wallet UX, field evaluation with municipal partners, for example; (5) *Policy extensions*: obligations, derived-data propagation [28]. Comparative evaluation against Consentio [2] (Fabric throughput), MA-CP-ABE [21] (policy expressiveness), and NuCypher (robustness) is essential before deployment.

Until these gaps are addressed, this architecture should be understood as a conceptual blueprint identifying design principles and research directions rather than a deployment-ready solution. Real-world deployment requires a multi-stakeholder consortium governing the permissioned blockchain, maintaining accredited authority registries, and overseeing transparent upgrade processes via on-chain governance. Integrating this socio-technical governance model with our technical architecture is essential for adoption and trust-building. This work paves the way for next-generation smart territories where data flows securely, transparently, and with user sovereignty at the core.

References

1. Elminaam, D.S.A., Abdel-Kader, H.M., Hadhoud, M.M.: Evaluating the performance of symmetric encryption algorithms. Int. J. Netw. Secur. **10**(3), 216–222 (2010)
2. Agarwal, S., D'Souza, M., Gharat, V., Shinde, S., Thakur, S., Vyas, T., Sharma, S.: Consentio: a blockchain-based consent management system. In: IEEE International Conference on Blockchain, pp. 1–8 (2020)
3. Alizadeh, T., Sharifi, A.: Toward a societal smart city: clarifying the social justice dimension of smart cities. Sustain. Cities Soc. **95**, 104599 (2023)
4. Ateniese, G., Fu, K., Green, M., Hohenberger, S.: Improved proxy re-encryption schemes with applications to secure distributed storage. ACM Trans. Inf. Syst. Secur. **9**(1), 1–30 (2006)

5. Simon Elias Bibri and John Krogstie: On the sustainability of smart and smarter cities in the era of big data: an interdisciplinary and transdisciplinary literature review. J. Big Data **6**(1), 25 (2019)
6. Blaze, M., Bleumer, G., Strauss, M.: Divertible protocols and atomic proxy cryptography. In: EUROCRYPT (1998)
7. Buterin, V.: Ethereum White Paper: A Next-Generation Smart Contract and Decentralized Application Platform (2013)
8. Chourabi, H, Nam, T., Walker, S., Gil-Garcia, J.R., Mellouli, S., Nahon, K., Pardo, T.A., Scholl, H.J.: Understanding smart cities: an integrative framework. In: 2012 45th Hawaii International Conference on System Sciences, pp. 2289–2297 (2012)
9. Christidis, K., Devetsikiotis, M.: Blockchains and smart contracts for the internet of things. IEEE Access **4**, 2292–2303 (2016)
10. Costan, V., Devadas, S.: Intel SQX Explained. Technical report, Cryptology ePrint Archive (2016)
11. Dinh, T.T.A., et al.: Privacyguard: enforcing private data usage with blockchain and attested off-chain contract execution. In: Data Privacy Management (2019)
12. Egorov, M., MacLane, D.: Nucypher KMS: Decentralized key management system. In: NuCypher Whitepaper (2020)
13. GDPR.: Regulation (EU) 2016/679 of the European Parliament and of the Council of 27 April 2016 on the Protection of Natural Persons with Regard to the Processing of Personal Data and on the Free Movement of Such Data, and Repealing Directive 95/46/EC (General Data Protection Regulation) (2016)
14. Groth, J.: On the size of pairing-based non-interactive arguments. In: Advances in Cryptology—EUROCRYPT 2016, pp. 305–326. Springer (2016)
15. Hoepman, J.-H.: Privacy design strategies. In: IFIP International Summer School on Privacy and Identity Management, pp. 446–459 (2014)
16. Intel Corporation.: Intel SQX: TCB Recovery. Technical Report, Intel Corporation, 2019. Documentation on SGX Trusted Computing Base Levels and Updates (2019)
17. Intel Integrated Performance Primitives Cryptography Library.: Intel Corporation, 2020. Constant-Time Cryptographic Primitives for SGX Enclaves. (). Intel Corporation
18. Jaiman, V., Urovi, V.: A consent model for blockchain-based health data sharing platforms. IEEE Access **8**, 143734–143745 (2020)
19. Kim, J., Park, S., Lee, K.: Secure electronic prescription system using proxy re-encryption and blockchain. Appl. Sci. **11**(24), 11679 (2021)
20. Kitchin, R., Dodge, M.: The (in)security of smart cities: vulnerabilities, risks, mitigation, and prevention. J. Urban Technol. **26**(2), 47–65 (2017)
21. Liu, Z.: Multi-authority ciphertext-policy attribute-based encryption with accountability. In: ASIACCS (2020)
22. Makhdoom, I., Zhou, I., Abolhasan, M., Lipman, J., Ni, W.: Privysharing: a blockchain-based framework for privacy-preserving and secure data sharing in smart cities. Comput. Secur. **88**, 101650 (2019)
23. Nakamoto, S.: Bitcoin: A Peer-to-Peer Electronic Cash System. https://bitcoin.org/bitcoin.pdf (2008). Accessed online 20 July 2023
24. Nam, T., Pardo, T.A., Conceptualizing smart city with dimensions of technology, people, and institutions. In: 12th Annual International Digital Government Research Conference, pp. 282–291 (2011)
25. Narayanan, A., Shmatikov, V. : Robust de-anonymization of large sparse datasets. In: 2008 IEEE Symposium on Security and Privacy (S&P). Seminal Work on Dataset Fingerprinting and Linkage Attacks, pp. 111–125 (2008)

26. Rantos, K., Drosatos, G., Demertzis, K., Ilioudis, C.: Advocate: a consent management platform for personal data processing. In: TrustCom/BigDataSE, pp. 1–8 (2018)
27. Szabo, N.: Formalizing and securing relationships on public networks. First Monday **2**(9) (1997)
28. Wang, C., Zhang, Y., Chen, L.: Dr.aid: tracking data disclosure via blockchain for enhanced data transparency and compliance. In: IEEE International Conference on Blockchain and Cryptocurrency (ICBC), pp. 1–9 (2021)
29. Woolley, J.P., Kirby, E., Leslie, J., Jeanson, F., Cabili, M.N., Rushton, G., Hazard, J.G., Lappalainen, I., Dugan, N.M., Zoubovsky, S., Knoppers, B.M.: Responsible sharing of biomedical data and biospecimens via the "automatable discovery and access matrix" (ada-m). NPJ Genom. Med. **3**, 1–7 (2018)
30. Zhang, Y., Deng, R.H., Shengmin, X., Sun, J., Li, Q., Zheng, D.: Attribute-based encryption for cloud computing access control: a survey. ACM Comput. Surv. **55**(8), 1–41 (2023)
31. Zyskind, G., Nathan, O.: Alex Pentland. Decentralizing privacy: using blockchain to protect personal data. In: IEEE Security and Privacy Workshops, pp. 180–184 (2015)

Performance Contributions of Traditional Lean and Lean-Smart in Manufacturing: A Metrics-Based Synthesis

Abir Rachidi[1(✉)], Driss Serrou[2], Hassan Mharzi[1], and Rabii El Gomri[3]

[1] Laboratory for Scientific Research and Innovation (ISA), ENSA, Ibn Tofail University, Kénitra, Morocco
{Abir.rachidi,h.mharzi}@uit.ac.ma
[2] Higher School of Technology, Kénitra, Morocco
[3] Transportation and Logistics Research Center (CRTL-ISTL), Casablanca, Morocco
Rabii.elgomri@istl.ma

Abstract. Lean has become a reference model for stabilising flow, reducing variability and embedding problem-solving routines in manufacturing, while Industry 4.0 technologies (IIoT, MES/APS, advanced analytics, computer vision) promise real-time visibility and predictive control. Their deliberate combination—often labelled Lean 4.0 or Lean-Smart—is widely promoted, but the respective contributions of Traditional Lean and digital layers remain insufficiently clarified. This paper focuses on manufacturing and examines under which conditions a Lean-Smart configuration adds measurable value beyond a well-implemented Lean system. Using a curated set of pre/post industrial studies, reported outcomes are organised into a unified measurement lens with four indicator families: flow and productivity, quality, cost and inventory, and sustainability/compliance. Changes are expressed as aligned percentage variations, and each improvement is attributed to its dominant lever (Lean or Lean-Smart) according to the underlying mechanism. The analysis shows that the first performance step—reductions in changeover and cycle time, stabilised OEE and improved FPY—is consistently delivered by plain Lean. Lean-Smart provides a second, selective step when losses are driven by information latency or coordination span, via e-Kanban/MES, IIoT-based maintenance and vision/traceability. A Lean 4.0 case from a mixed-model line in Tangier, where OEE rises from about 25% to over 90%, illustrates these mechanisms. The paper concludes with a decision frame for practitioners and perspectives for more robust empirical research.

Keywords: Lean manufacturing · Lean 4.0 · Lean-Smart · Industry 4.0 · overall equipment effectiveness (OEE) · performance metrics · discrete manufacturing

1 Introduction

In manufacturing, Lean is a reference approach for stabilising flow, reducing variability and embedding daily problem solving. Standardised work, SMED, Kanban and Jidoka, TPM and visual management are now widespread on shop floors. In parallel, Industry 4.0

Y. Benadada et al. (Eds.): GOL 2026, LNNS 2034, pp. 382–391, 2027.
https://doi.org/10.1007/978-3-032-29677-1_34

technologies—industrial sensing, IoT connectivity, MES/APS, data analytics and AI, computer vision and digital twins—enable real-time visibility, predictive control and tighter system-level coordination. Their deliberate combination, often labelled Lean-Smart or Lean 4.0, aims to preserve the human-centred strengths of Lean while closing feedback loops with digital signals and selected automation [1–3].

Reported manufacturing results are positive but uneven. Case studies and empirical analyses describe reductions in changeover and cycle times, better takt adherence, higher OEE and lower defect rates when established Lean practices (SMED, Kanban/Jidoka, standardised work, Lean Six Sigma) are combined with e-Kanban, MES, predictive maintenance and automated or vision-based inspection [4–8]. Several contributions also point to sustainability co-benefits—less scrap, improved energy and resource use, reduced paperwork and easier compliance—although these dimensions are measured inconsistently over time [9–12].

At the same time, results are fragmented by sub-sector and by how indicators are defined, making like-for-like comparison difficult [13, 14], and many accounts blur the boundary between what "plain" Lean achieves and what is truly incremental once a digital layer is added [15–17]. Under such conditions, it is hard for manufacturers, especially those facing skills, interoperability and change-management constraints, to decide when a robust Lean programme is sufficient and when investment in Lean-Smart capabilities is likely to generate a meaningful additional gain [18–20].

What is truly incremental once a digital layer is added [15–17]. Under such conditions, it is hard for manufacturers, especially those facing skills, interoperability and change-management constraints, to decide when a robust Lean programme is sufficient and when investment in Lean-Smart capabilities is likely to generate a meaningful additional gain [18–20].

This paper focuses on manufacturing and distinguishes the effects of Traditional Lean from those of Lean-Smart. Based on a curated set of pre/post studies, we group outcomes into four indicator families: (i) flow and productivity (changeover and cycle time, throughput, Overall Equipment Effectiveness, OEE), (ii) quality (First-Pass Yield, FPY; parts per million, PPM; scrap and rework), (iii) cost and inventory (unit cost, inventory levels and turns, overtime and downtime), and (iv) sustainability and compliance (energy use, material losses, traceability, paperless ratio). Changes are expressed as aligned percentage variations and linked to the dominant lever (Lean or Lean-Smart). The contribution is twofold: a compact comparison of Lean and Lean-Smart impacts on manufacturing performance, and a decision frame indicating when Lean alone is sufficient and when a targeted digital layer is warranted.

2 Literature Review

2.1 Lean and Lean Six Sigma in Manufacturing

Lean in manufacturing is essentially a system for stabilising flow and making problems visible. Standardised work, SMED, Kanban/Jidoka, TPM, 5S and visual management reduce variability, control WIP and expose abnormalities, which directly supports margins in discrete industries. Empirical studies show that well-implemented Lean programmes reduce changeover and cycle times, improve OEE and line balance, and cut defect rates through better process capability and fewer handling steps [4, 6, 17]. Lean Six Sigma (LSS) adds SPC, capability indices and DMAIC on top of these foundations, yielding higher FPY, less rework and scrap, and more stable processes, usually without heavy digital infrastructure.

2.2 Industry 4.0, Lean 4.0 and Smart-Sustainable Value Streams

Industry 4.0 brings IIoT/CPS, shop-floor connectivity, MES/APS, robotics, AI analytics, vision and digital twins. When combined deliberately with Lean, the literature refers to Lean 4.0 or Lean-Smart, aiming to reinforce Lean with real-time data, automated feedback and system-level coordination [1, 2, 8]. Integration studies extend classic Lean metrics (lead/cycle time, OEE, FPY, WIP, inventory turns) with digitalisation and traceability measures such as paperless ratio, data latency, connectivity and "smart VSM" scores, reporting faster reactions, earlier defect detection and more stable takt [4, 6, 21]. Smart-sustainable VSM adds environmental (and sometimes social) indicators—energy per unit, scrap, CO_2 proxies—on the same map as flow and quality, linking Lean 4.0 to ESG objectives [9–12].

2.3 Gaps and Limitations in Existing Studies

Three gaps recur. **Measurement:** defects are reported as %, ppm or DPMO; OEE often lacks A/P/Q breakdown; inventory and sustainability indicators use heterogeneous definitions, which hinders comparison [13, 14].

Attribution: many "digital" cases build on existing Lean foundations, making it unclear how much gain is due to Lean versus e-Kanban/MES, IIoT-PdM or vision/traceability, while other works understate the role of real-time data [15–17].

Design: most studies are single-site, before–after and short-horizon, with limited dispersion data and uneven sustainability coverage. These limitations motivate our use of a unified measurement lens, aligned percentage changes and explicit separation between Traditional Lean and Lean-Smart effects.

3 Problem Statement

On many shop floors, chronic performance losses stem from one recurring mechanism: variability enters the system faster than it can be seen and absorbed. Product mix and demand change, cycle times drift with micro-stoppages and rework, machines are intermittently unavailable, and information often arrives late on paper. The result is longer lead and cycle times, high WIP, defects discovered downstream, and stations oscillating between starvation and blockage. Traditional Lean stabilises work and makes problems visible. Standardised work, visual management, SMED, pull/leveling, TPM and jidoka/andon reduce cycle-time spread, compress changeovers, control WIP and allow safe stops at the point of cause, improving flow, OEE and FPY. Yet even in plants with a solid Lean backbone, two gaps persist: detection is still slow (problems surface at the end of batches or shifts) and synchronisation across cells remains fragile when coordination relies on manual signals and delayed reporting.

Lean-Smart configurations seek to close these gaps by accelerating what the system can see and how fast it can react. Digital pull and coordination (e-Kanban, MES/APS, connected boards) link consumption to replenishment, IIoT-based monitoring and simple predictive rules move maintenance to planned windows, and vision, torque/press-fit traces and digital poka-yokes push error detection upstream; smart-sustainable VSM extends this logic to energy, materials and documentation.

Despite numerous positive cases, manufacturers still lack a clear, comparable view of what Lean delivers on its own and what is genuinely incremental once digital layers are added. Reported gains are fragmented by context and metrics, the link between specific levers (SMED, TPM, e-Kanban, IIoT, vision, smart VSM, etc.) and observed changes is often implicit, and reviews warn that adding advanced technologies on unstable processes can overload teams and create new forms of waste[13, 14, 18, 20].

In this paper we therefore ask: under which conditions does a Lean-Smart configuration add measurable value beyond a well-implemented Lean system, and along which performance dimensions (flow and productivity, quality, cost and inventory, sustainability/compliance)?

4 Research Methodology

The study follows a two-step logic. First, we characterise losses on a representative mixed-model discrete line with frequent changeovers and demand variability. Across the literature, recurring mechanisms are arrival and mix variation, cycle-time spread due to micro-stoppages and rework, availability losses and unplanned downtime, late defect detection, gaps in material assurance, and information latency from paper-based records. On this basis, we reconstruct two like-for-like scenarios for the same line: a Lean baseline with standardised work, visual management, 5S, SMED, pull/leveling, TPM and jidoka/andon, and a Lean-Smart configuration that keeps this backbone but adds digital enablers that change decisions at the cadence of operations (e-Kanban or MES/APS, IIoT/PdM, e-Andon, vision/traceability). The process structure is held constant; only the speed and reliability of signals differ.

Second, we read reported results through a harmonised measurement lens with four indicator families: flow/productivity, quality, cost/inventory, and sustainability/compliance. For each metric we compute an aligned percentage change and invert "lower-is-better" indicators so that positive values always denote improvement. Each change is then tagged with its dominant lever—Lean when stabilisation tools plausibly explain the effect, Lean-Smart when faster digital signals or automated coordination are clearly involved. The empirical base is the curated set of reviews and primary cases that distinguish Lean from Lean-Smart and report at least one before/after metric. Given the prevalence of single-site, short-horizon before–after designs and heterogeneous definitions, we rely on aligned percentage changes, keep attribution conservative, and follow the sequencing principle explicit in several sources: stabilise with Lean, digitise where faster signals change decisions, then institutionalise the new cadence.

5 Case Study: Synthesis of Manufacturing Evidence

5.1 Mechanisms and Results in Discrete Manufacturing Lines

To interpret the evidence in mechanism terms, we concentrate on mixed-model discrete lines, which are the dominant context in Table 1. Across these cases, the recurring loss mechanisms are prolonged changeovers, cycle-time spread driven by micro-stoppages and rework, and late defect discovery. Where organisations deploy a disciplined Lean backbone—standardised work, visual management, SMED, pull/leveling, TPM, and jidoka/andon—the table repeatedly records shorter cycle and lead times, higher and more stable OEE (especially Availability and Performance), and better first-pass quality [6]. These Lean effects arise from exposing variability and stabilising the work: SMED compresses setup losses, pull dampens WIP oscillations, TPM reduces chronic downtime, and Jidoka prevents defective work from propagating. The same rows also report "classical" waste reductions—lower rework and scrap, improved layout and flow, and, in LSS-supported implementations, tighter process capability.

Table 1. Manufacturing results and dominant levers.

Metric family	Representative metric(s)	Direction / typical magnitude	Dominant lever-	Evidence hook (rows)
Flow & productivity	Changeover; lead/cycle; OEE (A/P/Q)	Changeover ↓ 30–50%; Lead/Cycle ↓ 15–35%; OEE ↑ + 5–15 pts	**Lean** (SMED, std. Work, pull, TPM)	[6, 15, 22]
Availability & stability	Unplanned downtime; Micro-stops	Downtime ↓ 25–45%; cycle spread tightens	**Lean-Smart** (IIoT PdM, e-Andon)	[5, 11, 17]
Quality	FPY; PPM/defects; Scrap	FPY ↑ + 2–6 pts (Lean); PPM/defects ↓ 30–70% (Lean-Smart)	**Lean → Lean-Smart** (vision, torque/press-fit traceability)	[10, 13]
Cost & inventory	Overtime/expedites; Turns	Expedites/OT ↓; Inventory turns ↑	**Lean**, reinforced by **e-Kanban/MES**	[22, 23]
Sustainability & compliance	kWh/unit; Paperless; Traceability latency	kWh/unit ↓ (where metered); Paperless → 100% (targeted forms); Traceability hours → minutes	**Lean-Smart** (digital VSM, traceability)	[5, 9]

The incremental step appears in entries where digital capabilities are deliberately added on top of this Lean baseline. Lines that close pull loops with e-Kanban, MES or lightweight APS and make the next job explicit at the station show steadier takt and reduced WIP; where condition monitoring and simple predictive models feed maintenance, unplanned downtime decreases and cycle-time distributions tighten; where computer vision, digital torque/press-fit traces or embedded poka-yokes shift error separation upstream, defect rates and rework fall further [5 , 10 , 17]. Several cases document substantial administrative gains through digitised traceability and shop-floor records: hours of document handling and investigation are removed, paper forms are eliminated, and KPI updates become automatic, which lowers compliance effort and accelerates response to quality or delivery incidents [5 , 9].

The ranges reported in Table 1 are not averages but spans of percentage changes observed across the manufacturing studies in the curated set, and each range is traceable to the sources listed in the "Evidence hook" column. Read through this lens, Table 1 summarises how Traditional Lean explains most of the improvements in changeover, lead/cycle time, OEE and FPY through stabilised methods and equipment, while Lean-Smart accounts for additional reductions in unplanned downtime, earlier error separation, higher traceability and paperless gains, all driven by faster, more reliable signals and automated coordination; programme-level studies further confirm this complementarity, as integrated Lean+I4.0 roadmaps achieve larger conversion-cost and flexibility gains than stand-alone Lean or stand-alone digital projects.

5.2 Industrial Case Study: Lean 4.0 in Tangier

The proposed reading of Lean and Lean-Smart effects is illustrated by an industrial case from a mixed-model discrete assembly line located in Tangier. For confidentiality reasons, the company and product family are anonymised; the line produces small series of assembled components on two shifts. Before the project, the line suffered from chronic micro-stoppages, long and unstable changeovers, frequent material shortages and extensive manual recording of production data.

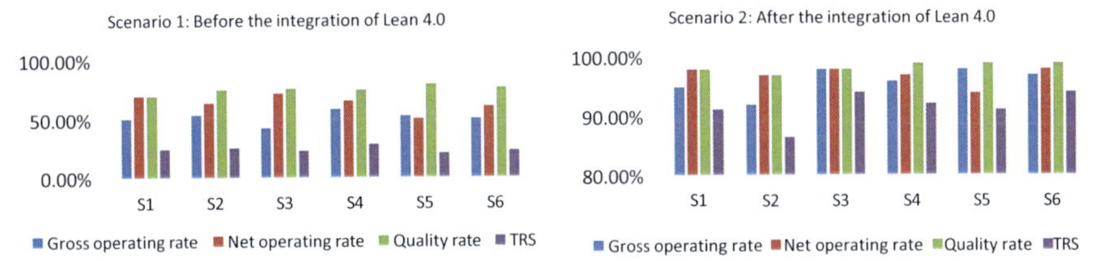

Fig. 1. OEE components and overall OEE before/after Lean 4.0 in Scenarios 1 and 2 (S1–S6).

OEE (TRS in the plant's terminology) was measured at line level using the standard decomposition: Availability (effective running time over planned time, excluding planned stops), Performance (actual output vs. theoretical output at nominal speed) and Quality (good pieces vs. total pieces). Six consecutive periods (S1–S6) were monitored in a diagnostic phase and again after a combined Lean and digital intervention. The Lean programme focused on standardised work, 5S, SMED on changeover tasks, TPM basics and visual management. The Lean-Smart layer then introduced e-Kanban for internal logistics, a simple MES for line-level monitoring, digital andon for escalation, and basic IIoT sensing on critical stations.

Table 2 and Fig. 1 summaries the results. In the diagnostic phase, chronic downtime and material shortages kept Availability and Performance low, leading to an OEE around 25 % despite a reasonable quality rate. After the Lean and Lean-Smart deployment, Availability and Performance improved sharply as setup overruns and recurrent stops were addressed, material assurance was stabilized and micro-stops became visible in real time. Across the six follow-up periods, OEE stabilized between 86% and 94%. This large change is therefore interpreted as a move from a severely under-performing line, with multiple structural issues, to a stabilized and digitally supervised line, rather than as a marginal optimisation on an already efficient system. The case remains a single-site before–after observation and is treated as illustrative, not as a statistical generalisation.

Table 2. Evolution of gross operating rate, net operating rate, quality rate and OEE over six periods.

Scenario 1: Before the integration of Lean 4.0						
Period	S1	S2	S3	S4	S5	S6
Gross operating rate	51.00%	54.00%	43.00%	59.00%	53.00%	51.00%
Net operating rate	70.00%	64.00%	72.00%	66.00%	51.00%	61.00%
Quality rate	70.00%	75.00%	76.00%	75.00%	80.00%	77.00%
OEE	24.99%	25.92%	23.53%	29.21%	21.62%	23.95%
Scenario 2: After the integration of Lean 4.0						
Period	S1	S2	S3	S4	S5	S6
Gross operating rate	95.00%	92.00%	98.00%	96.00%	98.00%	97.00%
Net operating rate	98.00%	97.00%	98.00%	97.00%	94.00%	98.00%
Quality rate	98.00%	97.00%	98.00%	99.00%	99.00%	99.00%
OEE	91.24%	86.56%	94.12%	92.19%	91.20%	94.11%

6 Discussion

The reviewed manufacturing studies and the synthesis in Table 1 show that the first performance step is still delivered by plain Lean. Standardised work, 5S, SMED, pull/leveling, TPM and jidoka/andon are consistently associated with shorter changeovers and cycle times, reduced lead time, more stable OEE and higher FPY, even with minimal digital support, by reducing task-time variation, compressing setup, controlling WIP and preventing defects from propagating [6, 22, 23]. Mechanistically, these tools reduce task-time variation, compress setup, control WIP and stop defects from propagating, making line behaviour more predictable.

A second, clearly incremental step appears when digital enablers are added on top of this Lean backbone. e-Kanban/MES, IIoT-based condition monitoring, e-Andon and vision/traceability reduce information latency and coordination gaps, leading to further reductions in unplanned downtime, tighter cycle-time distributions, additional FPY/defect improvements and lower documentation effort [1, 3, 5, 10, 13–15, 17]. Sustainability behaves similarly when smart-sustainable VSM or green/Lean-4.0 are used, with flow and quality gains often accompanied by lower energy and material intensity [9, 24]. From a managerial standpoint, the rule of thumb is simple: when losses are discipline-driven (unstable changeovers, inconsistent methods, micro-stops, WIP issues), priority should be given to consolidating Lean; when losses are clearly latency- or span-driven (late defect detection, slow escalation, availability shocks, weak synchronisation), targeted Lean-Smart levers are justified. Despite the limits of mostly single-site, short-horizon studies, the aligned percentage-change view and conservative attribution used here support a complementary, rather than competing, positioning of Lean and Lean-Smart.

7 Conclusions

This paper clarified what Traditional Lean delivers in manufacturing and what is genuinely incremental when Lean-Smart is adopted. Using a curated set of pre/post studies and a unified measurement lens (flow and productivity, quality, cost and inventory, sustainability/compliance), results were expressed as aligned percentage changes and each improvement was attributed to its dominant lever (Lean or Lean-Smart).

The synthesis shows that the first performance step is consistently provided by plain Lean: standardised work, 5S, SMED, pull/leveling, TPM and jidoka/andon reduce changeover and cycle times, stabilise OEE and improve FPY, even with limited digitalisation. Lean-Smart then adds a second, selective step when losses are driven by information latency or coordination span: e-Kanban/MES, IIoT/PdM, e-Andon and vision/traceability shorten detection-to-action time, cut unplanned downtime, move error detection upstream and reduce documentation effort, with similar patterns for sustainability when smart-sustainable VSM or green/Lean-4.0 frameworks are used.

For practitioners, a simple decision rule follows: prioritise Lean when losses are discipline-driven (unstable changeovers, inconsistent work, micro-stops, WIP issues), and justify Lean-Smart when losses are clearly latency- or span-driven (late defect detection, slow escalation, availability shocks, poor synchronisation), always respecting the sequencing "stabilise with Lean, then digitise where faster signals change decisions". For researchers, the main needs are stronger multi-site or quasi-experimental designs, harmonised indicator definitions and richer, longer-horizon sustainability data.

References

1. Skalli, D., Charkaoui, A., Cherrafi, A., Garza-Reyes, J.A., Antony, J., Shokri, A.: Industry 4.0 and lean six sigma integration in manufacturing: a literature review, an integrated framework and proposed research perspectives. Qual. Manage. J. **30**, 16–40 (2023). https://doi.org/10.1080/10686967.2022.2144784
2. Utiyama, M., et al.: Integration of lean manufacturing and factory physics in an industry 4.0 roadmap for augmenting operational performance. TQM **37**, 544–565 (2025). https://doi.org/10.1108/TQM-07-2023-0210

3. Mendes, D., Gaspar, P.D., Charrua-Santos, F., Navas, H.: Synergies between Lean and Industry 4.0 for enhanced maintenance management in sustainable operations: a model proposal. Processes **11**, 2691 (2023). https://doi.org/10.3390/pr11092691
4. Industrial Management and Plastics Forming Technology Team. Mechanics, Engineering and Innovation Laboratory, LM2I—ENSEM- Hassan II University Casablanca, Morocco; Ghouat, M., Benhadou, M., Benhadou, B., Haddout, A.: Assessment of the potential impact of Industry 4.0 technologies on the levers of lean manufacturing in manufacturing industries in Morocco. IJETAE **12**, 78–85 (2022). https://doi.org/10.46338/ijetae0722_08
5. Sundar, P.S., Chowdhury, C., Kamarthi, S.: Industrial internet of things enabled kata methodology of assembly line productivity improvement: insights from a case study. Processes **12**, 2611 (2024). https://doi.org/10.3390/pr12112611
6. Pozzi, R., Cannas, V.G., Rossi, T.: Data science supporting lean production: evidence from manufacturing companies. Systems **12**, 100 (2024). https://doi.org/10.3390/sys-tems12030100
7. Da Silva, I.B., et al.: Hot-stamped B-pillar for automotive driven by Lean 4.0. Proc. Inst. Mech. Eng., Part B: J. Eng. Manuf. **237**, 951–960 (2023). https://doi.org/10.1177/09544054221119328
8. Koteswarapavan, C., Pattanaik, L.N.: A novel tool-input-process-output (TIPO) framework for upgrading to lean 4.0. Int. J. Prod. Manage. Eng. **12**, 65–77 (2023). https://doi.org/10.4995/ijpme.2024.19723
9. Maldonado-Guzmán, G., Pinzón-Castro, S.Y., Garza-Reyes, J.A.: Does the integration of lean production and Industry 4.0 in green supply chains generate a better operational performance? JMTM **34**, 1120–1140 (2023). https://doi.org/10.1108/JMTM-02-2023-0034
10. Peralta-Abarca, J.D.C., Moreno-Bernal, P., Martínez-Bahena, B., Enríquez-Urbano, J., Bonilla-Sánchez, F.: Optimizing the rework area in an automotive parts supplier company by digital tools: a foundation for smart industry transformation. In: Nesmachnow, S., Hernández Callejo, L. (eds.) Smart Cities. Springer Nature Switzerland, Cham, pp. 77–91 (2024)
11. Braggio, L., Marinho, O., Soares, L., Lino, A., Rabelo, F., Muniz, J.: Integrating Agenda 2030 with Lean Practices and Industry 4.0 Technologies: Case Study in an Automotive Industry. Sao Paolo, Brazil, pp. 2024-36–0043 (2024)
12. Hossain, M.M., Purdy, G.: Integration of industry 4.0 into lean production systems: a systematic literature review. Manuf. Lett. **35**, 1347–1357 (2023). https://doi.org/10.1016/j.mfglet.2023.08.098
13. Saraswat, P., Agrawal, R., Rane, S.B.: Technological integration of lean manufacturing with Industry 4.0 toward lean automation: insights from the systematic review and further research directions. BIJ (2024). https://doi.org/10.1108/BIJ-05-2023-0316
14. Queiroz, G.A., Alves Junior, P.N., Mello, I.C.: Lean 4.0 deployment case studies in UK industrial companies: lessons learned. In: Silva, F.J.G., Ferreira, L.P., Sá, J.C., Pereira, M.T., Pinto, C.M.A. (eds.) Flexible Automation and Intelligent Manufacturing: Establishing Bridges for More Sustainable Manufacturing Systems. Springer Nature Switzerland, Cham, pp. 324–331 (2024)
15. Anbessia, M., Singh, B.: Optimization of industrial performance by lean 4.0. IJRASET **12**, 865–878 (2024). https://doi.org/10.22214/ijraset.2024.58948
16. Javaid, M., Haleem, A., Singh, R.P., Gupta, S.: Leveraging lean 4.0 technologies in healthcare: an exploration of its applications. Adv. Biomark. Sci. Technol. **6**, 138–151 (2024). https://doi.org/10.1016/j.abst.2024.08.001
17. Sah, B.P., Tanha, N.I., Sikder, M.A., Habibullah, S.M.: The integration of Industry 4.0 and lean technologies in manufacturing industries: a systematic literature review. GMJ **1**, 14–25 (2024)

18. Shafee, N.A.Q.M., Mohamad, E., Nordin, S.K.S., Abd Rahman, M.S., Zaidi, M.K.H.M., Ito, T., Abd Shukor, M.H.: Can Industry 4.0 work together with lean? An investigation from Malaysian lean manufacturing sector. Multidiscip. Sci. J. **7**, 2025055 (2024). https://doi.org/10.31893/multiscience.2025055
19. Sordan, J.E., et al.: Lean six sigma practices supported by industry 4.0 technologies: evidence from heavy vehicle manufacturers. JMTM **36**, 504–533 (2025). https://doi.org/10.1108/JMTM-10-2023-0471
20. Chen, J.C.: From Lean to Smart Manufacturing. Asian Productivity Organization (2024)
21. Silva, B.N., Khan, M., Han, K.: Integration of big data analytics embedded smart city architecture with RESTful web of things for efficient service provision and energy management. Futur. Gener. Comput. Syst. **107**, 975–987 (2020). https://doi.org/10.1016/j.future.2017.06.024
22. Gustavo Cunha, G., De Andrade Ussuna, G., Martins Coelho Trevisanuto, T., Antonio Gandolfo Rodrigues, M.: Estratégias Lean Six Sigma para melhoria contínua: um estudo setorial comparativo centrado nos setores alimentício e automotivo. fibinova **4** (2024). https://doi.org/10.59237/fibinova.v4i4.765
23. Alsadi, J., Maalouf, M., Goonetilleke, R.: Exploring the integration of lean six sigma and Industry 4.0 in the manufacturing industry: bibliometric analysis and systematic review. In: Proceedings of the International Conference on Industrial Engineering and Operations Management. IEOM Society International, Tokyo, Japan (2024)
24. Hossain, A., Khan, M.R., Islam, M.T., Islam, K.S.: Analyzing the impact of combining lean six sigma methodologies with sustainability goals. NHJ **1**, 123–144 (2024). https://doi.org/10.70008/jeser.v1i01.57

Author Index

Y. Benadada et al. (Eds.): GOL 2026, LNNS 2034, pp. 393–394, 2027.
https://doi.org/10.1007/978-3-032-29677-1

Kolibri 360 GmbH
Ferdinand-Jühlke-Straße 7
99095 Erfurt, Deutschland
produktsicherheit@kolibri360.de